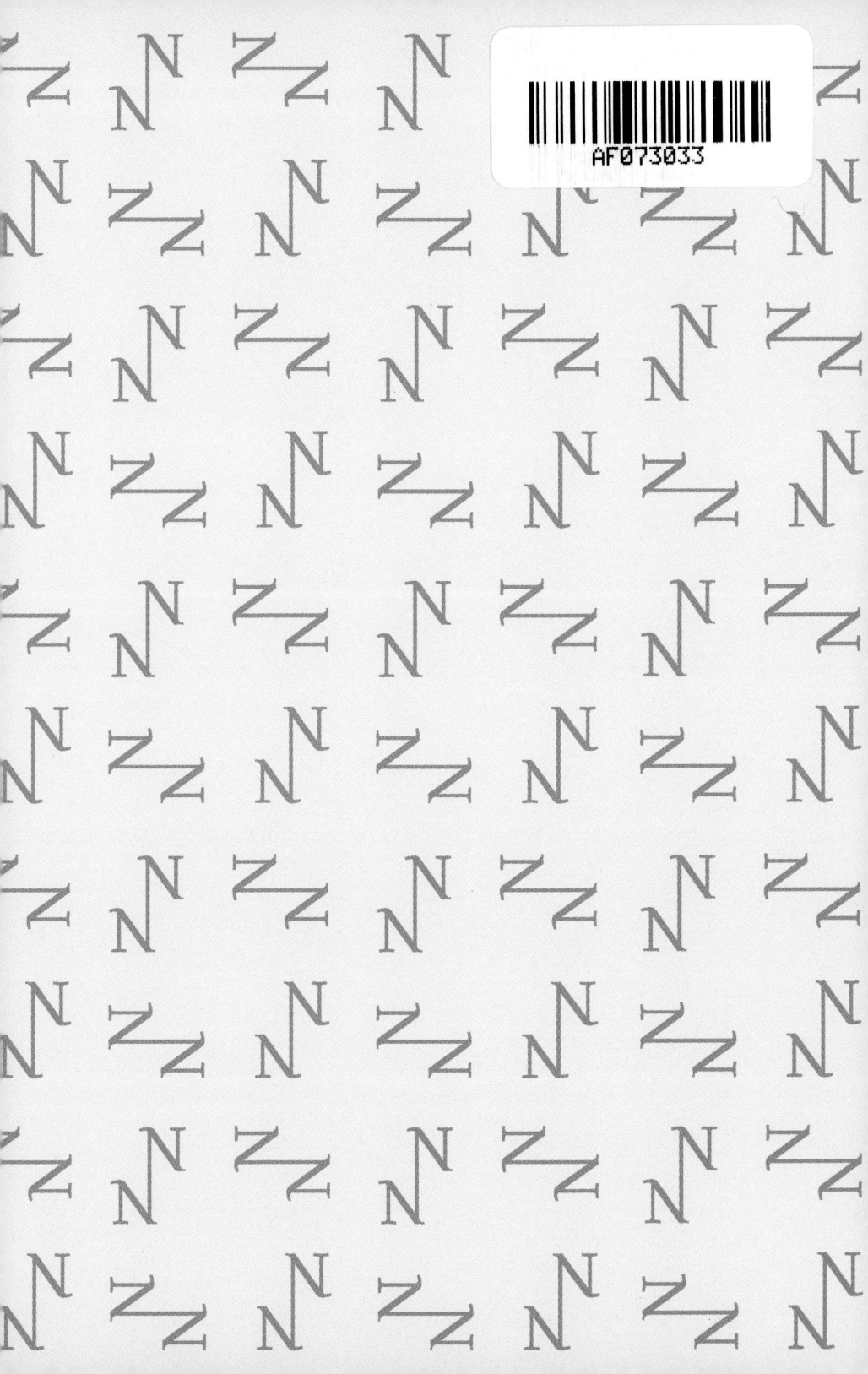

THE NEW NATURALIST LIBRARY
A SURVEY OF BRITISH NATURAL HISTORY
STOATS, WEASELS, MARTENS & POLECATS

EDITORS
SARAH A. CORBET, ScD
DAVID STREETER, MBE, FRSB
JIM FLEGG, OBE, FIHort
Prof. JONATHAN SILVERTOWN
Prof. BRIAN SHORT

*

The aim of this series is to interest the general reader in the wildlife of Britain by recapturing the enquiring spirit of the old naturalists. The editors believe that the natural pride of the British public in the native flora and fauna, to which must be added concern for their conservation, is best fostered by maintaining a high standard of accuracy combined with clarity of exposition in presenting the results of modern scientific research.

THE NEW NATURALIST LIBRARY

STOATS, WEASELS, MARTENS & POLECATS

A Natural History of British & Irish Small Mustelids

JENNY MACPHERSON

This edition published in 2024 by William Collins,
an imprint of HarperCollins*Publishers*

HarperCollins*Publishers*
1 London Bridge Street
London SE1 9GF

WilliamCollinsBooks.com

HarperCollins*Publishers*
Macken House, 39/40 Mayor Street Upper
Dublin 1, D01 C9W8, Ireland

First published 2024

© Jenny MacPherson, 2024
Photographs © individual photographers. See Picture Credits, p. 372

All rights reserved. No part of this publication may be reproduced, stored in a retrieval system or transmitted in any form or by any means, electronic, mechanical, photocopying, recording or otherwise, without the prior written permission of the copyright owner.

A CIP catalogue record for this book is available
from the British Library.

Set in Nexus Serif Pro and Nexus Mix Pro

Edited and designed by
D & N Publishing
Baydon, Wiltshire

Printed in Bosnia-Herzegovina by GPS Group

ISBN 978-0-00-833493-2

All reasonable efforts have been made by the author to trace the copyright owners of the material quoted in this book and of any images reproduced in this book. In the event that the author or publisher is notified of any mistakes or omissions by copyright owners after publication of this book, the author and the publisher will endeavour to rectify the position accordingly for any subsequent printing.

Contents

Editors' Preface   vii
Author's Foreword and Acknowledgements   ix

1   Meet the Weasels: An Introduction to the Mustelid Family   1
2   On the Origin of Weasels: Taxonomy and Evolution   15
3   Differences and Similarities, Coexistence and Competition   27
4   From a Richness of Martens to a Paucity of Predators:
    A History of Exploitation and Persecution   47
5   Polecats and Pine Martens: The Bumpy Road to Recovery   79
6   Pine Marten Reintroductions to Wales and Beyond   103
7   Where Weasels Aren't Wanted: Small Mustelids as Invasive
    Non-native Species   147
8   Mapping Phantoms: The Challenges of Surveying and
    Monitoring Small Mustelids   185
9   Mustelid Menus: What Do Weasels Eat?   209
10 Of Mice and Mustelids: Predator and Prey Interactions   235
11 Pine Martens and Squirrels: Inference and Implications   259
12 Mythology, Monarchy and Mustelidae   275
13 Mustelid Mysteries: What We Know and Don't Know   295

References   323
Species Index   344
General Index   357
Picture Credits   372

Editors' Preface

For most British and Irish naturalists, the term 'small mustelids' amounts to Stoats and Weasels, as they are the two species most frequently encountered in the field. Alongside these in this book are the red-letter species the Pine Marten and the Polecat. There is a family likeness – the rather pointed snout, powerful jaws and sharp fangs, and long, sinuous, slender body with short legs. Mustelids are possessed of dense fur, once of substantial international economic and more particularly social value. Black Stoat tails against white fur (ermine) are an essential accoutrement of the British peerage, and the number of black tail tips featured is jealously guarded. At one time in the distant past, high-class ladies' slippers were made from mustelid body fur called vair: some clumsy corruption of this to the French for 'glass' is a possible origin of the otherwise improbable and impractical glass slippers supposedly worn by Cinderella to the ball.

The mustelids are the largest family within the mammalian order Carnivora, all of which are predators, depicted as especially fearsome killers in the form of the 'big cats', notably lions and tigers. As big as these are, on a power-to-weight basis, Stoats and Weasels are extremely skilful predators, capable of bringing down and killing prey far larger than themselves, a Stoat tackling an adult Rabbit being a striking example. The slender body of the Weasel is supremely adapted to the pursuit of small rodents down burrows only marginally wider than the Weasel's skull, and so flexible is its skeleton it can reverse in the tunnel by in essence performing a somersault.

This delightfully written text, by Jenny MacPherson, accompanied by many stunning photographs, is wide-ranging and packed full of information, particularly in the subtexts on genetics. It admirably fulfils the aims of the New Naturalist series of interesting the general reader in the wildlife of Britain by recapturing the enquiring spirit of the old naturalists. The editors believe that the natural pride of the British public in native fauna and flora, to which must be added concern for their conservation, is best fostered by maintaining a high standard of accuracy combined with clarity of exposition when presenting the results of modern scientific research.

This absolutely fascinating book on small mustelids does just this and fills a void in the available zoology texts, making it an excellent addition to the New Naturalist library. It really is a book that, once started, you will not want to put down.

FIG 1. Pine Marten *Martes martes* with the characteristic apricot bib.

Mustela nivalis have, in turn, been reviled and revered, and some of the folklore associated with these little mustelids seems bizarre now, to say the least. At various times, Weasels have been said to conceive young through their mouths, to give birth through their ears, and to be able to cure other animals with herbs (more about that later). Now, having once almost eliminated Pine Martens from England and Wales, we want them back to rid us of Grey Squirrels *Sciurus carolinensis* so, as you can see, we have had a long and complicated relationship with the mustelid family.

Mustelids make up the largest family within the mammalian order of carnivores, with 59 species in 22 subgroups (or genera). Members of the *Mustelidae* include weasels, stoats, polecats, martens, otters, mink, Fishers *Pikania pennanti*, Wolverines *Gulo gulo*, badgers and others. Among them are species with behavioural and physical adaptations for a range of different lifestyles. Examples of this diversity can be seen in British mustelids alone, whose habits range from the digging, highly social European Badgers to the tree-climbing, solitary Pine Martens, the semiaquatic Eurasian Otters and the small and slender Weasels, perfectly formed for hunting rodent prey in their burrows. What all the mustelids have in common is their well-developed anal scent glands, which produce secretions that are often rather pungent. The scent from these glands is used for

CHAPTER 1

Meet the Weasels: An Introduction to the Mustelid Family

My fascination with small mustelids began many years ago, when I was out badger-watching in Scotland one evening. As I sat on a bank above the sett at dusk with my binoculars poised, waiting for some European Badgers *Meles meles* to emerge, the most beautiful animal I'd ever seen crossed my field of view. It was a Pine Marten *Martes martes*. It stopped, stood up on its hind legs and sniffed the air, and then went on its way. From that moment on I was hooked, and I'm sorry to say that the badgers, when they came out a little later, could not compete. Since then, I have been involved with research and conservation of Pine Martens and other small mustelids and have been fortunate enough to have had many similar sightings, but the thrill never gets old.

What is it about these little carnivores that is so intriguing? Mustelids are interesting because, in spite of their small size, they are supremely adapted and fearsome predators: some kill prey much larger than themselves – even Lions *Panthera leo* don't do that alone – and no one who has ever witnessed a Stoat *Mustela erminea* pursuing and taking down a European Rabbit *Oryctolagus cuniculus* twice its size could fail to be impressed. Their dense furs have been valued for fashion and for practical clothing and as a result prompted major changes in mammal distribution (e.g. American Mink *Neovison vison*) and abundance (as in the case of Sea Otters *Enhydra lutris*, the value of which also led to Russia once owning Alaska). It seems difficult to believe now, but the hunting of Eurasian Otters *Lutra lutra* was also once a key part of British rural life. 'Polecat' has long been a term of abuse because of that little mustelid's unmistakable smell, with prominent scent glands being characteristic of the family. Weasels

FIG 2. Five-toed Pine Marten *Martes martes* footprint.

communication, territory marking and defence. Mustelids have five toes on all feet, leaving characteristic five-toed footprints (a useful field sign to look out for), and often have strong claws.

Most of the mustelids have the long, slender body shape typified by weasels, polecats and otters. They usually have short legs, small, rounded ears and a long, flattened head. Badgers and Wolverines have broader, stockier bodies but they still retain the bounding movement that is typical of the family. There is a huge difference in size between the smallest and largest mustelid species. The Weasel is not only the smallest mustelid but also the world's smallest carnivore. It may weigh as little as 31 g when fully grown. This is a huge contrast with the North Pacific Sea Otter, the largest member of the family, which can reach 45 kg.

PRESENT-DAY SMALL MUSTELIDS IN BRITAIN AND IRELAND

Most mammalian predators in Britain are mustelids and seven species of them now live in the British Isles (Fig. 3). The Weasel, Stoat, European Polecat *Mustela putorius*, Pine Marten, Eurasian Otter and European Badger are native to

Great Britain. Feral American Mink are not native but are a recent addition to the assemblage, following escapes and releases from fur farms. Of these seven species, only five (Stoat, Pine Marten, Eurasian Otter, European Badger and American Mink) are present in Ireland and the reasons for this are discussed later on. Confusingly, the Irish Stoat *Mustela erminea hibernica* is often referred to colloquially as the weasel, but hereafter whenever the name 'weasel' is used, it refers solely to *Mustela nivalis*. Apart from the European Badger and the Eurasian Otter, the native British and Irish mustelids are small or medium-sized and, by convention, referred to collectively as the 'small mustelids'. All belong to the genus *Mustela* (Stoat, Weasel and European Polecat) or *Martes* (Pine Marten).

FIG 3. The seven species of mustelid in Britain. From the top: European Badger *Meles meles*, Eurasian Otter *Lutra lutra*, Pine Marten *Martes martes*, European Polecat *Mustela putorius*, American Mink *Neovison vison*, Stoat *Mustela erminea* and Weasel *Mustela nivalis*.

FIG 5. (a) A Weasel *Mustela nivalis* is caught by a Grey Heron *Ardea cinerea* and (b) tries to escape. (c) The Weasel demonstrates flexibility and fights to free itself. *continued overleaf*

FIG 5 *continued*. (d) The weasel gets free of the heron and (e) bites the bird on the cheek before escaping.

pine marten recovery project: Henry Schofield, David Bavin, Lizzie Croose, Josie Bridges, Steve Carter, Huw Denman, Cat McNicol, Alastair Willcox, Alexandra Tomlinson and Alice Bacon.

We stand, as they say, on the shoulders of giants and that is particularly true here. Much of what we know about stoats and weasels is due to the years of research by Carolyn King and Roger Powell. Others, whose work on mustelids I have also called upon and cited copiously throughout this book, include Johnny Birks, Robbie McDonald and many more.

Huge thanks to all of the talented and generous photographers whose beautiful images of these animals illustrate my descriptions of them and their lives. A picture truly does paint a thousand words, especially those by Robert Cruickshanks, Mark Strachan, Gary Clarke, Jason Hornblow, Colin Smith, John Dellow, Mark Lockett, Frank Greenaway, Hugo Fourdin, Jean-Michel Bompar, Nick Upton and Ruth Hanniffy.

Last but definitely not least, I want to thank my fantastic family, Don, Ella and Adam, for their love and support and for enduring my long absences over the years on mustelid-related fieldwork.

Author's Foreword and Acknowledgements

In this book you will see that I have focussed on Britain's small mustelids, which, largely thanks to Kenneth Grahame, have much more of an image problem than their larger cousins, the beloved badger and otter. Since the publication of Grahame's classic, *The Wind in the Willows*, generations of small children have grown up accepting that author's characterisations of wise old badger and cheery, helpful otter whilst believing that 'weasels – and stoats ... well, you can't really trust them, and that's the fact'.

In addition to their shared vilification over the years, the stoats, weasels, pine martens and polecats have much more in common with each other than with the highly social, burrowing badgers or the otters in their watery world. I have shone a light here on the small mustelids in the context of their history, ecology and the circumstances in which they have managed to rub along with us humans for all of this time, often to their detriment. I hope I manage to whet your appetite to learn more and appreciate what they have gone through in their shared history with us as well as their unique qualities, of which we should be far more appreciative.

ACKNOWLEDGEMENTS

First of all, I would like to thank my dear friend and mentor, Pat Morris, who first suggested that I write this book. His encouragement and helpful comments were invaluable, as always. I owe heartfelt thanks to my friends and colleagues at Vincent Wildlife Trust who, over the years, have not only encouraged my obsession with small mustelids (especially pine martens) but actively shared in it. Particularly worthy of mention are my compatriots and co-conspirators on the

They are characterized by their tubular body shape, which enables them to follow their prey down small tunnels and burrows. However, because of this similarity in body shape, they can be difficult to distinguish from each other. This is particularly true when, as is usually the case, you catch only a fleeting glimpse of one in poor light. Though superficially similar, there are some distinct differences in both appearance and ecology that distinguish the small mustelid species from each other. These differences have evolved over a long period of time and are what makes it possible for them to live alongside each other.

WEASELS AND STOATS

Weasels and Stoats are generally much smaller than the rest of the mustelids. The other distinguishing feature is the sharp contrast between the reddish-brown coat colour on their back and sides and the creamy white fur on the throat, chest and belly. This is not found in any of the larger members of the weasel family. Both Stoats and Weasels can be active during the day and are sometimes seen bounding across a road or peering out from features that provide cover, such as dry-stone walls and wood piles.

What is often referred to in Britain as the Common Weasel is properly named the Least Weasel as it is the smallest member of the weasel family. It was said that a Weasel's skull would fit through a wedding ring, which, given a small Weasel and a very large wedding ring, may be true but it wouldn't go through mine. It may be the smallest of all the carnivores, but apart from its diminutive size, there is nothing else 'least' about the Weasel. Make no mistake, this is a fierce little predator. It has short legs and a slender body (13–23 cm). The fur is chestnut brown on the back and head, while that on the belly is creamy white. The division between brown and cream is irregular and spotted (Fig. 4a). This irregular pattern and the 'gular spots', tiny patches of brown fur on each side of the white throat, are different for each animal so can be used to identify individuals. The Weasel's tail is short (3–6 cm) relative to its body (Fig. 4b) and is a uniform chestnut brown colour with no black tip. Although there is quite a difference in size between Weasels and Stoats, the black tip to the Stoat's tail (which is always present, even in babies or in the white winter coat) is usually the only visible difference when an animal is seen at a distance.

If seen moving, the Weasel's gait is often quicker, flatter and less bounding than that of the Stoat. Someone once described it to me as being 'like a rocket-powered sausage'. Weasels are diurnal, probably to avoid predation by owls, but

FIG 4. Weasel *Mustela nivalis*, showing (a) the irregular division between back and belly fur and (b) the relatively short tail.

can be active during both day and night and are found in a variety of habitats, including forest, coastal dunes, meadows and farmland. They will hunt in dense grassland where there is an abundance of small mammals such as Field Voles *Microtus agrestis* but avoid being out in the open where they themselves are vulnerable to predation from raptors and other larger carnivores. In encounters with Weasels though, predators often underestimate the magnitude of their diminutive opponent's bravery. No predator, regardless of size, can assume to have the upper hand when taking on a Weasel. An example of this was recounted to me recently by Mark Strachan who, while walking on the seafront near his home in Peterhead, witnessed an epic battle between a Weasel and an unwitting Grey Heron *Ardea cinerea* (Fig. 5). Mark had stopped to watch a couple of herons hunting in some long grass behind nearby rocks. The pair were having some success diving in and catching voles when one of the herons grabbed a Weasel instead. The bird was taken by surprise when the Weasel fought back ferociously, twisting and screaming loudly (much to the alarm of passers-by, who described the noise as 'like a baby crying'). After an acrobatic struggle that lasted for some minutes, the indignant Weasel bit its assailant on the face and managed to escape. Mark saw the same heron on subsequent occasions, easily recognisable by the scar on its left cheek, still hunting in the grass for voles but probably a bit more warily.

FIG 6. Stoat *Mustela erminea*, showing the straight division between cream and brown fur on the body, and the black tail tip.

The feisty Weasel would have been using the long grass for hunting and the nearby rocks for cover. In common with other small mustelids, Weasels also use linear features such as hedgerows and stone walls for shelter and will den in tree roots, hollow logs and the abandoned burrows and dens of their prey.

The Stoat is slightly larger (20–30 cm) than the Weasel and has a longer tail (5.5–13 cm) with a distinctive black tip. Actually, although it is always referred to as a 'tip', it makes up about one-third of the tail's total length. The Stoat's body is sandy brown in colour on the back and head with a cream belly, and the division between brown and cream fur along the flank forms a straight line (Fig. 6). The black tail tip is a key identifying feature, and if the live animal is seen, the gait is a characteristically arched-back, bounding movement.

In Scotland and the north of England and Wales, the Stoat turns white in winter, known as being 'in ermine', and in fact the Stoat is often referred to as the ermine when sporting its winter coat. This white fur was highly prized as a symbol of status and moral purity as, according to legend, the ermine would give itself up to the hunter and be killed rather than soil its white coat in the mud while trying to escape. In the milder south and in Ireland, Stoats stay brown in winter, although parts of the body may go white, and elsewhere in England and Wales, Stoats with piebald or mottled winter coats are regularly seen.

Being larger than Weasels, Stoats are more often seen above ground, which puts them at risk from birds of prey. However, the Stoat's black tail tip, which is present all year round, may be an effective defence against raptors.

Predator-deflection marks
Many small animals have contrasting spots of colour on their rear ends, which are believed to attract the eye of a predator and deflect its attack away from more

vulnerable body parts such as the head and neck. Roger Powell suggested that the black tail tip of Stoats might have the same function. The black tip is always conspicuous in Stoats, but especially in the winter, when it remains black in stark contrast to the rest of the coat, which turns white. In experiments with captive hawks and model stoats and weasels, he tested whether the black tail tip could be an impediment to birds of prey when hunting Stoats. He presented each hawk with a series of these models in turn, some of which had a black spot on their back, some on their tail, and some of which were plain. What he found was that the model stoats with tail spots and the model weasels with no tail spots (both of which were most like the real thing) were missed by the hawks much more often than the other models. It seemed that if the black spots were placed on the body, the hawks focussed their attacks on the spots, with the result that they nearly always caught either the stoat- or weasel-sized model. However, if the spot was placed on the long, thin tail of the model stoat, the hawks failed to grasp hold of it, and they also sometimes checked their attack at the last moment, as if they had not seen the rest of the model until then.

On the other hand, when the spot was placed on the short tail of the smaller, weasel-sized model, the hawks usually caught it because the rest of the body was close enough to be within reach of the birds' talons. The hawks took fractionally longer to notice and react to the model weasels with no spots, and often missed them. Powell concluded, therefore, that the black tail tip on Stoats is a classic predator-deflection mark, and that the smaller Weasels do not have it because their tails are too short for the mark to be far enough away from the body (Powell 1982). This begs the question of why Weasels do not have longer tails, so that a black tip would be a benefit. Powell suggests that they may be too small to keep a longer tail warm during the cold northern winters experienced across much of their range. Another explanation could be that Weasels are less exposed to raptor predation than Stoats because they spend so much of their time underground or under snow cover.

PINE MARTENS, POLECATS AND MINK

Pine Martens, European Polecats, Polecat-Ferrets *Mustela putorius* × *Mustela putorius furo* and American Mink all have the classic long, sinuous, mustelid body shape but are much larger and heavier than Stoats. They are also more strictly nocturnal, making it difficult to see and identify live animals with certainty. The smell that polecats can emit from their anal scent glands when threatened is particularly powerful and unpleasant.

FIG 7. The European Polecat *Mustela putorius* has dark brown guard hairs over buff-coloured underfur over most of its body, giving it a two-tone appearance, a dark facial 'bandit' mask, white fur on the muzzle and white ear margins. This pattern of coloration distinguishes it from the similar-sized American Mink *Neovison vison*, which is usually uniformly dark (although light-coloured variants were also bred for their fur and are occasionally seen in the wild).

The characteristic black and white (aposematic) markings warn predators that polecats are smelly and probably taste pretty repulsive. Anyone who has ever kept Ferrets *Mustela putorius furo* will be familiar with the scent. According to genetic (mitochondrial) analyses, Ferrets were domesticated from wild European Polecats around 2,500 years ago and are still sufficiently closely related for the two to interbreed and produce hybrid offspring (Davison *et al.* 1999). Consequently, polecats can look very similar to feral Ferrets, or to Polecat-Ferret hybrids (Fig. 8). However, in true European Polecats, the dark fur on the face extends to the nose, and there is clear contrast between the pale cheek patches and frontal band and the dark facial mask. Polecats have dark fur on their paws and dark guard hairs over the body, so even though the creamy underfur shows on the flanks, there are no white hairs present on the body. Polecats are absent from Ireland but there are feral Ferrets across the island, some of which look very like polecats.

FIG 8. The Polecat-Ferret hybrid *Mustela putorius* × *Mustela putorius furo* can appear very similar to the European Polecat *Mustela putorius*, but the presence of one or more of the following characteristics can be used to distinguish them to some extent: pale feet, indistinct facial band and white patches on the throat and chest. However, dark Polecat-Ferret hybrids can be difficult to separate from polecats in the field and the only certain way of distinguishing them is by DNA analysis.

FIG 9. Pine Marten *Martes martes*, showing the apricot bib.

The Pine Marten, at about the size of a small domestic cat, is the largest of the small mustelids. It is a chestnut brown, cat-like animal with large, rounded ears fringed with pale fur, a long, bushy tail and a distinctive creamy yellow to apricot chest and throat patch. The pattern of the chest patch, or 'bib', is unique to each Pine Marten and so this can be used to identify individuals.

The Pine Marten has relatively longer legs and a longer tail than either the American Mink or European Polecat and its movement when on the ground is characteristically loping or bounding. When startled or investigating, martens will sit upright on their hind legs like Meerkats *Suricata suricatta*. It is the species most likely to be seen in trees, although Pine Martens also spend much of their time on the ground. Although the large, creamy yellow to apricot throat patch is diagnostic, smaller, whiter patches are found on the throat of some American Mink and Polecat-Ferrets. American Mink and Stoats can also climb trees fairly well but are not normally arboreal, whereas Pine Martens are agile climbers.

FIG 10. American Mink *Neovison vison* are a uniform dark chocolate brown which may appear almost black at times. There is often a white chin patch present and sometimes white patches also on the chest, belly and groin. The tail is slightly bushy and approximately half the body length.

FIG 11. The Eurasian Otter *Lutra lutra* is semiaquatic and can be found in freshwater and coastal habitats. Otters have a thick, dark brown coat, short legs and a long, thick tail, tapering off to a point. The head and body of a male otter can measure up to 90 cm, with a further 40 cm-long tail. The females are slightly smaller, at up to 80 cm, with a 35 cm-long tail.

They rely on this agility to escape from predators by scooting up the nearest tree so they don't need to emit an off-putting smell. This has led to their alternative name of the 'sweet mart' as distinct from the European Polecat, which was called the 'foul mart' (among other things!).

The American Mink (Fig. 10) is a non-native species that has become widely established throughout Britain and Ireland as a result of escapes and some deliberate releases from fur farms (see Chapter 7). Mink are semiaquatic and usually found near water, where they may be mistaken for Eurasian Otters (Fig. 11). However, American Mink are substantially smaller than otters with a much thinner and shorter tail.

SMALL MUSTELID DISTRIBUTION

Although the focus of this book is the small mustelids found in the British Isles, it is worth taking some time to consider the family as a whole and where they came from. Mustelids are the most widely distributed carnivore family and can be found on every continent except Antarctica and Australia, although Stoats and Weasels have been introduced by humans to New Zealand with disastrous consequences for many native species, as discussed in Chapter 7. Worldwide, mustelids occupy an impressive range of habitats, from tundra and boreal forests through to deserts, grasslands and tropical forests.

All mustelids are carnivorous; however, the diets of different species vary highly from those that specialise on rodents, such as the weasels, through to the otters, whose diets consist of mainly aquatic prey, including fish, crustaceans and other aquatic invertebrates. At the other end of the spectrum are the generalists, such as the martens, whose seasonal diet can include high proportions of fruit when it is available. Adaptation to their varied habitats and diets can be seen in the shape and structure of the teeth and skeletons of different species. This diversification is a good example of adaptive radiation, the process by which species rapidly evolve from a common ancestor into a multitude of new forms filling different ecological niches. However, as a result of the wide diversity within the family, it has proved challenging to resolve the evolutionary relationships and history of the mustelids.

CHAPTER 2

On the Origin of Weasels: Taxonomy and Evolution

In the early attempts at classification, back in 1735, weasels were grouped together with martens, polecats and civets on the basis that they were all small carnivores with a similar slender body shape. However, it was not until 1817 that the German naturalist Johann Gotthelf Fischer von Waldheim formally named the Mustelidae (Fischer 1817). Since the early part of the twentieth century, taxonomists have sorted mustelids into various subfamilies using physical characteristics and similarities among species. In 1945, George Gaylord Simpson identified five subfamilies, which were the Lutrinae (otters), Melinae (badgers), Mellivorinae (honey badgers), Mephitinae (skunks) and Mustelinae (martens and weasels). This taxonomic scheme was not necessarily intended to reflect the evolutionary history (phylogeny) or relationships within the group, but it was used for almost 60 years as a framework to interpret mustelid biology and evolution.

With the development of molecular and genetic methods, studies on mustelid phylogeny based on DNA sequences began to be published. These found that skunks and stink badgers weren't mustelids at all, despite being similarly pungent. This resulted in them being elevated to a family of their own, the Mephitidae. In recent years, using multiple different methods, the mustelid family tree has been resolved into eight primary lineages (clades) that define eight subfamilies (Fig. 12). The resulting phylogenetic trees have transformed our understanding of mustelid evolutionary history and relationships among the family.

It is now agreed that modern mustelid lineages originated in Eurasia, which is consistent with the fact that the earliest-known fossil specimens of ancestral mustelids have been found in Eurasian deposits from the Late Oligocene, about

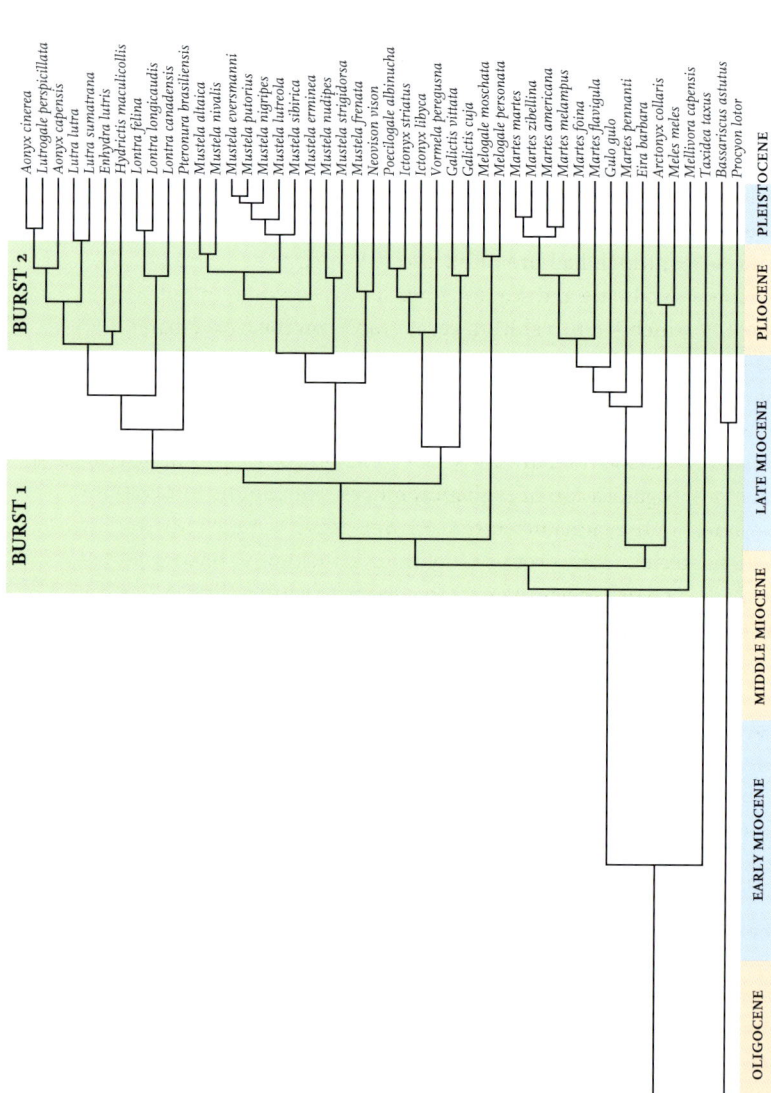

FIG 12. Chronogram of the Mustelidae. Vertical green bars indicate two bursts of diversification. Geological timescale is shown at the bottom. Redrawn from Koepfli et al. (2008).

33 million years ago (mya). A large majority of the mustelid diversity in North and South America is thought to have stemmed from lineages that repeatedly migrated out of Eurasia when it was still connected to other landmasses. The exact number and sequence of dispersal events, however, has been difficult to decipher because of gaps in the fossil record.

EVOLUTIONARY DIVERGENCE

There seem to have been two major bursts of evolutionary divergence within the mustelids that may have been driven by changes in global climate during the Miocene Epoch, which lasted from roughly 24 mya to about 5.3 mya. During the Middle Miocene, global warming created temperate climates throughout much of the world, which had profound effects on plants and animals. Further changes in climate and sea level increased overall terrestrial aridity and seasonality, which led to a shift in vegetation from the closed habitats of tropical and subtropical forests to more open woodland and grassland habitats. These changes in vegetation also had an impact on animal communities and may have led to diversification in a variety of groups, including mustelids, as a result of geographic isolation, divergent selection between different habitats and the creation or reorganisation of ecological niches. The Middle Miocene saw a major turnover in the fauna of western Europe (termed the Middle Vallesian 'crisis'), which affected many groups of mammals, including the carnivores. Almost half of carnivore species known to have gone extinct during this turnover were mustelids, and turnover in mustelids remained high throughout the Late Miocene. The combined effects of changes in habitats and extinction of earlier lineages may have created ecological opportunities that promoted the initial burst of diversification of modern mustelids. The onset of a cooler, drier climate during the Pleistocene (2.6–0.01 mya) then led to a dramatic expansion of low-biomass vegetation. In Eurasia and North America, this included grasslands, steppe and taiga. Diversification of four of the five *Martes* species (*M. americana*, *M. martes*, *M. melampus* and *M. zibellina*), which are all associated with taiga forest, occurred at the same time that this habitat was expanding. Linked to these environmental changes was a diversification of rodents and passerine birds. As these prey species diversified to exploit the new habitats, this also provided new niches for predators. It has been suggested that the burst of diversification of mustelids in the Pliocene may have been promoted by the diversification of prey species, as species in genera such as *Martes* and *Mustela* became specialised in hunting small rodent prey. It is thought that the evolution of small body size in *Mustela* was partly driven by adaptation to exploit the abundant resources provided by rodent diversification during the Pliocene.

HISTORY OF SMALL MUSTELIDS IN THE BRITISH ISLES

In the British Isles, as elsewhere across northern latitudes, a series of severe 'ice ages' (collectively called the Pleistocene) was interspersed with warmer intervals, the 'interglacials'. In Britain, there were four major glaciations: the Beestonian, Anglian, Wolstonian and Devensian. The interglacials are respectively known as the Cromerian, Hoxnian, Ipswichian and Flandrian, which is the one we are currently in. Mustelids are represented in the fossil record at various intervals in Britain. One of the earliest is the Clawless Otter *Aonyx reevei*, found in Pliocene/Early Pleistocene deposits in East Anglia. In West Runton, in deposits from the Cromerian interglacial, several modern species of mammal appear in Britain for the first time. These include a Weasel, a Pine Marten and a Eurasian Otter (Yalden 1999). The pollen found in the same deposits indicates that Norfolk at this time (approximately 563,000–478,000 years ago) was a wooded landscape but that it was open woodland rather than closed forest habitat. Pine Marten was also found at the later (Hoxnian) interglacial site at Swanscombe in Kent. Stepping forward to the Wolstonian glaciation, a cold-adapted mammal fauna is found that includes Woolly Mammoth *Mammuthus primigenius* and Reindeer *Rangifer tarandus* but also one mustelid, the Wolverine.

During the last (Devensian) glaciation, most of Britain and Ireland was once again covered in ice. Even in the extreme south, beyond the ice sheet, was bare

TABLE 1. The reappearance of extant mustelid species in Britain at successive interglacial periods.

	Cromerian	Hoxnian	Pre-Ipswichian 1	Pre-Ipswichian 2	Ipswichian	Flandrian (present)
Eurasian Otter	✓		✓	✓		✓
European Badger		✓			✓	✓
Pine Marten	✓	✓	✓			✓
Stoat		✓			✓	✓
Weasel	✓	✓				✓

Derived from Yalden (1999).

tundra with sparse vegetation. Very few of the mammal species found in the British Isles today could have survived these glacial conditions; probably only Weasel and Stoat, if there were lemmings to hunt, as well as Mountain Hares *Lepus timidus*.

This period was the last one in which Woolly Mammoths roamed Britain. There was also a giant polecat in Britain at this time. Originally regarded as a full species *Mustela robusta*, it is now thought to be simply a large, Late Pleistocene form of the more familiar *M. putorius*. First recognised in Germany, this was described in Britain for the first time from a site in Kent. The skull was about 12 per cent larger than modern European Polecats so these animals would have been able to feed on larger prey, such as the European Sousliks (or European Ground Squirrels) *Spermophilus citellus* that were also present in Britain at that time.

A WARMING WORLD

The Pleistocene ended about 12,000 years ago, after the Devensian glaciation, and the climate rapidly changed from an arctic coolness to the warm, temperate climate of the Mesolithic, or Middle Stone Age. In Britain, the land was essentially tree-covered, and the sparse human population now present survived largely by hunting and fishing. It has been suggested that much of lowland Britain at this time would have resembled the remnant primeval forest of Białowieża National Park in eastern Poland. Maroo and Yalden (2000) combined the results of studies of sub-fossil mammals and pollen analysis in Britain with recent work in Białowieża to estimate the likely nature of the Mesolithic mammal fauna of Britain. The resulting assessment is that there would have been about 22,000 Eurasian Otters, 110,000 European Polecats and perhaps 66,000 Stoats. They suggest that the Weasel would have been the most common carnivore at that time, numbering approximately 480,000, with around 147,000 Pine Martens, making that species the second most numerous. This is often cited to demonstrate the change in fortunes of the Pine Marten, which is now the second-rarest carnivore in Britain after the Wildcat *Felis silvestris*. (It is thought that in Mesolithic Britain, there were more than 66,000 Wildcats, putting them at fifth place in the league table of carnivore numbers.)

During the Late Glacial Period, Britain was still part of continental Europe since sea level was greatly lowered because of the quantity of water locked in the Earth's ice caps. However, as the climate warmed, so the ice melted, and sea levels rose. It is estimated that by about 8,000 years ago, Britain would have been cut off from Europe as the last link via the southern part of the North Sea disappeared

(Montgomery *et al.* 2014). This is quite soon after the ice had retreated and consequently there was relatively little time for animals of warmer climates and conditions to colonise or return to Britain. There are several species that apparently failed to reach Britain naturally or return to Britain under their own steam, so to speak, before they became isolated from continental Europe by rising sea levels. One of these, the Stone Marten *Martes foina*, has been the subject of some debate.

A tale of two martens

The Stone, or Beech, Marten is a common mustelid found throughout much of western Europe, where it occurs alongside the Pine Marten. The Stone Marten appears to be more tolerant of humans and is frequently found near (or even in) villages and houses, where it can cause problems for the human inhabitants to varying degrees. Stone Martens damage cars by chewing through cables and, twice (in April and November 2016), caused the Large Hadron Collider at CERN (European Council for Nuclear Research) to be shut down after climbing on electrical transformers located above ground. (Neither animal lived to tell the tale, but the second can now be seen in the Rotterdam Natural History Museum, where it was stuffed and put on display.)

There is still some debate as to whether the Stone Marten did once occur in Britain but was exterminated at some point in history by our ancestors. Pine Martens and Stone Martens are very similar in size, shape and skull morphology; however, there are subtle differences in their teeth (specifically the third upper premolar, the external face of which is concave in Pine Martens but convex in Stone Martens). The Pine Marten usually has a creamy yellow to orange bib, whereas that of the Stone Marten is whiter, but the colour does fade on older taxidermy specimens exposed to daylight. Bib colour is the most obvious distinguishing feature, but there are other physical and behavioural differences between the two martens. This is hardly surprising given that, according to genetic analyses, they diverged as separate species sometime between 2 and 3 million years ago.

One of the Pine Marten's physical adaptations to colder climates is that its paws are well furred on the underside and in between the pads (Fig. 15). This helps with warmth in winter and in snowy conditions. By contrast, the underside of the Stone Marten's paws is not furry. Another difference is that the exposed skin of the nose (the rhinarium) is dark brown in Pine Martens but pink in Stone Martens. In addition to the bib of the Stone Marten being white, so is the underfur of the coat (Fig. 16), whereas that of the Pine Marten is dark grey (Fig. 14), giving it a richer, more luxuriant appearance.

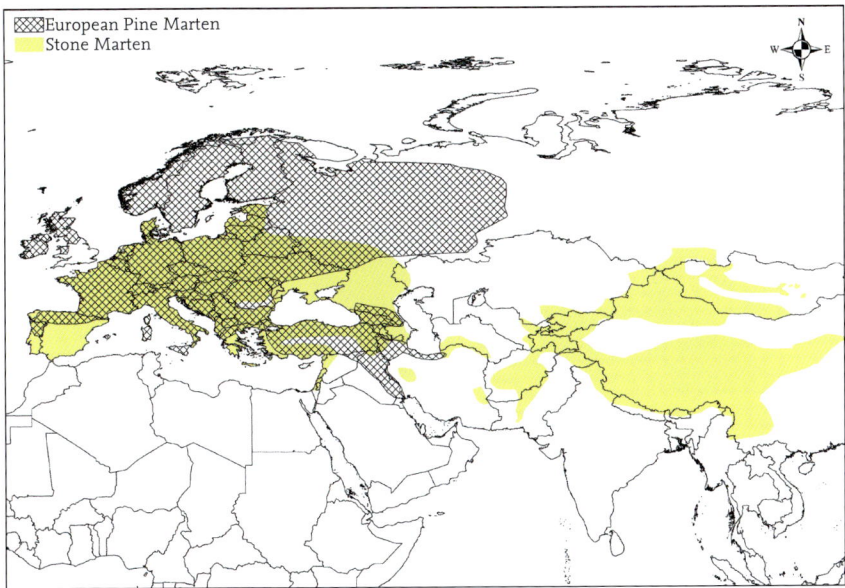

FIG 13. Native distributional ranges of Pine Marten *Martes martes* and Stone Marten *M. foina*. The distribution of the cold-adapted Pine Marten extends much further north than that of the thermophilic Stone Marten, which can be found as far south as Asia. International Union for Conservation of Nature (2022). Red List of Threatened Species (version 2022–23).

FIG 14. Pine Marten *Martes martes* in snow.

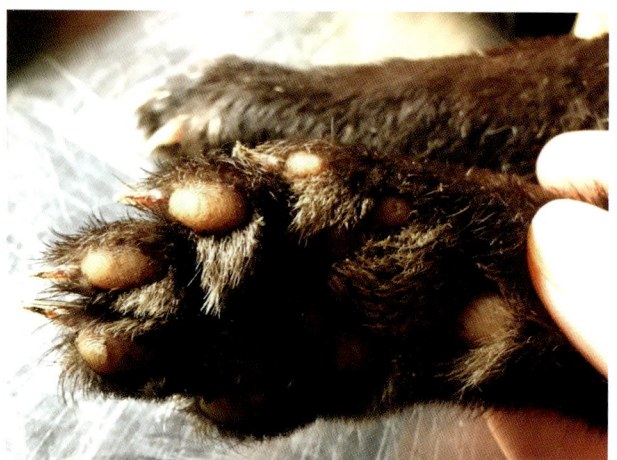

FIG 15. The underside of a Pine Marten *Martes martes* foot, showing fur between the pads.

The two species are sympatric (occur together) across much of continental Europe but, despite having similar diets of small mammals, birds and fruit, they have different habitat preferences. Pine Martens generally occupy extensive woodland and den in tree cavities, whereas, being much more tolerant of human disturbance, Stone Martens can be found in farmland, towns and villages, where they will often den in buildings with high levels of human disturbance.

Prior to the 1800s, naturalists recognised two types of marten in Britain, and described them as the 'white-breasted' and 'yellow-breasted' (Fig. 17) marten, suggesting the presence of both Stone and Pine Marten. There are a number of credible written references to the historical presence in Britain of two types of marten (including by Pennant in 1768, Fleming in 1828, MacGillivray in 1833–43, Goldsmith in 1840 and Bell in 1874). All of these accounts describe recognisable differences between the two species. Thomas Pennant, a Welsh naturalist writing in 1768, referred to the white-breasted 'martin' [sic] as *foina* and *huhss marder* (house marten) and described the yellow-breasted or 'wood' martin, *bela goed* (Welsh), as being much darker and with greatly superior fur. In 1828, John Fleming noted that the Pine Marten prefers wild situations, while the Stone Marten would approach houses and was easily tamed. In Poland's (1892) reference book about the international fur trade at the end of the nineteenth century, he suggested that although both species of marten had been present in Britain, the Stone Marten was by then probably extinct.

The idea that two separate species of marten were present in the British Isles persisted until 1879, when a paper was published which argued that sub-fossil evidence ruled out the presence of the Stone Marten (Alston 1879). The following

FIG 16. Stone Marten *Martes foina*, showing white bib and underfur.

FIG 17. Yellow-breasted Pine Marten *Martes martes*.

year, a Stone Marten was recorded from Craygord Woods in Kent, but this was dismissed as being undoubtedly an escaped animal. It has to be conceded that there are not many known sub-fossil specimens of British *Martes*; however, skulls from cave deposits in Britain all appear to be from Pine Martens, which are recognisable from the slight differences in dentition. Alston argued that the earlier accounts of white-breasted animals could be explained as faded museum specimens of Pine Martens or aged animals whose bibs had lost their colour. Taxidermy martens do fade over time, but this is not something that is seen in live animals.

Imported from Rome?
If Stone Martens were indeed once present in Britain, how could they have got here? Natural colonisation by this species is highly unlikely. The evidence suggests that Stone Martens persisted in a southeast Eurasian refuge during the last glaciation (Yalden 1999), and by the time they spread back into western Europe, the land bridge connecting Britain to the continent had gone. However, it cannot be ruled out that our forebears, with their proclivity for moving species around, imported some Stone Martens much later on. A couple of possible scenarios have been suggested which could explain how they might have arrived in Britain (Burton *et al.* 2018). One is that the species was imported during Roman times. This would coincide with the large-scale movement of other domestic and wild mammals into Britain which occurred from this period onwards. Stone Martens are allegedly easy to tame, and it is reported that both the ancient Greeks and Romans used them for rodent control before the domestication of cats. If some semi-tame Stone Martens were brought to Britain for this reason, subsequent escapes could have resulted in the establishment of the species in the wild. An alternative scenario relates to the high value of marten pelts to the early fur trade. By the fourteenth century, demand was such that the Pine Marten was becoming scarce in England, and skins were having to be imported from Scotland and elsewhere to meet demand. It is not beyond the realms of possibility that some landowners may have thought it prudent to import Stone Martens to release on their land to establish populations for harvesting. Both of these scenarios are plausible, but we shall never know for certain if either or both are true and if, as it is argued, Stone Martens were here but became extinct at some point during the mid-1800s.

THE FAUNA OF IRELAND

When we look at Ireland, it seems that even fewer species made it across before the land connection was submerged. Britain and Ireland are neighbouring

FIG 18. Irish Stoat *Mustela erminea hibernica*, showing irregular division between white and brown fur.

islands, but there are significant differences in their native fauna and flora. Britain has a set of species that, although restricted, is broadly similar to that of nearby parts of continental Europe. In contrast, Ireland is very species poor (47 terrestrial mammals and bats native to mainland Britain compared with only 36 in Ireland). However, Ireland does have some unexpected forms among those present. As is the case in Britain, few of the species currently present in Ireland would have been there during the Last Glacial Maximum (LGM) when both landmasses were predominantly covered by ice, so most of the colonisation has occurred since that time. This colonisation to mainland Britain was possible when it was still linked to the continent, as discussed above; however, the restricted native fauna of Ireland suggests that there was not such a long-lasting land connection between Ireland and either Britain or continental Europe. While lowered sea levels mean there would have been a land connection between Ireland, Britain and continental Europe up until about 16,000 years ago (the end of the LGM), there are doubts whether Ireland was connected to either landmass afterwards, which would explain the restricted biota in Ireland. There are no European Moles *Talpa europaea*, Common Shrews *Sorex araneus*, Water Shrews *Neomys fodiens*, Roe Deer *Capreolus capreolus*, Field Voles *Microtus agrestis*,

TABLE 2. Differences in the range of sizes between Stoats *Mustela erminea* in Britain and Ireland (means given with range in brackets).

Species	Sex	Head and body length (mm)	Tail length (mm)	Weight (g)
Irish Stoat	Male			
	County Down	252 (240–274)	85 (72–104)	233 (194–293)
	County Waterford	278 (267–288)	95 (88–105)	335 (302–369)
	Female			
	County Down	209 (184–221)	66 (57–73)	124 (100–161)
	County Waterford	230 (212–245)	70 (64–77)	165 (117–197)
Stoat (Britain)	Male	291 (260–318)	102 (67–119)	367 (252–471)
	Female	262 (244–278)	87 (69–100)	242 (180–303)

Data from Harris & Yalden (2008).

Water Voles *Arvicola amphibius* or dormice native to Ireland. From the mustelid family, there is no fossil evidence that either European Polecats or Weasels reached Ireland before the land bridge disappeared, and neither species is part of the present-day fauna. Of the mammals that do occur, the Irish Hare *Lepus timidus hibernicus*, a distinct form of the Mountain Hare, is present. The Irish Stoat is a smaller subspecies, and the division between the dorsal brown and ventral white fur is irregular, unlike the straight line of British Stoats.

These species, which are distinct enough to be recognised as subspecies, have presumably been isolated in Ireland sufficiently long enough for this to occur. Single fossil specimens of the Irish Hare and Irish Stoat have been dated to the Late Glacial Period. Given that both species are cold tolerant, they may have been able to survive conditions in Ireland at that time. Stoats are found over a wide range of temperature conditions, from warm to arctic (King & Powell 2007). They are currently found in the High Arctic of Greenland and Canada, where they feed on lemmings. It is known from fossils that lemmings survived in Ireland during the LGM, so there would have been a food supply for Stoats at that time. It has been suggested that Stoats are sufficiently cold tolerant to have survived close to the British–Irish ice sheet at the LGM and therefore they were present on the exposed landmass of Ireland, having colonised at that time or earlier. The Stoats in Ireland would then have become isolated as sea levels rose.

CHAPTER 3

Differences and Similarities, Coexistence and Competition

On the island of Ireland, where the contemporary fauna includes only two small mustelids, the Irish Stoat and the Pine Marten, competition for resources between the two species is less likely than in mainland Britain, where the small-carnivore guild is much more crowded.

However, despite the fact that Britain is home to Weasels, Stoats, European Polecats and Pine Martens, due to the variety of feeding and habitat specialisations across the different species, direct competition is avoided, for the most part. Whilst the small mustelids in Britain are superficially very similar, with the same long, thin body shape and carnivorous diet, they have coevolved

FIG 19. The Irish Stoat *Mustela erminea hibernica* has fewer small carnivore competitors in Ireland than Stoats *Mustela erminea* on mainland Britain.

TABLE 3. Measurements and weights of small mustelid species in Britain (means given with range in brackets).

Species	Sex	Head and body length (mm)	Tail length (mm)	Weight (g)
American Mink	Male	397 (330–450)	193 (150–220)	1,153 (850–1,805)
	Female	338 (320–360)	168 (135–190)	619 (450–810)
European Polecat	Male	398 (330–450)	149 (125–165)	1,111 (800–1,710)
	Female	367 (335–385)	133 (125–145)	689 (530–915)
Pine Marten	Male	494 (480–520)	248 (225–270)	1,890 (1,600–2,150)
	Female	440 (410–460)	231 (220–240)	1,330 (1,100–1,450)
Stoat	Male	291 (260–318)	102 (67–119)	367 (252–471)
	Female	262 (244–278)	87 (69–100)	242 (180–303)
Weasel	Male	216 (195–248)	49 (32–62)	125 (81–85)
	Female	184 (175–194)	39 (35–46)	68 (48–107)

Data from Harris & Yalden (2008).

with sufficient differences between their ways of life to allow them to coexist, despite apparent overlaps in these ecological niches.

It would be expected that there is some competition between these small mustelids as they have broadly similar diets (small mammals and birds) and there is size overlap between at least some pairs of species in Britain.

SPECIALIST PREDATORS

If you take the example of Stoat and Weasel coexistence, the Weasel, being the smaller of the two species, has an advantage under certain scenarios. Weasels are better able to hunt small rodents in their burrows, even when they are scarce, and can breed rapidly when there is a glut of prey. Stoats, however, as a result of their larger size, can win in aggressive encounters or direct competition with Weasels. They also have the advantage of being able to take bigger alternative prey when rodents are scarce. Variation in the environment means that the balance of these advantages (and therefore which species has the upper hand) is constantly changing.

Stoats and Weasels are the two most similar-looking species and the pair in which interspecific competition and coexistence have been widely studied and discussed. Both species are highly variable in body size and, as do most mustelids, show pronounced sexual dimorphism, with males being approximately a third

FIG 20. Weasel *Mustela nivalis* x-ray showing the flexibility of the spine.

larger than females of the same species. In Britain, male Stoats are always larger than male Weasels, but male Weasels and female Stoats can be very similar sizes, as shown from the range of measurements in Table 3. Both Stoats and Weasels evolved as specialist predators of small mammals and all aspects of their ecology and behaviour are linked to their abilities to exploit this plentiful but unstable resource. However, Weasels are the supreme rodent specialists: their small body size means that they can easily pursue rodents down tunnels and their breeding strategy enables them to take full advantage of a temporary abundance of prey, such as cyclical peaks in vole numbers.

Mice and voles feature in the diets of all of the small mustelids in varying frequencies, but the Weasels and Stoats, in particular, are impressively efficient rodent predators. Indeed, their genus name of *Mustela* translates literally as 'mouse' (*mus*) 'spears' (*tela*), but this is exemplified in Weasels '*Mustela nivalis*' (mouse spears of the snow). Everything about the Weasel is designed for hunting small prey in dark, confined spaces. Whereas in other carnivores the legs are about the same length as the body, a Weasel's legs are only a third to half its body length. This means that they can swing through their full range of movement in a very small space and do not have to be folded, which would hinder their movement. The spine is flexible, with thin spinal processes that enable the vertebrae to articulate in such a way that the Weasel can turn in a tight space by rolling over and then walking back over its own rear quarters. Quite the party trick! In her (excellent) book, Carolyn King describes having seen a weasel 'leap into a hole and then look out again in a single fluid action so fast the tail was not in before the nose came out again' (King & Powell 2007). A Weasel is small and light and built in such a way that the shoulders and hips do not have to carry any

FIG 21. Weasel *Mustela nivalis* carrying a vole with ease.

great weight. This adds to the sinuous, streamlined character of its skeleton. The neck is long, so prey can be carried in the Weasel's mouth without getting in the way of its front feet (Fig. 21), and the skull is the widest part of the skeleton, so that if a Weasel's head will fit through a space, then so will the rest of it.

The skull is strong and relatively heavy, providing a large surface area for attaching the formidable jaw muscles. The jaws are short, which maximises the amount of force from the powerful temporal muscle. The canine teeth are long and sharply pointed to concentrate that force into a small area for a deadly killing bite; truly, as Carolyn King calls them, 'hair trigger mouse traps with teeth' (King & Powell 2007). Weasels (and Stoats) have a relatively higher bite force, pound for pound, than Lions, bears and other much larger predators. Scaled up to the size of a Tiger *Panthera tigris*, a Weasel would be a truly terrifying prospect.

It is thought that the early Weasels descended relatively recently from larger, marten-like mustelids, becoming smaller over time to better exploit the new niche as predators of voles, mice and lemmings. Their smaller size and other adaptations for getting into the runways and burrows of rodents meant that during cold phases of the Pleistocene, the Weasels were readily able to burrow through snow, to find shelter from the extreme temperatures at the surface, and gain access to the underground tunnels of their prey. There is evidence, both

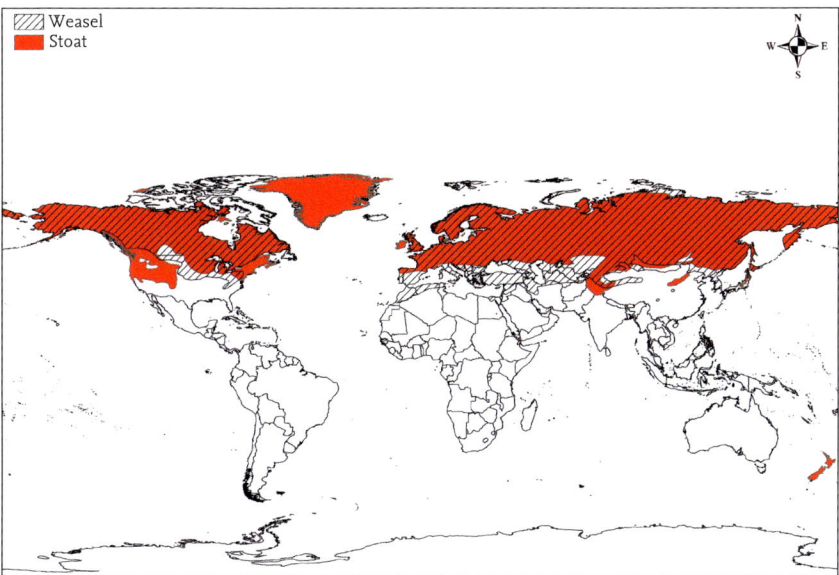

FIG 22. Global distribution of Weasel *Mustela nivalis* and Stoat *M. erminea* (n.b. both are introduced into New Zealand). Much of the huge modern-day geographic ranges of the Weasel and Stoat are within climatic zones which are characterised by severe winters with prolonged periods of snow cover. International Union for Conservation of Nature (2016). Red List of Threatened Species (version 2022–23).

from past and contemporary Weasels and Stoats, that small size is advantageous to these animals in cold climates. Fossil Stoats *Mustela palerminea*, the ancestor from which modern Stoats are descended, show variation in size between the colder periods of the Middle Pleistocene, when they were smaller, and warmer phases, when they were larger. Fossil Weasels found in Poland show a similar trend, with those dated from the last glaciation being smaller and those from the Eemian interglacial (c.120,000 years before present) larger.

THE WEASEL WAY OF LIFE

Paradoxically, there are serious consequences of being a small animal in a cold climate that are not entirely compensated for by the advantage of being able to shelter under the snow. Animals have to achieve a balance between the energy they gain from their food and the energetic costs of getting it. Mammals have to use energy just to maintain a constant body temperature, and heat is lost across their surface. Small animals and long, thin animals have a relatively large surface area to volume ratio, and therefore both the size and shape of a Weasel are hugely inefficient in physiological terms, which imposes a real cost to living the Weasel way of life. Their long, thin shape is ideal for hunting through small spaces and burrows, but the result is that Weasels are very susceptible to the cold. Their body temperature is 39–40 °C, and their body heat escapes easily, not just because of their shape, but also because they cannot afford to insulate themselves with a thick coat or too much fat. Layers of fur or fat might impede their sinuous movement or restrict their access to smaller vole runways and burrows. Weasels moult twice a year, in spring and autumn, but their winter coat is not much warmer than their summer coat, and fat has to be confined to any dips in the body outline. Even a well-fed Weasel in peak body condition manages to retain an enviably sleek figure. The cold northern winters that are a feature across much of the Weasel's range are periods of real energetic stress: just maintaining body heat at rest uses twice as much energy as in summer. In winter, Weasels rely on having a thickly insulated nest, which they often take over from recent prey and, adding insult to injury, improve by lining with the fur from that prey. The thickness of the lining is a good indication of how long the Weasel has been in residence.

But however safe and warm a Weasel's nest may be, eventually hunger will mean it has to venture out into the cold again and use up more energy for hunting. Weasels have large appetites to satisfy the huge amounts of energy they need: weight for weight, weasels have one of the highest food requirements of all carnivores. Weasels in captivity eat between a quarter and a third of their body weight each day, and it is likely that more active wild ones need even

FIG 23. Weasel *Mustela nivalis* in a burrow.

more. Compare this with Red Foxes *Vulpes vulpes*, which only need to eat about 8 per cent of their body weight each day. To meet this energy requirement, Weasels have to eat often, around 5–10 times a day, and so need to find food at frequent intervals. In cold temperatures, a significant proportion of their food intake is burnt up just to keep warm. A Weasel is a solitary hunter, darting from one patch of cover to another, criss-crossing irregularly, investigating every small hole or tunnel it comes across and occasionally pausing to stand up on its hind legs and look around from a vantage point. These alert, rapid movements are a physical manifestation of the animal's constant hunger and 100-mile-an-hour metabolism. A Weasel's resting pulse rate is about 400–500 beats per minute, compared with the Stoat's 360–390 beats per minute (and a human's of between 60 and 100). This places a great load on the heart, which in Weasels is large relative to their body weight. A Weasel's digestive system is adapted to deal with food coming in large, infrequent packages, and meals pass quickly through the short gut. Studies using dyed bait fed to Weasels show that this reappears in 2–4 hours and the defecation rate is high, averaging 19 scats (faeces) per 24 hours (Gillingham 1984). Weasels cannot afford the luxury of taking a long nap after a meal.

BREEDING STRATEGIES OF WEASELS AND STOATS

The Weasel's 'live fast, die young' strategy works well for an opportunist species that can exploit a wide range of habitats and whose main rodent prey can be extremely plentiful in some years. One of the keys to success for such species is that they have the capacity to respond quickly to any local improvement in environmental conditions. Populations of many small mammal species fluctuate more or less regularly, with cyclic peaks and troughs of different frequency and amplitude. When there is a sudden increase in the population density of voles, adult Weasels are rapidly able to produce more young. The characteristics of Weasels that make this possible and contribute most to a high rate of population increase are early maturity, more than one litter per year, large litter size and more than one litter per adult female's lifetime. Female Weasels are mature at 3–4 months old and can breed in their first year. Mating takes place in February, and kits are born just over a month later (after a gestation of around 36 days). There can be up to eight kits in each litter and once they are weaned, usually around May, the female can breed again before the autumn, as can her daughters. While prey is abundant, this increases the young Weasels' chances of surviving to become adults, and the adults' chances of surviving long enough to breed

FIG 24. The Stoat's *Mustela erminea* spine is similarly flexible to that of a Weasel.

again. Weasels can adjust their breeding effort almost immediately, and so their breeding success is very closely linked to the density of rodents. When prey availability is high, Weasel population densities can reach 0.2–1 per ha.

Small rodents are also prey for Stoats although, being larger, Stoats are excluded from the smallest burrows; however, they are also able to switch to bigger alternative prey when small rodents are scarce. Stoats have one disadvantage when it comes to exploiting a temporary glut of rodents and that is their breeding cycle includes a period of delayed implantation, or embryonic diapause. Mating in Stoats takes place between April and July, and then the fertilised egg is held dormant in the uterus until the following spring. Only then does the egg implant into the uterine wall and begin to develop into a pregnancy. This means that productivity in Stoats varies according to the availability of prey in the year of implantation, and not the year of fertilisation. Stoats only have one litter per year whereas, in years when small rodents are abundant, adult female Weasels can breed a second time. Young female Weasels are also able to breed in the season of their birth. This means that the potential productivity (and rate of population increase) of Weasels is considerably higher than that of Stoats. By the end of a good breeding season in Britain, the number of offspring produced by a single female Weasel could be about double the maximum of a female Stoat.

This lower annual productivity of Stoats is compensated for, to some extent, by having a slightly longer lifespan than Weasels. Although in the wild Stoats will rarely survive more than a year and a half, they can live up to 8–9 years. Weasels have a short lifespan of up to three years at most, but the great majority of them will die well before the end of their first year. In addition, although male Stoats are not sexually mature until their second year, females are able to be mated at 2–3 weeks old, while they are still blind, deaf and hairless in the nest. This has huge

FIG 25. In spite of its short legs, a Weasel *Mustela nivalis* can carry large rodent prey with ease.

benefits for any adult male that can get access to them, as he has the opportunity to mate with several females at the same time – the adult and her daughters. The advantage for the females is that they are fertilised before they disperse in midsummer. High population turnover and a lack of pair bonding probably reduces the chances of an adult male Stoat mating with his own offspring. This is a remarkably efficient breeding system and means that the young females are assured of a litter the following season even if there has been unusually high mortality, or a female has colonised a new area and there are few or no prospective mates around at that time. This (and their surprising abilities as swimmers) is one reason why Stoats have been highly successful at colonising islands.

HUNTING LARGER PREY

While Weasels of both sexes can go down all but the very smallest vole burrows, which measure 22–28 mm in diameter, the larger Stoat is more limited, which is why it is a less effective rodent hunter than the Weasel. Nonetheless, by virtue of their size, Stoats have a much more flexible diet than Weasels, and so they are able to adapt in times and/or habitats where small mammal prey is relatively scarce. Stoats are thought to have originally specialised in feeding on larger rodents such as Water Voles, and still feed mainly on these where they occur.

FIG 26. Stoat *Mustela erminea* with a recent European Rabbit *Oryctolagus cuniculus* kill, showing the relative size difference between predator and prey.

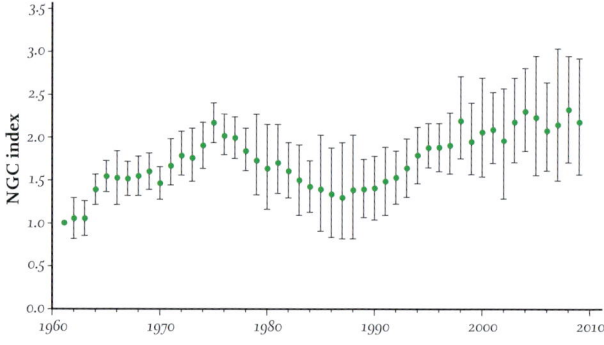

FIG 27. Index of bag density for Stoat *Mustela erminea* from 1961 to 2009. Redrawn from the Game and Wildlife Conservation Trust's National Gamebag Census.

Some years ago, I observed this directly while carrying out some research on causes of mortality in Water Voles at Shapwick Heath and Ham Wall nature reserves in Somerset (MacPherson & Bright 2010). These are non-linear wetland sites (a reedbed and a grazing marsh), where Water Voles persist despite the long-term presence of American Mink. Stoats were seen regularly, including on one memorable occasion when a Stoat swam across a ditch, after one of the Water Voles we were radio tracking. The Water Vole was found dead, minutes later, having been killed by the characteristic small mustelid bite to the back of the neck – what you might call a 'smoking gun'. Water Voles were once much more common in Britain than they, sadly, are now. However, in Britain and elsewhere where Water Voles are scarce or absent, rabbits are often the major prey of Stoats (Fig. 26). A Stoat hunting a rabbit is an impressive sight. Those who have watched it say that a Stoat will often appear to select one particular rabbit and will then follow that individual relentlessly until the kill. It is reported that sometimes the rabbit will appear to give up and freeze, like the classic 'rabbit in the headlights', in which case it is possible that the rabbit is literally frightened to death, before any physical injury occurs. This is supported by examination of a number of European Rabbits killed by Stoats, which showed no evidence of serious injury, leading their examiners to the conclusion that the rabbits had died of fright. The Stoat's apparent ability to frighten to death an animal much bigger than itself is remarkable and, considering that an adult rabbit can weigh 1,200–2,000 g, whereas a Stoat will weigh a maximum of around 445 g, this tactic might help to reduce the risk of it being injured by the struggles of the larger prey.

As a result of their reliance on rabbits, Stoat numbers were severely reduced following the introduction of myxomatosis in rabbits in the 1950s. There are no national surveys carried out for Stoats and Weasels, so data are from the Game and Wildlife Conservation Trust's (formerly Game and Wildlife Conservancy) records of numbers killed each year on game estates, but these showed a severe decline in numbers of Stoats killed in the years after the initial outbreak of the disease (Tapper 1992). In contrast to Stoats, Weasel abundance increased during and after myxomatosis, probably as a result of reduced rabbit grazing, increased rough grassland and increased abundance of Field Voles. Indices of the numbers of Stoats killed then increased steadily from the 1960s alongside rabbit recovery, although there was another dip in the 1980s (Fig. 27).

Striking a balance

The coexistence of similar species like Stoats and Weasels is possible because the ecological overlap between them is counterbalanced in this way, so that one does not always displace the other. In Britain, both Stoats and Weasels live in almost any habitat and are active at any time, but these different breeding and feeding strategies associated with the difference in their body size reduce competition between them. In summary, Weasels are more efficient at exploiting small rodent prey and are able to breed rapidly to take immediate advantage of temporary peaks in rodent numbers. However, this specialisation makes them vulnerable to local extinction during rodent declines. The Stoat has a more generalised diet and, being larger, is probably the dominant species in direct competition, but is limited by delayed implantation to producing only one litter a year. The Weasel therefore has the advantage in exploitation competition, where the prey resource is used directly and is no longer available, and the Stoat has the upper hand when it comes to interference competition, where the larger species physically prevents others from accessing prey. There are published studies showing that both direct aggression and scent communication from the larger and socially dominant Stoat can exclude Weasels from some habitats with high prey densities. This interference occasionally takes the form of predation, confirmed by evidence of Weasel remains in Stoat scats.

All of these factors mediate competition between Stoats and Weasels, so that they can coexist, but fluctuations in population sizes of the two species are common. Studies show that each species in turn can become locally extinct for various periods of time. Niche overlap between species only results in competition when resources are scarce, and therefore Stoats and Weasels can coexist when prey populations are high, but when prey populations decrease, it is likely that one or the other species will gain the upper hand. Initial population sizes and predation abilities as well as the availability of different prey determine whether it is Stoats or Weasels that go extinct locally, whereas it is the size and connectivity of habitat and proximity to the nearest population that dictate if or when recolonisation by the eliminated species occurs. In large, heterogeneous environments with a range of fluctuating prey species, Stoats survive by switching between prey sizes. Weasels do not usually continue to hunt prey at low densities unless they have no alternative, so local declines of voles can lead to the emigration of Weasels. On small islands, where the choice of habitats is limited and emigration is not an option, this can be a problem. On Terschelling, a small (680 km^2) island off the coast of the Netherlands, over 100 Weasels and fewer than 10 Stoats were introduced in 1931 with the aim of controlling Water Voles, which had reached such high numbers that they were damaging trees and gardens.

FIG 28. Weasels *Mustela nivalis* are supremely adapted rodent predators.

Despite the fact that many more Weasels than Stoats were released, by 1934, the Weasels had gone extinct while the Stoats had increased in number to at least 180. By 1937, there was still a fluctuating but substantial population of Stoats present, but the Water Voles were extinct (Van Wijngaarden & Morzer Bruijns 1961). There was plenty of small prey for Weasels on Terschelling but, on such a small island, it is thought that interference competition from Stoats was unavoidable. Without the Water Voles, the Stoats could not survive either and they, too, are now extinct on Terschelling (Mulder 1990). Ecological theory suggests that strict carnivores, such as Stoats and Weasels, are only able to establish permanent populations on relatively large islands and that ecologically similar species are unlikely to be found together on islands below a size threshold.

ISLAND POPULATIONS OF STOATS AND WEASELS

The effect of island size is reflected in the distribution of Stoats and Weasels on the offshore islands around Britain. Stoats are established on Islay, Jura, the Isle of Wight, Mull, Sheppey, Skye, Anglesey (Ynys Mon) and Holy Island, Raasay and Bute. Stoats were introduced as early as the seventeenth century to Mainland Shetland, Muckle Roe, Burra and Trondra. There are also unconfirmed records from Lismore and Luing, as well as recent records from Scalpay (Skye) and Canna. There are old or occasional records from Great Cumbrae, Handa, Iona, Orkney, Rhum, Staffin and Colsay, Haaf Gruney, Whalsay and Yell (Shetland).

The Irish Stoat occurs on the Isle of Man as well as across the island of Ireland. The Stoat is not recorded on Rathlin, Arran, Coll, Colonsay, Gigha, Scarba, Tiree or the Outer Hebrides.

Weasels are established only on Sheppey, Skye, Anglesey (all of which have bridges to the mainland), the Isle of Wight and Bute. There have also been recent records from Scalpay. There are old records from Mull, Jura, Raasay and Islay but Weasels have not been seen on these islands recently, suggesting that they may now be extinct. There have been occasional sightings on Brownsea, Ramsey, Hilbre, Holy Island and Tiree but these are most likely of 'visiting' animals (in the case of Holy Island, which has a causeway at low tide) or individuals that were accidentally introduced.

In 1979, King and Moors looked at the distribution of all the small mustelids on the islands around Britain and concluded that the approximate minimum conditions required for a British island to support a persistent population of mustelids were at least four species of mammalian prey and an area larger than 90 km^2 (King & Moors 1979a).

TABLE 4. Islands around Britain on which Stoats *Mustela erminea* and Weasels *M. nivalis* have been recorded.

	Area (km^2)	Stoats	Weasels
Anglesey (Ynys Mon)	714	Established	Established
Brownsea	2		Possible visiting
Burra and Trondra, Shetland	2.75	Introduced	
Bute	122	Established	Established
Canna	11.3	Recent records	
Hilbre	0.05		Possible visiting
Holy Island	5		Possible visiting
Islay	620	Established	Old records
Isle of Man	572	Established	
Isle of Wight	381	Established	Established
Jura	367	Established	Old records
Lismore	24	Unconfirmed	
Luing	143	Unconfirmed	
Mainland Shetland	969	Introduced	
Muckle Roe, Shetland	18	Introduced	
Mull	875	Established	Old records
Raasay	64	Established	Old records
Ramsey	2.6		Possible visiting
Scalpay (Skye)	6.53	Recent records	
Sheppey	94	Established	Established
Skye	1,656	Established	Established
Tiree	78.34		Possible visiting

Raasay and Scalpay are smaller than Skye but are in close proximity to it. Raasay is a little over 900 m from Skye at the narrowest point of the Sound of Raasay, and Scalpay is a swim of only around 300 m from Skye. Similarly, Muckle Roe, Burra and Trondra are small islands, but all are connected by bridge to Mainland Shetland.

Both Stoats and Weasels have been recorded on some of the smaller islands around Britain in the past, but records suggest that they do not persist for long. There are historical records of Stoats on Whalsay (20 km^2), Iona (10 km^2), Handa (less than 1 km²) and Colonsay (less than 1 km²), and Weasels were once present on Islay, Jura, Mull, Raasay and Bute (Corbet 1966, King & Moors 1979a, Harris & Yalden 2008).

On islands with no voles, it seems that Weasels fail to establish, but even when voles are present, the irregular fluctuations in their populations and therefore availability as prey mean that Weasels are vulnerable to extinction in low vole years. The Isle of Wight, at 380 km², is the smallest of the British islands with both Stoats and Weasels. Ireland is considerably larger than the Isle of Wight, and yet there are no Weasels in Ireland. This was described by Corbet (1966) as 'something of a zoogeographical mystery'. The explanation for this is based on the assumption that following the last glaciation (c.12,000–8,000 years before present), when the British Isles were recolonised by temperate species, Stoats arrived first and were therefore able to reach Ireland and some of the Scottish islands before rising sea levels cut them off. It is inferred from its absence that the Weasel missed this opportunity and arrived much later. However, King and Moors (1979) point out that there is no reason why the Weasel, a species equally as well adapted to the cold as the Stoat, should not have returned to Britain at the same time as the rest of the arctic fauna. There is currently no evidence of Weasels from the fossil record of arctic animals that were present in Britain during the last glaciation in the Late Devensian. However, mustelid fossils are notoriously rare, and absence of evidence is not necessarily evidence of absence. Weasels and Stoats were both present at that time in Poland, which is at the same latitude as Britain and would have been connected to it via Germany, Holland and the bed of the North Sea. Therefore, King and Moors (1979) suggested that Weasels may have been present in Ireland in the early postglacial period, along with Stoats and the other cold-adapted species such as lemmings. Both Field Voles and Bank Voles *Myodes glareolus* are less cold-adapted species than lemmings and did not arrive in Britain until after it was cut off from Ireland. Consequently, when lemmings went extinct in Ireland, they were not replaced by voles. If there were Weasels in Ireland at this point, they would then have lost their competitive advantage over Stoats and probably disappeared shortly after. However, there is no evidence for this anywhere in the fossil record (Yalden 1999).

MINK, POLECATS AND MARTENS

Even when resources are not limited, coexistence of small mustelids can break down in the face of a non-native competitor such as American Mink. Mink can have a significant impact on populations of Water Voles, a preferred prey of Stoats, and also eat rabbits where they are present, so are an additional competitor for both Stoats and European Polecats. One study in Poland showed that numbers and distribution of Stoats declined following the invasion of the American Mink into a forested area of northeastern Belarus (Sidorovich & Macdonald 2001). This was observed over a long period of time so was not likely to be merely a temporary response to a short-term decrease in prey. Since the

FIG 29. American Mink *Neovison vison* can take a wide range of prey, such as European Rabbits *Oryctolagus cuniculus*, when they are available.

FIG 30. European Rabbits *Oryctolagus cuniculus* are also an important prey for European Polecats *Mustela putorius* in Britain.

Stoat population declined, Weasels began to populate marshland habitats, from which, it was thought, they had previously been excluded by Stoats.

In common with Stoats, European Polecats in Britain are also very reliant on rabbits as prey. Polecats will also often make use of rabbit burrows as den sites. In Britain, they utilise a wide range of habitats from lowland farmland and woodland to wetlands, marshland and other riparian habitats. They show a preference for dense cover and shelter, although habitats that support high densities of rabbits or rats are favoured. Elsewhere in Europe, they are described as semiaquatic and are predominantly a wetland species with a diet that is high in amphibians. In Britain, seasonal variation in polecat diet tends to reflect the changing availability of prey. Amphibians are consumed most often in spring, with lagomorphs (mainly rabbits) in the summer months and small mammals in autumn and winter.

European Polecats are bigger than Stoats, yet slightly smaller than Pine Martens. Polecats are usually nocturnal although they are occasionally active during daylight hours. Their eyesight is not their most well-developed sense, but they make good use of their ears and nose to search for prey, sniffing the ground as they amble along. They are not good climbers, and the polecat is not known for its speed over open ground, but it can hunt very efficiently underground, where it is perfectly adapted for exploring the tunnels and chambers of a rabbit burrow. Using its finely honed sense of smell, it follows scent trails through the darkness towards the occupants, and, once upon them, the polecat has the element of surprise. Where rabbits are common, polecats spend a lot of time searching underground in rabbit burrows. The burrows are also used as convenient daytime resting sites. Once a polecat has killed a rabbit, it will often stay underground for a few days while it eats the kill in the warmth and safety of the burrow. One assumes that the other inhabitants would seek alternative accommodation during this time. Polecats can live up to 14 years and both sexes can breed from their second year. Mating takes place in early spring (February–April), with ovulation induced by mating. Like the slightly larger Pine Martens, polecats have only one litter a year, between May and June, but unlike Pine Martens, polecats have large litters (commonly between five and ten kits) and no delayed implantation.

The Pine Marten is the only small mustelid in the genus *Martes* that occurs in Britain and Ireland and it is differentiated from the other mustelids by aspects of its behavioural ecology, morphology and life history. Pine Martens have evolved to occupy a semi-arboreal (tree-dwelling) niche and they have specific adaptations which enable them to exploit a woodland environment. These include long, powerful forelimbs with strong muscles for climbing, and proportionally large feet. Each of the five toes ends in a markedly curved claw for gripping onto tree bark. These claws are sharp and semi-retractable like those of cats, to protect

TABLE 5. Mating and breeding cycles of small mustelids.

Species	Mating occurs	Births	Delayed implantation (days)	Litters/year	Kits/litter	Gestation (days)
American Mink	Feb–Mar	May	✓ (40–70)	1	4–6	28–31
European Polecat	Feb–Apr	May–Jul	✗	1	5–10	40–43
Irish Stoat	May–Jul	Early Apr	✓ (c.270)	1	5–12	c.28
Pine Marten	Jul–Aug	Feb–Mar	✓ (c.230)	1	1–4	c.30
Stoat	Apr–Jul	Apr–May	✓ (c.270)	1	6–9	c.28
Weasel	Apr–Jul	Apr	✗	1–3	1–6	34–37

FIG 31. Pine Marten *Martes martes*, showing ankle rotated through 180 degrees.

them from wear whilst running on the ground or anywhere else that might blunt them. Their ankles, in common with other tree-climbing species such as squirrels and dormice, can rotate through a full 180 degrees (Fig. 31), which enables martens to maintain their grip from all four feet when climbing downwards. Pine Martens have excellent vision, essential for navigating through a complex, three-dimensional environment and finding prey in the trees as well as on the ground.

In common with the other mustelids, Pine Martens are carnivores. They are, however, highly omnivorous and will eat whatever is locally and seasonally most abundant and easy to obtain. For most of the year, this is small mammals, which, in Britain, include Field Voles, Bank Voles and Wood Mice *Apodemus sylvaticus*. Although not a specialist predator of birds, they will also take adult birds, nestlings and eggs if they come across them. Dietary studies have shown that these tend to be those of the more common woodland species like Wood Pigeon *Columba palumbus*, Blue Tit *Cyanistes caeruleus* and Great Tit *Parus major*. Large-bodied invertebrates can also make up a large proportion of the diet at certain times of the year, and in autumn, when fruit is available, there is evidence that martens will preferentially switch to eating berries such as Bilberry *Vaccinium myrtillus*, Rowan *Sorbus aucuparia*, Blackberry *Rubus fruticosus* and Wild Cherry *Prunus avium*.

Pine Martens are slow breeders but are potentially one of the longest-lived of the British and Irish mustelids. I knew one captive-bred female, called Poppy, who lived for around 20 years in captivity, but it is likely that a marten would count itself very lucky if it reached the age of 10 years in the wild. Pine Martens do not usually breed until their third year. Males are not sexually mature until they are three years old. Females can mate in their second year, when they are approximately 18 months old, but, like Stoats, the reproductive cycle of Pine Martens involves a lengthy embryonic diapause. This means that while mating takes place in late summer, births do not occur until early the following spring (March–April). Fertilised eggs are dormant over winter and then around February they start to develop. Each fertilised egg becomes a rapidly dividing ball of cells, known as a blastocyst, and implants into the wall of the uterus where gestation (pregnancy) begins. In Pine Martens, gestation lasts for approximately 30 days.

DELAYED IMPLANTATION: MATING THEN WAITING

Delayed implantation (DI) is found in some other mammals, but it is particularly prevalent among the mustelid family. Around 34 per cent of mustelid species have DI, compared with fewer than 0.05 per cent of mammals overall. One potential advantage of DI is that it means the timings of mating and birth can be uncoupled. This frees animals from having to time mating and parturition to coincide with optimal external conditions, and each can take place at independently favourable times of year. This uncoupling effect of DI enables young to be born earlier in the season, giving them the maximum amount of time to develop before the challenge of their first winter. This is supported by studies which show that species with DI tend to give birth earlier than those without (Table 5). This is likely because increasing day length, the trigger that stimulates implantation, is also what initiates the mating season in many species, and so delayed blastocysts implant at the same time as mating begins in species with direct development. The exception to this is the Pine Marten, where mating occurs in late summer. This would have the greatest benefit in strongly seasonal environments and, indeed, a study by Thom *et al.* (2004) found that DI was significantly more common at greater distances from the equator. This suggests that DI could have evolved to avoid the timing restrictions imposed on breeding by long winters, but the trait is now adaptive only at extreme latitudes. Another potential explanation is that DI allows mating to occur at the optimum time, which may be when females have the greatest opportunity to choose high-quality mates. One example of this is in the American Mink. DI enables the timing of births to be synchronised following a protracted period over which the female mates with several males, resulting in multiple paternity of litters. The delay is short and variable (40–70 days) to facilitate this. The rut can occur at a time when male mink are in prime body condition and the females are then able to give birth at a time of year when food is abundant.

Researchers at Oxford (Thom *et al.* 2004) concluded that DI was likely to be a trait of the common ancestor of the Mustelidae, but one which has changed state several times since. As it is a physiological trait under hormonal control, the mechanism may not be especially complex, and the trait could be latent in non-delaying species. The fact that blastocysts of European Polecat, a non-delaying species, survive when delay is artificially induced, along with the variable delayed implantation of American Mink, suggests that, in these species at least, the endocrine pathways show some flexibility.

Despite the potential advantages of DI, there is also a cost. DI effectively increases the minimum time between each reproductive event, so it should be more prevalent in longer-lived species, which are able to spread their reproductive investment over several years. Longevity is positively associated with the presence of DI, supporting the hypothesis that the time cost imposed by DI prohibits its occurrence in short-lived species. However, the Stoat is a notable exception to this, which raises the question of why. Stoats need to produce their young as early in the season as possible, so they have plenty of time to grow before the winter. However, the adults need to avoid mating too early in spring, when conditions may be severe. This is particularly true in some of the more northerly parts of their global range, where winters are very harsh. Delayed implantation allows both mating and birth to occur at the optimal times. It also accommodates the pre-dispersal mating of female Stoats which, while guaranteeing successful fertilisation, requires DI in order for the females to reach maturity before they give birth.

The potential for strong competition among the sympatric small mustelids in Britain results in a need for different strategies to reduce ecological overlaps. Each species has a different way in which it has adapted to its environment, ranging from that of the slow-breeding, relatively long-lived Pine Marten, which has adopted what is a very effective strategy in a stable (i.e. slow-growing forest) environment, to the 'live (and breed) fast, die young' approach of the Weasels, which enables them to thrive in almost any setting. These differences between the life histories and, in particular, the reproductive strategies of the small mustelids have had a huge impact on how each species has fared in the face of changing fortunes in Britain and elsewhere.

CHAPTER 4

From a Richness of Martens to a Paucity of Predators: A History of Exploitation and Persecution

The small mustelids, in common with all the mammalian carnivores native to Great Britain, have had mixed fortunes at different times, depending on whether they were valued for fur, reviled as vermin, used for sport or appreciated as rodent-catchers throughout their shared histories with humans. The collective noun for Pine Martens is 'a richness' (or *richesse*) although, as these animals are fiercely solitary for the most part, it is difficult to imagine a scenario where such a noun would be required. However, along with that of many mustelids, the luxuriant fur of martens was highly valued in the past, and so it is likely that the term was coined to refer to a collection of pelts rather than to any mass gathering of live Pine Martens ever seen. Humans have always killed animals for meat, as well as for fur, and this includes mustelids. Fossil Pine Marten remains found with human bones in cave sites in Britain suggest that humans were already killing Pine Martens, probably for their fur, as early as the Mesolithic. At this time, postglacial tree cover was at its maximum, providing near-continuous woodland habitat, and it has been estimated that there were nearly 150,000 Pine Martens in Britain (Maroo & Yalden 2000). A richness indeed. Only the Weasel is thought to have been more numerous. However, this heyday for predators was not to last. Starting as far back as the Neolithic, humans began shaping the landscape to suit their own ends. This included clearing areas of woodland to create better grazing for their livestock and to grow crops, which would have reduced some of the habitat for

FIG 32. The arboreally adapted Pine Marten *Martes martes* would have enjoyed extensive forested habitat in Mesolithic Britain.

forest-dwelling animals such as the Pine Marten. Animals that had not previously been a problem were now seen as competitors. Predators that were seen as a threat to domesticated cattle, sheep and goats were hunted, and the numbers of Eurasian Wolf *Canis lupus lupus* and Eurasian Lynx *Lynx lynx* slowly started to decline across this period. Nonetheless, up until the early Middle Ages, there was still an abundance of wildlife to be found in Britain. Animal fur was used for clothing or as goods with which to trade, and this reached a peak in Britain from the twelfth century right through until the middle of the sixteenth century.

The coats of furbearing animals evolved as a means of insulating them and maintaining their body temperature, whatever the climate. The length and density of hairs in the coat (pelage), as well as the ratio of longer, glossy guard hairs in the outer coat to that of the short, insulating underfur, vary

FIG 33. The luxurious fur of Sea Otter *Enhydra lutris*.

between species but are also variable within species depending on time of year and climate of the location where individuals are found. The most luxuriant come from animals trapped in the depths of winter in regions with the coldest climates. By the Middle Ages, the trade in fine furs from the forests of northern Europe was already well established. As far back as Roman times there had been a market within the empire for Swedish traders to sell black fox skins (Veale 1966). The pelts of wild-caught mink and otter were important to the economies of the northern lands and, later on, the incentive for much of the early exploration of North America was the hunting and trapping of furbearing mammals. As local habitats were overexploited and hunted out, fur companies were motivated to find new hunting grounds (Lightfoot *et al.* 2013). In the eighteenth and nineteenth centuries, wild mustelids constituted a substantial proportion of the furs traded in North America.

MUSTELIDS A LA MODE IN THE MIDDLE AGES

In the Middle Ages, the Sable *Martes zibellina* was among the most highly prized of the mustelids found in Europe, because of its rich, dark brown to black fur. Sable skins were imported into Britain from the far north of Russia and Scandinavia. Stone Marten fur, referred to as *foynes*, after its French name of *fouine*, also came in from Europe, but it is possible to imagine that live animals as well as pelts were imported. If this was the case, then escapees or releases could have been a part of the British fauna for a while, as suggested by naturalists' descriptions up until the 1800s of the white-breasted marten (see Chapter 2).

Of the native British mustelids, the Pine Marten was much sought after. Other members of the family that were commonly used in the medieval fur trade were Stoats, which were harvested in winter for their white ermine winter coats. Polecat fur, called *fitch* or *fitchew*, was less popular but was also used, as was the fur of Weasels, although the most commonly used Weasels were those from more northerly parts of Europe, whose coats also turn white in winter. These white Weasel skins were known as *lettice* and were a cheaper alternative to ermine.

For those who could afford them, heavy, fur-lined clothes played a vital part in the daily life of men and women in the Middle Ages. They were worn indoors as well as outside, at a time when there were few other ways of keeping warm in the almost year-round cold of the British climate. The wealthy, in their draughty great halls and stone castles, would have needed several layers of warm clothing on all but the hottest summer days. However, fur garments were not just valued for their warmth but also as statements of style, wealth and social standing. In the Middle Ages, the furs that an individual could wear were clearly set out in a series of laws known as the sumptuary legislation of the fourteenth century. In 1337, the first of these statutes dictated that only the royal family, prelates, earls, barons, knights and clerks with at least £100 could wear furs. This was superseded in 1363, with the substitution listing the furs that could appropriately be worn by each class in society. This ranged from the nobility, who could wear ermine, Baltic squirrel (the soft blue/grey and white winter fur of Eurasian squirrels from the coldest parts of northern Europe) and other fine imported skins, through to citizens and craftsmen with sufficient income, who could only wear lamb, rabbit, cat or fox (Veale 1966). Progression through the ranks, as denoted by these outward displays of status, was the aspiration of anyone with social ambition and, as a result, the fur trade flourished.

FIG 34. John the Fearless (1371–1419), second Duke of Burgundy, wearing robes trimmed with Sable or Pine Marten fur.

Fashion, as we know it, first emerged in the Middle Ages and was firmly linked with wealth. The fur trade since then has always been particularly susceptible to changing fashions, and this was as true in the Middle Ages as at any other time. The pace of change may have been slower, but the exotic and the exclusive have always been highly coveted. In the late twelfth century, sable and ermine, as the most expensive skins available, were particularly admired. King John paid £2 for a single sable skin and £5 for a lining of ermine – the equivalent of more than three years' wages for most, at a time when a carpenter was paid threepence a day. Large quantities of smaller skins such as ermine were required to make a single garment. One of Henry IV's robes was made of 80 ermine and 12,000 squirrel skins.

In the thirteenth and fourteenth centuries, *minever* and *gris* (fine squirrel skins from the north) were the height of fashion, inferred both from the volume of these in cargoes unloaded in London and the large quantities bought for the king and his court. Nonetheless, by the fifteenth century, squirrel was rarely to be found in the wardrobes of the fashionable and appears to have been replaced in popularity by sable and marten skins. The brightly coloured fine woollen cloths that were worn in the thirteenth and fourteenth centuries were well matched by small, light furs. However, by the end of the fourteenth century, fashions had changed, and dark, rich velvets, damasks and brocades were increasingly popular. These were best set off with the fuller deep chestnut and brown to black furs of marten and sable. Between 1413 and 1418, Henry V bought 625 pelts of sable, 2 sable fur linings, 20,000 marten skins and 113 linings of marten (Veale 1966). By 1440, furs for Henry VI's own use were largely of marten. Despite the considerable expense involved, men of high social standing at court were quick to follow the example set by their monarch. Even though marten and sable skins are twice the size of squirrel and ermine, once these larger skins became more fashionable, huge numbers were still needed. It took 250 'backs of marten' to line a purple velvet gown for Henry VI, and the sleeves of the same garment were lined with 68 marten belly skins. The price of such skins was such that whereas in 1407 a lining of the best-quality *gris* might have cost just over £4, a lining of comparable size fashioned from marten fur would have cost £8–10 (and one of sable probably at least £13). In spite of these costs, most of the gentry, whether prominent or not, were soon wearing marten.

FIG 35. King Henry VI in one of his many marten-trimmed garments.

Throughout the Middle Ages, those who could afford it spent lavishly on finery and furs for themselves and their households. Some even went so far as to ornament the liveries of their horses with fur. In 1471, 115 ermine skins were used just for the trappings of Edvard IV's horses. Even footwear was made with fur. Henry VII had boots lined with marten skins to keep him warm, each pair needing 15 skins (Veale 1966). Slippers were often lined with *vair*, made up of the grey back and white belly fur of Baltic squirrel sewn together in an alternating pattern (named *vair* from the Latin *varius*, or 'variegated'). *Vair* as a shortened form of 'vairy' (fairy) was also a colloquial name for Weasel in some areas (Williams & Jones 1873, Wright 1898–1905). As an interesting aside, in 1697, when Charles Perrault was transcribing the story of Cinderella in French, he wrote down that her *vair* slipper was instead made of *verre* (glass), so the poor actress playing Cinderella in countless dramatisations ever since has had to suffer the discomfort of wearing slippers of glass, instead of the softest fur of either squirrel or Weasel, whichever the original referred to.

Hunting and trapping furbearing animals would have been comparatively easy in the thirteenth and fourteenth centuries, and markets were supplied by a variety of sources, including local villagers. Following the introduction of European Rabbits to the English mainland by the Normans, rabbit fur quickly became popular, and the skins of Mountain Hare, Red Fox, Wildcat and Eurasian Otter were still plentiful. However, it is not known how readily the more highly prized skins of marten were to be found in England by that time. In her

FIG 36. Illustration showing King Edward IV, his queen, Elizabeth Woodville, and his heir, the future Edward V, all wearing ermine.

book on the history of the fur trade in Britain, Veale (1966) concluded that it is probable that the numbers of Pine Martens were greatly reduced in the Middle Ages as the area of settlement spread, and that by the thirteenth and fourteenth centuries, they were found only in the more remote parts of England and in the rest of the British Isles. Even in Wales, Pine Martens were sufficiently rare for their price to have been set at some time in or before the thirteenth century at 24 times the price of sheepskin. Even earlier, in a twelfth-century charter, the citizens of Swansea had been forbidden to hunt the marten, and, judging from the scarcity of later references, it seems unlikely that during the later Middle Ages martens could have been hunted in England in any great quantities. Pine Martens fared better in the wilder countryside of Ireland and Scotland. Both countries supported a wide variety of furbearing animals in the later Middle Ages and were famous for their exports of skins. This can be seen in the lists of tolls for Scotland, which show a flourishing trade not just in Pine Marten, but also Wildcat, Red Fox, Red Squirrel, Eurasian Beaver and Eurasian Otter skins. At one time, Inverness was renowned for its trade in Scottish *mertick* (marten), and there was also a considerable export through Leith. These Scottish skins were of such fine quality that they could attract even German merchants. A customs levy of fourpence per timber (30 skins) was imposed from around 1324, rising to sixpence per skin by 1424. The notes or writings of early naturalists give us a good account of the exploitation of Pine Martens as far back as the fourteenth century. Marten skins from the colder climate in northern Scotland were greatly valued, and martens were evidently being killed in high numbers. However, the population could not sustain the continual drain on its numbers. Pine Martens were recorded as being hunted in the wild on the (virtually treeless) Outer Hebrides from 1549. It is thought likely that they were introduced by humans so as to have a ready supply of the martens' prized fur. However, overharvesting probably led to their extinction on the islands by the 1880s (Birks 2017). This pattern of decline was paralleled on the mainland, although the demise of martens in the lowlands is attributed more to the fact that it was eliminated there as a predator of poultry and other domestic fowl, rather than for the value of its fur.

The popularity of furs lasted for several centuries and, as a result, hundreds of thousands, possibly millions, of native mammals in Britain were killed for their skins and many of these were small mustelids. By the mid-sixteenth century, the high fashion for furs had begun to wane as the wealthy spent their money instead on the splendid fabrics that were becoming available. Precious metals, stones and ornate jewellery were also worn as symbols of wealth and status. Quilted and padded styles of dress came into vogue, and these were already warm without the need for the addition of fur lining.

FIG 37. Medieval illustration depicting men hunting for Eurasian Otter *Lutra lutra*.

By the sixteenth century, glass windows were also in more general use, and the growing output of coal and more frequent provision of chimneys meant that it was possible to develop more efficient methods of heating the large houses of the rich. It is possible, too, that the sumptuary legislation of the sixteenth century played some part in the demise of a fashion for fur which was already dying. In the sixteenth century, these acts were very detailed and their aims far more protectionist than they had previously been. In an act of 1509, one significant change worth noting states, for the first time, that none under the degree of gentleman, with only a few exceptions, were to 'use or wear any furs whereof there is no like kind growing in this land of England, Ireland or Wales'.

It may also be true that the constant hunting of furbearing animals was a significant factor in the decline of the medieval fur trade in Britain. We can see that even before the implementation of rewards for killing these animals as vermin, there had already been a long period of sustained killing for an entirely different purpose. There are no records to verify the extent of this, but it is highly likely that it had already had an impact on the populations of several species, particularly the Pine Marten. By the end of the Middle Ages, fur traders were complaining of scarcity and high prices. The kings of Scotland had forbidden the export of marten skins in 1424 and of hides in 1485. An act of 1592, confirmed in 1593, which further prohibited their export, explicitly mentions the 'exorbitant dearth which prevailed' by that time. European Polecats, Stoats and Weasels were not so affected but for different reasons. Polecat fur was not as sought after as that of Pine Marten, whilst the Stoats and Weasels, with their greater population densities and higher rates of reproduction, were better able to withstand the levels of harvesting to which they were subjected for their pelts.

VERMIN IN ERMINE: THE TUDOR VERMIN ACTS AND THEIR IMPACT ON THE SMALL MUSTELIDS

The first legislation relating to vermin control was enacted in 1532 under the monarchy of Henry VIII and remained on the statute book up until around 1800. This made provision for organised vermin control to be carried out in the parishes of England and Wales for financial reward. It may have seemed ironic to those villagers who were now being paid to kill as vermin many of the same species that they had formerly sold to markets for their valuable fur. However, when viewed within the social context of the time, the rationale for this war that was declared on many wildlife species can be better appreciated.

The sixteenth century was characterised by a rapidly expanding human population, serious deprivation and a series of failed harvests (Hoskins 1964). Against this backdrop, the introduction of parliamentary acts to initiate campaigns of vermin destruction can be seen as reasonable measures aimed at increasing food production. Following successive outbreaks of the plague in the fourteenth century, the population of England and Wales had declined from 3.7 million to around just 2.5 million

FIG 38. Portrait of King Henry VIII, under whose reign the first of many vermin acts was added to the statute book.

by the last quarter of the century. Population recovery was remarkably slow, with the result that by 1525, there were still only 2.6 million people in England, half a million in Scotland and just 210,000 in Wales. In England and Wales, the population at this time was still almost exclusively rural, London being the only major urban area, with around 50,000 inhabitants (Lovegrove 2007). After 1525, however, the population began to increase again, steeply at first but then slumping again in the 1550s, following a series of disastrous harvests and outbreaks of disease. After that there was a steady rise in population which continued to the end of the sixteenth century and into the early decades of the next. In England especially, this growing population brought with it a growing set of problems. Prices of agricultural produce increased as a result of increasing demand, and landlords were quick to put up land rents which, in many areas, put unsustainable pressure on leaseholders. Prices that could be asked for produce in the increasing urban markets were higher than those which could be achieved by selling to local villages and communities; therefore, they frequently had to go without many basic necessities. Demand often outstripped supply, especially in years of poor harvests, and in the latter half of the sixteenth century there was a whole succession of disastrous harvests. These resulted in serious food shortages and led to widespread malnutrition, disease and death (Morris 2002). This background of population growth and agricultural disasters undoubtedly motivated the succession of vermin acts across this period. Wildlife was seen as an enemy if it was thought to compete with humans in any respect, but particularly in relation to agriculture or any other food production.

From our modern perspective of relative abundance, it is easy to condemn the war of attrition on many native species that began during this period. However, for the majority of the poorer classes at the time, both in towns and the country, life was a daily struggle against hunger, squalor, disease, unemployment and death. Any competition for food resources was unacceptable, so any and all species of bird or mammal that were known (or even suspected) to impact on crops or stock were seen as a legitimate target and one that should be destroyed.

In England and Wales, it was Henry VIII who, in 1532, introduced the first of a series of acts which became known familiarly as the Vermin Acts. These were followed by further acts during the reign of his daughter, Elizabeth I, the most important of which was that of 1566. Some of the crucial changes in Elizabeth's (1566) version of the act include defining 'vermin' as a legal category and expanding the species covered by the term to include a wide range of mammals. The 'Acte for the preservation of grayne' listed all the species of birds and mammals that were considered as vermin and therefore deserved to be eradicated. This included all of the native mustelids: Pine Martens, European Polecats, Weasels, Stoats,

European Badgers and Eurasian Otters, as well as Wildcats, Red Foxes, European Hedgehogs *Erinaceus europaeus*, mice, rats and European Moles. Property owners were required to pay a small tax to their respective parish each year from which statutory bounties would be paid by the church wardens for 'the dystructyon of noyful fowells and vermin'. Although the list was primarily aimed at the protection of agricultural crops, it went further than that to include a range of native species, such as Wildcat and Common Kingfisher *Alcedo atthis*, which had nothing to do with agriculture but were presumably considered a nuisance in other respects. Paradoxically, some of those species listed are predators of genuine agricultural pests and so were actually beneficial to agriculture. The 1566 list remained the guideline for continuing persecution through the parishes for several hundred years, and the act was not repealed until 1863. Almost all of the native species that are still regarded as vermin today were present on the Tudor list. There was a hierarchy of bounties (literally the price on their heads) for species listed in the act. One penny was paid 'for the heads of every fitchou, polcatte, wessel, stote, fayre bade [marten] or wilde catte'. This was at a time when the average wage of an agricultural labourer was fourpence a day, so it was quite an incentive.

The act set an important precedent that has had two significant effects. Firstly, it established by legal statute that a specific group of animals (mostly carnivores) act against the wellbeing of society and therefore deserve to be exterminated. This has contributed to the negative relationship between humans and many animal species that continues today. Secondly, community participation was required to carry this out, and a lack of active participation in 'vermin control' was a punishable offence. In this way, the category of vermin became a tangible focus and a scapegoat for other larger and more abstract problems.

FIG 39. European Polecats *Mustela putorius*, along with all the native mustelids, were named as vermin in the Acte for the Preservation of Grayne 1566.

FIG 40. European Polecats *Mustela putorius* can have large litters, which may be why, despite being targeted as vermin, they managed to persist in most parts of Britain until the early nineteenth century.

Whilst all of the mustelids were considered as vermin of varying degrees of nuisance, the polecat was seen as one of the worst offenders. The name polecat is thought to be derived from the French *poule chat* (chicken cat), reflecting the animal's reputation as a poultry killer. At a time when every smallholder kept chickens, as well as ducks and geese, polecats were seen as a serious threat because of the difficulty of keeping them out of the henhouse, before strong wire netting was available. The only option was to eradicate them. In addition to poultry, since around the fourteenth century, rabbits had become an important sustainable source of meat for many. Following their introduction to Britain by the Normans in the late eleventh or early twelfth century, numerous warrens were constructed in England, Scotland and Wales. A successful warren was a very valuable asset for the landowner, assuring a healthy profit each year and also providing permanent employment for a warrener and someone to make and maintain the nets and snares used for trapping and ferreting. As a consequence of the importance of rabbits, both commercially and as a basic food source for many rural communities, lethal control of polecats was a high priority. Ironically, the characteristics of the polecat which make it such an efficient predator of rabbits have been exploited and selectively bred in Ferrets, a domesticated form of the polecat, for centuries (Davison *et al.* 1999). However, these very qualities made wild polecats a serious pest of rabbit warrens. The rates of payment that were agreed in counties throughout England and Wales and implemented by the church wardens give some indication of the value placed on controlling polecats. In the original statute of 1566, the specified rate was a penny per head. However, by the beginning of the seventeenth century, this had risen to fourpence, twice the amount that was paid for Stoats or Weasels. Remarkably, despite all this, polecats appear to have remained widespread in Britain until the 1800s, although there were some earlier local extinctions. However, with the rising popularity of game shooting in the nineteenth century, there was an increase in predator control to protect game birds, and this further incentivised killing of polecats.

Knowledge about the ways of destroying polecats and other animals classed as vermin was probably common throughout the Middle Ages, but during the

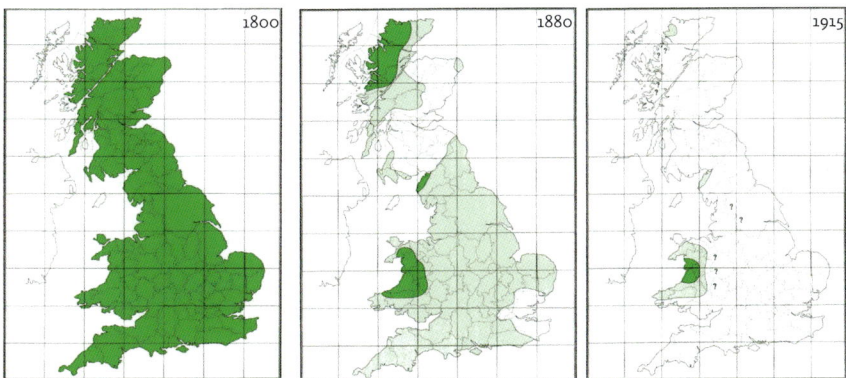

FIG 41. Changing European Polecat *Mustela putorius* distribution, 1800–1915. With the exception of the Red Fox *Vulpes vulpes*, no mammal was killed in the name of vermin control more widely or in such great numbers in England and Wales between the seventeenth and nineteenth centuries than the much-maligned polecat. Light-green shading indicates areas where Polecats were rare, declining or localised; dark green where common or widely distributed. Redrawn from Langley & Yalden (1977).

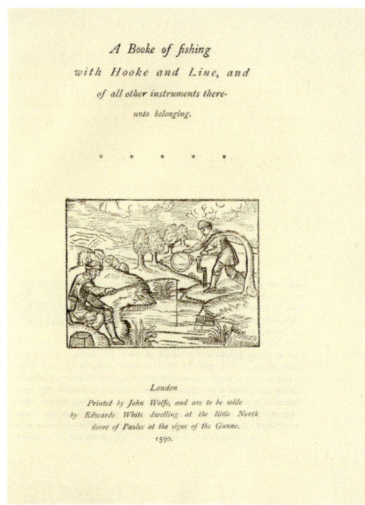

FIG 42. One of many early manuals on predator control.

early modern period this was increasingly disseminated in the form of the manual. A number of these were in print at the turn of the seventeenth century, the most popular of which was snappily titled *A booke of fishing with hooke and line, and of all other instruments thereunto belonging. Another of sundrie engines and trappes to take polcats, buzards, rattes, mice and all other kindes of vermine and beasts whatsoeuer, most profitable for all warriners, and such as delight in this kinde of sport and pastime.* This was written by Leonard Mascall, clerk of the kitchen to the Archbishop of Canterbury. Mascall's manual for destroying vermin includes detailed descriptions of 34 traps and 9 recipes for poison.

By the time of the Tudor Vermin Acts, the Pine Marten, although not common, was still relatively widespread in Britain. In England, counties such as Cornwall, Devon, Somerset, Dorset, Hampshire, Kent, East Sussex and Westmorland were still well wooded and supported good populations of marten. Churchwardens' records from these counties show steady but low levels of killing in some parishes and more sporadic payments in others. There can be little doubt that by this period, loss of woodland habitat had already left martens much more patchily distributed than the other small mustelids, so it is unsurprising that the numbers killed as vermin were also fewer. Historical records of marten bounty payments paint a picture of a sporadic, opportunistic campaign of attrition. Roger Lovegrove's analysis of records from churchwardens' accounts shows the earliest record in 1664 and the last one in 1824, with the majority occurring between 1710 and 1780 (Lovegrove 2007). This is consistent with Langley and Yalden's (1977) conclusions about several counties, from which they suggest Pine Martens had gone completely by the mid-1800s.

We only know a little about the methods that were used to trap Pine Martens before guns were widely available for vermin control. One opportunistic method was to follow the marten's tracks in snow back to its hiding place, usually in the hollow of a tree. From there it was driven or smoked out (by burning grass and heather or sometimes tobacco) into a bag. Martens were also known to be vulnerable to spring traps and baited deadfall traps. In his 1892 book, *A vertebrate fauna of Lakeland*, Macpherson (no relation to the author) gives a detailed account of the good sport to be had in winter hunting the 'sweet mart' with foxhounds. He writes that 'The "foil" of the Mart is sweet, but all dogs will follow it', and follow it they do for the best part of a day until eventually it tries to outrun the hounds across open ground for about a mile before being caught and killed by the dogs (Macpherson 1892).

In the early nineteenth century, Knapp (1829) writes that the local value of marten skins in Gloucestershire was two shillings and sixpence, although he comments that furs from the colder regions were better and more readily obtained. This represented a far better return than Pine Marten heads taken to the churchwarden for payment, which were only paid at fourpence. This may be another explanation for the relatively low numbers of martens in many parish records. Macpherson makes a similar point that at Kendal in the mid-nineteenth century, a dog marten pelt would fetch six shillings and sixpence and the smaller 'bitch' one five shillings and sixpence. In addition, one caught 'wick' (alive) could easily be sold in Whitehaven and elsewhere in the west for as much as 10 shillings to be used for baiting with dogs. The poor marten was considered game enough to show good sport even though it could not fight as ferociously as the 'foumart' (polecat). In further comparisons between the marten and the polecat, Macpherson reflects on the differences in ease of trapping the two species. Pine Martens were, apparently, easily trapped as 'they do not seem so suspicious of traps as some wild animals'. He suggests that the cause of the sudden decrease of martens in some areas at the time was probably due to their lack of wariness of a baited trap, going on to say that martens are very fond of rabbits and 'where these are numerous, and are systematically trapped, if there are any martens about, they are almost certain to get accidentally into the rabbit traps' (Macpherson 1892).

The Stoat, along with the European Polecat, was also a known predator of

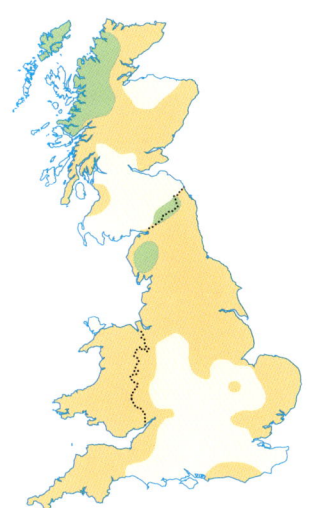

FIG 43. Pine Marten *Martes martes* distribution in 1850. Pine martens were already extinct (white) or rare and declining (gold) in much of Britain by this time. Areas in green show where they were still thought to be common. Adapted from Langley & Yalden (1977).

FIG 44. A young man with a Pine Marten *Martes martes* which he has clubbed to death after catching it in the trap near his feet (late 1800s).

poultry and invader of rabbit warrens. One of the old colloquial names for the Stoat was Cain, after the murderous brother in the Bible, giving a clear indication of how Stoats were perceived. Despite this apparent image problem and whatever the feelings expressed about the Stoat, it was nowhere near as reviled as the polecat, nor, according to the records, were Stoats killed in significant numbers (Lovegrove 2007). Across the majority of parishes in England and Wales, it seemed that Stoats were not a cause of great concern in the seventeenth or eighteenth century.

When it came to the Weasel, it seems that we have had an uncertain love–hate relationship with this diminutive but impressively efficient little predator. Paradoxically, while it is viewed with some suspicion, and has a reputation as both a bloodthirsty villain and a ferocious killer, it has also long been recognised as a potential ally. This is essentially because its main prey species are small rodents which are greater nuisance pests, particularly to farmers. Much of the prejudice against Weasels is unjustified. Though they will certainly kill chicks or ducklings if the opportunity arises, this is far outweighed by the numbers of rodents that they take. After going through churchwardens' accounts of vermin payments from the sixteenth to nineteenth centuries, Lovegrove (2007) reports being left with the strong impression that the killing of Weasels was not encouraged in England and Wales throughout the period. This is despite the fact that they were specifically listed in the Vermin Act of 1655. Weasels appear to have been targeted in only one parish out of more than 1,500. At Arksey, in Yorkshire, twopence per head was paid for 685 Weasels between 1722 and 1767, with a maximum of 60 killed in a single year. This level of killing would have had a negligible effect on the local Weasel population, as is evident from the continuing numbers killed each successive year. In parishes elsewhere, the numbers of Weasels killed through the late seventeenth and early eighteenth centuries were small to negligible, suggesting that they may often have been a result of by-catch in traps targeting other species (such as European Polecat and Stoat). It seems that across the country, Weasels were seen as beneficial around houses and farms and, as a result, their presence was not only tolerated but probably even encouraged. They were almost certainly abundant in most areas at a time when there was plenty of ideal habitat and a profusion of small rodents. In the days when corn ricks and stackyards were ubiquitous on farms, such sites would have been full of rodent prey and perfect hunting grounds for Weasels.

Variations in vermin payments from parish records analysed by Roger Lovegrove in his book *Silent Fields* illustrate how the emphasis of vermin control reflected shifts in agricultural demand – for example, showing an onslaught on Bullfinches *Pyrrhula pyrrhula* that was recorded when there was a marked increase in orchard planting and demand for fresh fruit, and targeting species suspected of preying on farm poultry or other livestock at times when grain was in good supply.

THE KILLING FIELDS: ENCLOSURE AND THE RISE IN GAMEKEEPING

The next significant development in the fortunes of predators began with the Parliamentary Enclosure Acts. Between 1760 and 1870, around 4,000 enclosure acts resulted in the change of c.2,800,000 ha (approximately a sixth of the total area of England) from common land to enclosed land (Fairlie 2009). These changed the nature of the countryside, resulting in far-reaching social consequences. From the sixteenth to the eighteenth centuries, much of rural Britain had been unenclosed and farmed under the medieval open field system. Native woodlands were already seriously depleted, the best timber having been taken for use in the shipping and building trades, with much of the rest cleared for firewood and charcoal, so that by the end of the nineteenth century, woodland cover in England was less than 5 per cent (Smith & Gilbert 2001). The main purpose of the Enclosure Acts was to rationalise landholding and increase agricultural productivity. The great open fields of the Middle Ages, as well as many of the commons and 'wastes', were broken up and fenced in. In addition to the physical changes that were manifested, the ultimate social effect of enclosure in England and Wales was to concentrate ownership of the majority of land into the hands of a small number of individuals. A class of prosperous capitalist tenant farmers was created, and many small freeholders were effectively dispossessed of their former living. These were forced to either accept low-wage employment as farm labourers or migrate to new urban areas in search of other prospects. The gap between rich and poor widened, as a rigid three-tier class system emerged, comprised of landowner, tenant farmer and labourer. It is said to have been the most dramatic agricultural, landscape and social change that had occurred in rural Britain for a thousand years or more, and it also had significant implications for wildlife and its control. The enclosures made possible the huge farming improvements that took place during the agricultural revolution, but the process completely disrupted the medieval landscape pattern. Much of the land enclosed was common land and, by 1795, almost a million hectares of this had been hedged in. As enclosures spread onto the cultivated land, new crops and new farming systems evolved. Advances in agriculture meant a growth in farm profits which, in turn, led to higher land rents. This new wealth provided the basis for the developing increase in game management.

By the early decades of the nineteenth century, the small mustelids and other species had been subject to persecution of varying intensity across the parishes for at least 300 years. However, this was now to accelerate, as 'vermin' (and especially predator) control was gradually taken over by an increasing number of highly motivated and well-armed gamekeepers, tasked with protecting the shooting interests of landowners and improving the production of game birds. At the end of the nineteenth century, the reshaped countryside, particularly arable farmland, was a near perfect habitat for partridge and pheasant. The new miles of hedges provided sheltered and well-drained nesting cover, while cereal crops formed tall, dry habitat for foraging broods. The patchwork crop rotation ensured feeding areas in winter and summer for these birds and other game such as Brown Hares *Lepus europaeus* in lowland farmland. The approach that had transformed farming was now applied to game management, supported by the wealth of the new sporting landowners. All that was needed to release this potential was systematic management instigated and funded by the wealth of the sporting landowners.

Shooting has always been the prerogative of landowners, and laws were designed to prevent anybody but the wealthy landowner from having access to game. Villagers without a means to feed their livestock through the winter had to resort to hunting fresh meat from the wider countryside. Low-level poaching was a long-established necessity, and even more so after enclosure when many villagers were no longer able to produce enough food for their families. They simply could not understand that the wild animals, which had previously been fair game for anyone in the open countryside, were now the sole property of a single individual. As protection against this illegal competition for his game, it became necessary for the landowner to employ men to defend it. It is not certain when the employment of gamekeepers by our modern understanding of the job first began, but the gamekeeper's role was clearly established by the early nineteenth century. In addition to deterring poachers, the gamekeeper was employed to rear and protect the maximum numbers of game birds possible.

FIG 45. In the unfinished painting, *Polecat Fur* (Klimt, 1916), the woman is depicted wearing a European Polecat *Mustela putorius* fur coat, an incredible luxury at the time that shows her wealth and high class.

In the latter years of the nineteenth century, many estates had been bought up by wealthy businessmen whose money came from industry or commerce. They were keen to acquire the social standing of having a country estate and to adopt the sporting lifestyle that came with it. Not being dependent on rental incomes

FIG 46. A royal shooting party with King Edward VII (third from the right), 1909.

from their tenant farmers, they had little interest in agricultural improvements, concentrating rather on developing their pheasant and partridge shoots. Up until around 1840, shotguns were usually sold as single guns, and most game was taken during walked-up shooting. However, during this decade, coinciding with the refinement and increased reliability of the percussion cap firing system, matching pairs of shotguns were made, giving rise to the increased popularity of driven shooting. As shotgun design and manufacturing improved, so did the attractiveness of game shooting, especially amongst royalty and the rich landowners (Fig. 46). Game shooting reached the height of its popularity in late Victorian Britain, by which time shooting was a fashionable and highly competitive pastime, and one that was enjoyed by the Prince of Wales and other prominent members of society. This encouraged a spirit of rivalry, judged by the size of an individual's 'bag' (the number of birds shot) at the end of the day. Estates were managed to maximise the numbers of birds for shooting, and drives were organised to ensure that a continuous supply of game was made to fly over the guns.

One of the key factors in producing a large surplus of wild game birds on any shoot has always been predator control. Trapping, shooting and, in the past, poisoning have been the main methods by which gamekeepers have achieved this. Over the years, new trap designs and approaches to predator control have become available, and changes have been imposed by improved animal welfare legislation and increased legal protection for some species, such as the Pine Marten. In the early 1900s, when labour was cheap and gamekeepers were employed across most of Britain, their main strategy was a blanket approach to predator elimination that encompassed most carnivorous birds and mammals.

A HISTORY OF EXPLOITATION AND PERSECUTION · 67

FIG 47. The gin trap was designed so that any weight placed on the flat piece of metal between the jaws would trigger the trap, and the jaws would snap shut.

Traps and poisoned baits were set indiscriminately, the whole year round. Up until it was outlawed, in 1958, the gin trap (an abbreviation of *engin*, the French for 'engine') (Fig. 47) had been the mainstay of gamekeepers for generations. The gin trap was simple and devastatingly effective. It was a mechanical device which, once triggered, was designed to catch an animal by the leg using a pair of spring-operated jaws, either with or without a serrated edge, or teeth.

The gin trap came in a variety of shapes and sizes and was capable of catching, killing or maiming a wide range of mammals and birds. It was widely used for killing 'ground vermin', including Stoats and Weasels, to which end huge numbers of gin traps were deployed by keepers around all the hedges and woodland on their estates.

In contrast to the apparently laissez faire attitude of most parishes towards Weasels and Stoats during the seventeenth and eighteenth centuries, the focus on game preservation in the nineteenth century resulted in a marked change in attitude towards both species. The Weasel's usefulness in controlling rodents was quickly forgotten as it became perceived as an enemy of game interests, along with many other predators. The primary aim of producing maximum game bags rapidly ensured that Stoats and Weasels were high on the list of undesirable vermin. The Badminton Library book on game rearing at around this time condemns the Stoat in unashamedly subjective terms, and says that, 'A more lithe, active, crafty and bloodthirsty foe to all other animals has never been constructed' (Walsingham & Payne-Gallwey 1887). Once gamekeepers were active on shooting estates and equipped with gin traps and guns, Stoats were killed in huge numbers.

FIG 48. Gamekeepers often displayed their kills on 'gibbets', to demonstrate their prowess. Many early photos illustrate the high numbers of small mustelids that were frequently killed.

Up until the early nineteenth century, while shooting was available throughout most counties of England, the uplands of Scotland remained largely unvisited by English sporting tourists, due to the difficulties of getting there from London. However, after rail links were built in the 1840s, this was no longer an issue. Stalking, shooting and fishing on the large estates in Scotland quickly gained popularity, with the grouse season becoming a key feature in the social calendar of the wealthy and ambitious.

In Scotland, the earliest estate vermin records show the steady beginnings of systematic Stoat and Weasel control, but many records refer to the two species simply as 'weasels' so the exact numbers of each are not known. In the 14 years from 1782 to 1796, a relatively modest 45 Stoats and Weasels were killed by John Sinclair at Taymouth, along with a similar number of Pine Martens. (It is worth noting that six times as many European Polecats were killed on the same estate during this period.) However, in just 10 years, from 1818 to 1828, 2,526 Stoats and Weasels were killed on the Buccleuch shooting estates in southwest Scotland. At two estates in Perthshire, 'weasels' were killed at an even more impressive rate (7,198 and 9,852, respectively, in just 11 years). These combined Stoat and Weasel figures show that on Scottish estates in the nineteenth century, fairly heavy control was exercised in almost equal measure but with a slight bias towards Stoats as, on shoots where the two species were differentiated, Stoats killed outnumbered Weasels but by fewer than 2:1 (Lovegrove 2007).

In the early part of the twentieth century, the gamekeeper really had only two duties: catching poachers and destroying vermin. He was able to devote most of his time to the latter task, so it was carried out with ruthless efficiency and with the goal of eliminating any species of animal or bird which might take the occasional game bird chick.

The effect of this extent and scale of predator control on carnivorous birds and

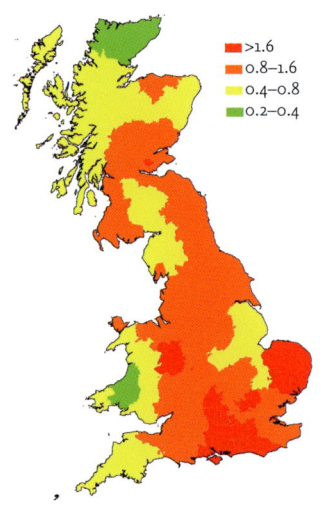

FIG 49. Gamekeeper density per 1,000 ha in each county in 1911. Between 1871 and 1911, the number of gamekeepers steadily increased, reaching a peak of 23,056. Most of Britain was covered with a contiguous network of well-keepered estates, which meant that virtually every inch of the countryside was subject to this continuous annihilation of vermin. Redrawn from Tapper (1992).

mammals was devastating. In their paper 'The decline of the rarer carnivores in Great Britain during the nineteenth century', Langley and Yalden (1977) showed that the distributions of the marten and polecat were reduced by the early nineteenth century to those small areas of Britain where gamekeeping intensity (numbers of keepers) was lowest. The fact that these species have recovered as gamekeeping continues to decline lends strong support to the argument that it was intensive and indiscriminate predator control which brought them to the brink of extinction a hundred years ago.

Following the onslaught on them for their precious fur in the Middle Ages, Pine Martens had already disappeared from much of southern Britain. They survived through the subsequent campaigns against predators only in those parts of Britain (and Ireland) where pressure from humans was at its lowest. In the remote uplands of Wales and northern Britain, there was little competition from people for resources, and martens were able to survive here in small numbers. However, it was in the far northwest of Scotland where the largest populations remained, during the species' nadir from the late eighteenth to early twentieth centuries. In the western Highlands, the rocky crags and mountainsides provided a surrogate for the three-dimensional habitat of woodlands that martens are adapted to exploit. Nonetheless, these upland habitats are nowhere near as rich in prey and other resources so it would have been a much harsher existence for martens in the low-density populations here. It has been estimated that by the

FIG 50. European Polecats *Mustela putorius* in Wales where gamekeeper densities were low throughout the eighteenth and nineteenth centuries.

FIG 51. The Red Kite *Milvus milvus* also found refuge in Wales at the height of its persecution elsewhere in Britain.

early 1900s, the British marten population had declined by more than 98 per cent from its Mesolithic 'richness' of around 150,000 to a mere 2,000. This was echoed in Ireland, where martens similarly retreated to the rockier, more inhospitable habitats in the western counties. The strongholds in Ireland were in Galway and among the limestone crags of the Burren in County Clare, as well as smaller populations in the south around Counties Waterford, Cork, Kerry and Louth. Martens declined in Ireland at a later point than in Britain, largely because there was a lower intensity of pressure from gamekeepers. However, it seems that the increased use of strychnine poisoning to control predators, associated with intensified sheep farming, is correlated with the disappearance of martens from many parts of Ireland in the 1950s and 1960s.

The pattern of decline for the polecat shows that, unlike the marten, which was already scarce by 1800, polecats were still common and widespread at that time. Polecat fur was not as prized as that of ermine or marten, and the polecat's rate of reproduction was able to keep pace with the early levels of killing as vermin. As a result, it appears that polecats were still relatively widespread in most of Britain until the early nineteenth century. However, during the 1800s and concurrent with increases in gamekeeping, populations of polecats showed marked declines. These were documented in published reports by county naturalists throughout the nineteenth century. By 1850, polecats had become rare in large parts of southern and central England, as well as in the Scottish Borders. By 1880, the only areas where polecats were still common were in mid-Wales, northwest Cumbria and the northwest Highlands of Scotland. Local extinctions of polecats appear to have peaked between 1870 and 1910. As gamekeeper numbers climbed, reaching a maximum in 1911, polecats disappeared from 57 counties across England, Scotland and Wales. The polecat found refuge in west Wales with its stronghold in an area of approximately 70 km radius around Aberdyfi (Aberdovey) (as shown in Figure 41). Data from the national census show that in 1911, this was a region of sparse human habitation and one with very low levels of gamekeeping (fewer than one keeper per 1,000 ha). Polecats (and, notably, Red Kites *Milvus milvus*) were able to survive here through a period when they were eradicated from most other parts of Britain.

THE TWENTIETH CENTURY: A HUNDRED YEARS OF CHANGE

The outbreak of the First World War, in 1914, followed by the decline of traditional sporting activities and the loss of an entire generation of gamekeepers, led to a vast reduction in predator control. Nonetheless, by this time, it appears that only Stoat and Weasel populations, along with Red Foxes, had been unscathed by predator control. After the so-called 'Great War' ended in 1918, efforts were made to restore game management, albeit at a lower level, on the larger estates, but the Second World War following on so soon after, in 1939, took a further toll. By 1946, there was very little game shooting, and the sport was slow to recover. The 1951 census recorded just 4,391 gamekeepers but this had dwindled to an estimated 2,500 by the 1980s (Tapper 1992). To quote P. G. Wodehouse, 'the fascination of shooting as a sport depends almost wholly on whether you are at the right or wrong end of the gun'. Nonetheless, shooting remains a popular pastime and contributes around £2 billion per year to the UK economy (PACEC 2014). However, wild game management is no longer widespread, with more than 80 per cent of shoots now relying on releasing hand-reared game into the countryside. Approximately 30–40 million pheasants are reared and then released for shooting each year. Conversely, since the 1970s, the numbers of grouse killed each year, as well as the area of land used for grouse shooting, have both declined significantly. Land management for game, particularly in the lowlands, is fragmented and often separated by large areas with no gamekeeper.

Gamekeeping practice and methods of predator control have also changed considerably in the past hundred years. Poisoning was outlawed as a means of predator control in the 1911 Protection of Animals Act, although it is easy to lay poison baits without being caught, and so the practice continued to be used illegally by some (Tapper 1992). The Pests Act of 1954 made it an offence to use any spring trap, other than those approved by the appropriate ministry (currently the Department for Environment, Food and Rural Affairs, or DEFRA), for the purpose of killing or taking animals. The Spring Traps Approval Order was introduced in 1957, and in 1958 approval for use of the gin trap in England and Wales was removed on the grounds that it was considered inhumane and indiscriminate. Gin traps continued to be legally used in Scotland until as recently as 1971 and are still occasionally found in illegal use (https://raptorpersecutionscotland. wordpress.com/ 2018/11/30/illegal-gin-trap-found-set-near-nairn-highlands/ [accessed September 2020]). It is important to note that the authorities in England, Wales, Scotland and Northern Ireland all publish their own versions of the approval order which, although generally very similar, are often published at different times and so can become out of step. The order is not only very specific about traps but also about which species each trap can be used to catch and the manner in which the trap may be used. The Fenn 'humane trap', introduced in the 1950s, was a popular replacement for the gin for controlling smaller mammalian predators such as Stoats and Weasels. Even though a report as far back as 1961, documenting the results of 130 captures of 6 species, suggested that the Fenn should not be regarded as the last word in humane traps (Bateman 1988), its low cost and ease of use meant that it was widely used up until a recent change in the law.

In April 2020, the UK implemented new restrictions as part of the Agreement on International Humane Trapping Standards (AIHTS), to which it is a signatory. The AIHTS requires participating countries to prohibit traps for furbearing species that do not meet the specified standards for humaneness. Of the species to which the AIHTS applies, only the Stoat, Pine Marten, Eurasian Otter, European Badger and Eurasian Beaver *Castor fiber* occur in the UK, and the Stoat is the only one of these species without full legal protection. A number of spring traps that were widely used in the UK (including the Fenn, as well as the Springer, Solway, Magnum, Conibear and Kania traps) did not meet the AIHTS humaneness criteria for Stoats. Prior to April 2020, Stoats could be legally trapped or killed in the UK without a licence, but since then, general licences have been introduced to authorise the control of Stoats in England, Wales and Scotland provided the terms of the licence are met. These specify the traps that may be used. It should be noted that since 1980, the Irish Stoat is a protected species, and it is an offence to trap or kill Stoats in Ireland under the Wildlife Act 1976 (Republic of Ireland) and the Wildlife (Northern Ireland) Order 1985.

One condition of the use of approved spring traps is that they have to be set in a suitable tunnel to prevent accidental capture of non-target animals. Most of the Stoats and Weasels killed by gamekeepers in Britain now are killed in these tunnel traps (Fig. 52), though Stoats are sometimes shot after being 'squeaked up' by someone mimicking the sound of an injured rabbit. On upland moors and on shoots where wild Grey Partridge *Perdix perdix* is the main quarry, gamekeepers still regard Stoats as serious predators that will take nesting game birds, eggs or poults if they have the opportunity. On pheasant shoots, where the majority of birds are captive-bred and released, Stoats are seen as less of a problem, but a Stoat would still do serious damage if it broke into a pheasant rearing pen.

FIG 52. A tunnel trap on upland moor in Scotland.

However, predator control is time-consuming and expensive and so the objective nowadays is not to eliminate every predator, as it was in the past, but to reduce predation during the critical period when birds are nesting and rearing chicks, in order to maximise breeding success. It is common practice to set a network of tunnel traps at a density of around 100 traps per keeper per 400 ha to reduce ground predators like Stoats and Weasels, but tunnel trapping is usually now concentrated in spring and early summer when it is of most benefit, rather than being carried out all year.

HOW STOATS AND WEASELS SURVIVED THE ONSLAUGHT

Stoats and Weasels, by virtue of their life histories and high rates of reproduction, are extremely resilient in the face of efforts to control them (King & Moors 1979b). Both species have very variable reproductive success and high rates of natural mortality. Births and deaths are rarely in equilibrium, which means that local variation in density is normal, and local extinctions are common. However, Stoat and Weasel populations are very resilient against total extinction because it only requires a small number to survive in a favourable patch nearby for recolonisation to occur as soon as conditions are suitable. McDonald and Harris (2002) carried out a study in 2002 that made use of demographic data from a large number of Stoat (822) and Weasel (458) carcasses collected from gamekeepers carrying out predator control on game estates in Britain. The data were used to construct population models for each species and to test the effects of changes in demographic parameters on population growth rates. Their collection strategy meant that they were able to compare the effects of culling on Stoat and Weasel populations sampled in the same areas and by the same means. The same population model was used for both Weasels and Stoats, with different parameter values to reflect the different life histories and reproduction biology of the two species. The assumptions of the model were that survival was independent of density and that there was no immigration or emigration (a closed population). In spite of this necessary oversimplification, the model is extremely useful in demonstrating the differences and similarities between the two, often coexisting, species.

A population growth rate was estimated for both Stoats and Weasels on British game estates. Values for the rate of increase (r) that are greater than 0 mean that the number of new animals born and recruited into the population is exceeding the number of deaths, so the population will increase, whereas if r is less than 0, then the rate of mortality is greater than recruitment, and the

FIG 53. Stoat *Mustela erminea* with young.

population will decline. During the years sampled by McDonald and Harris, the population growth rates averaged −0.05 for Stoats and 0.3 for Weasels. The model was then adjusted to assess how sensitive the population growth rates were to changes in demographic parameters, including survival of each age class, fertility and probability of a late-summer litter (for Weasels). For both Weasels and Stoats, the single factor with the most influence on *r* was survival of the first-year age class. If the modelled survival rate of newborn Weasels in the first three months was cut from 0.88 to less than 0.56, then the modelled population would decline. This probably simulates conditions in years or habitat patches where food is very short. However, when the values for newborn survival and probability of second litters were both set high in the model, the Weasel population growth rate increased. Both of these parameters are strongly influenced by the availability of food (prey), and so this gives an indication of how Weasel populations are able to quickly respond in years of high rodent abundance.

For Stoats, one of the critical factors that influenced *r* was the survival rate of second-year females from April–June. This is when they produce and nurse their first litters and is a critical period in the survival of those young. If the female dies before her young are weaned and able to hunt for themselves (at approximately 10–12 weeks old), it is highly unlikely that they will survive. This means that trapping female Stoats during spring to early summer has a disproportionately large effect on the population. When the survival rate for that group of females was increased only slightly in the model (from 45 to 54 per cent), the population growth rate went from negative to zero. This means that when, or where, the survival of this group is high, r becomes positive, and the population increases. There is plenty of evidence that the value of r does frequently oscillate from positive to negative for Stoat populations, resulting in the unstable numbers that are typically observed everywhere. Clearly, local Stoat populations can sometimes withstand high rates of mortality, whether these are natural or imposed. Patchy environments provide a mosaic of breeding opportunities for Stoats, and the high dispersal capabilities of their offspring mean that there is always a supply of young to migrate across the landscape and recolonise vacant patches after every local extinction.

Weasel populations are also unstable, showing marked fluctuations in response to small rodent availability. Species in which r fluctuates between high and low values tend to be very resistant to control, such as that carried out on game estates. Their 'live fast, die young' strategy is such that they are capable of very rapid, albeit temporary, variations in population density. For control measures to have an impact on these populations, more individuals must be removed than can be replaced, at least at the local level, which means that a huge proportion of the target population has to be removed. For Stoats and Weasels, it has been shown that this proportion must be at least 80 per cent, but even then, those that are removed will be replaced very quickly. However, even at this rate of removal there will be no effect on the density of the resident population if it only replaces, rather than adds to, natural mortality (compensatory rather than additive mortality). For example, when natural mortality is high (as it is in winter), trapping can remove large numbers of animals which would otherwise have died anyway, therefore having little effect on local density. When trapping is carried out intensively, it is possible to reduce Stoat and Weasel densities temporarily and for as long as trapping continues; however, the animals that are removed can be quickly replaced by immigration from surrounding untrapped areas. As a result, Stoats and Weasels are not thought to have undergone significant nineteenth-century declines.

There are no national surveys for Stoats or Weasels and so, ironically, the only data on their populations in Britain are from the numbers killed on shooting estates, reported in the Game and Wildlife Conservation Trust's (GWCT) National Gamebag Census (NGC). These suggest that Stoats and Weasels are still common and widespread throughout mainland Britain. Analysis of trends in NGC data from 1966 to 2016 (Aebischer 2019) showed that Stoat captures had increased by 50 per cent, mostly in the last 25 years of the dataset. Weasel bags had changed

FIG 54. Weasel *Mustela nivalis* in grassland.

very little overall, but when the data were split into shorter time series, there had been a halving and then doubling over time. Earlier declines were most marked in East Anglia and the East Midlands (Tapper 1992). During and after myxomatosis in European Rabbits, there was a significant increase in Weasel abundance. This was likely to have been as a result of an increase in Field Voles as areas, previously grazed down by rabbits, grew into suitable rough grassland habitat for voles.

The NGC data showed a decline in the number of Weasels killed (per unit area) on game estates from the 1960s onwards. This decline is unlikely to be the result of trapping by gamekeepers for the reasons already discussed. Weasel populations fluctuate with vole abundance, so it is possible that the recovery of rabbits has had a negative effect on Field Vole populations and, consequently, Weasels. Indices of Weasel numbers from the NGC started to increase again from the 1990s but are still below those recorded in the 1960s. In the 1990s,

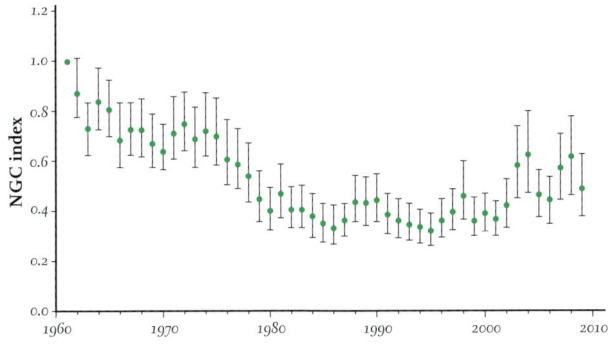

FIG 55. Index of bag density for Weasel *Mustela nivalis* from 1961 to 2009. Redrawn from the Game and Wildlife Conservation Trust's National Gamebag Census.

Harris *et al.* (1995) estimated a British population of 450,000. A more recent review of the status of British mammals (Matthews *et al.* 2018) concluded that it was not possible to provide an updated figure because no further evidence was available. The only indicator of trends from the NGC (Aebischer *et al.* 2011) suggests a 51 per cent increase between 1995 and 2009, but this trend is not adjusted for trapping effort. For these reasons, Weasel is listed as data deficient and the future prospects for the status of both its range and population currently unknown.

For Stoats, the annual NGC records from 1961 to 2001 suggest that there was an overall decline in numbers in the first 20 years, after which the populations stabilised. Data from one estate in Suffolk showed a greater than tenfold reduction in the numbers of Stoats killed after myxomatosis decimated the rabbit population, but this was followed by a recovery as rabbits began to increase again. In 1995, Harris *et al.* (1995) estimated the total population of Stoats in Britain to be 462,000, This was based on the authors' expert opinion of population density estimates for all habitat types. There has since been a 28 per cent increase in the numbers of Stoats culled on game estates reporting to the NGC between 1995 and 2009, but this does not account for trapping effort, so it may reflect an actual population size increase or an increase in trapping effort. The latest estimate for the total population of Stoats in Britain is 438,000 (Mathews *et al.* 2018). This is based on density estimates from Harris *et al.* (1995), so differences in the two estimates are the result of changes between the two reviews in the estimated availability of key habitat types. As is the case for Weasels, the 2018 review concluded that Stoats are data deficient with respect to population size and that the future prospects for both population and range of the species are unknown.

We can see that small mustelids have been hunted, trapped and killed for a variety of reasons throughout their shared history with humans. This reached a peak during the nineteenth century and into the early twentieth century, a period when populations of many carnivores in Great Britain underwent severe declines in range and numbers. Of the small mustelids, it seems that only Stoats and Weasels came through this period relatively unscathed, while European Polecats and Pine Martens were only able to persist in refugia where trapping pressure was the least intensive.

Since the 1970s, several changes have reduced some of the anthropogenic pressures on Britain's carnivores. Land management practices have changed, and legal protections have been put in place for conservation and animal welfare. As a result, martens and polecats have been recovering and recolonising some of their former ranges.

CHAPTER 5

Polecats and Pine Martens: The Bumpy Road to Recovery

After the loss of almost an entire generation of gamekeepers to the First World War and the consequent alleviation of predator control in the early twentieth century, the European Polecat and Pine Marten populations in Britain slowly started to recover. At the end of the war, in 1918, there had followed a brief revival in game shooting but then, with the collapse of agricultural prices in 1921, many estate owners were forced to abandon such extravagances as game preservation and in many cases to sell up altogether. Almost a quarter of the land in England and Wales was sold between 1914 and 1927. The majority of new owners did not have the financial resources to maintain game shoots on the scale seen before the war. By the 1930s, only around 40 per cent of agricultural land in Britain was subject to some form of gamekeeping, and the number of full-time gamekeepers had almost halved to around 14,000 (Tapper 1992).

POLECATS AND RABBITS – A COMPLICATED HISTORY

One of the consequences of the intensive predator culling that had previously been carried out by gamekeepers was its benefit to Britain's European Rabbit population. The Weasels, Stoats and Red Foxes which were controlled to protect pheasants and partridge on shooting estates were also the rabbit's natural predators. As game preservation declined, rabbit numbers continued increasing to what was described as plague proportions. By the 1930s, the rabbit population was estimated to be 60–100 million. A rapid fall in agricultural prices at this time encouraged the adoption of low-input, more economical methods of extensive

FIG 56. Poster promoting the benefits of keeping European Rabbits *Oryctolagus cuniculus* for meat during the Second World War (designed by Frederick H. K. Henrion, 1941).

production, with barbed-wire fencing as a cheaper alternative to maintaining hedges. The result was an increase in scrubbiness that also favoured the rabbit population. This abundance of their preferred prey should have been an enormous benefit for polecats. However, once again it put them into direct competition with humans.

Rabbits had long been regarded as 'poor man's chicken', a cheap meat that was mainly eaten out of necessity by those with low incomes. The Great Depression of the 1930s, followed by the lean years of the 1939–45 Second World War, increased this market and led to a thriving trade in rabbits for their meat and fur. It is estimated that at least 36 million wild rabbits were killed each year in Britain in the 1930s. Across the countryside, thousands of wire snares were set each night so the snared rabbits could be gathered in the following day and carcasses sent to markets. Predators such as European Polecats, Stoats and Red Foxes were routinely removed in areas where rabbits were trapped to prevent them from taking or damaging any snared rabbits and to protect the income from sales.

With the outbreak of the Second World War, domestic food production in Britain had to be increased to compensate for the reduction in food that could be imported. Homegrown crops were a staple component of wartime diet, and agricultural damage by rabbits could not be tolerated. The Ministry of Agriculture initiated campaigns with the aim of eradicating rabbits by any means necessary. This included trapping, snaring, ferreting and gassing (although this last method was unpopular with farmers as there were concerns about whether gassed rabbit meat was fit for human consumption). Whilst eradication was not achieved, these wartime measures did bring about a significant reduction in rabbits. Meat shortages during the war and for some years after meant that rabbit became an attractive meat, for which there was a thriving black market. This intensified pressure on the rabbit population. At the same time, postwar changes in agricultural practices meant a mass grubbing up of hedgerows and scrub clearance, which also had an adverse effect on rabbits. In addition, the outlawing of the gin trap in the 1950s benefited predators such as Stoats and Weasels. However, it was myxomatosis that had the most widespread and devastating impact on rabbit numbers from the 1950s.

Myxomatosis cuniculi was first seen in South America in 1898 and then in California in 1930. In 1933, the Australian authorities asked Sir Charles Martin, an internationally renowned physiologist and pathologist, to carry out a series of experiments on the virus and potential methods of transmission. These showed that in an enclosed paddock area of 500 yd^2 (approximately 420 m^2), myxomatosis was 100 per cent fatal to wild rabbits. However, further experiments on the rabbit-infested island of Skokholm, off the Pembrokeshire coast, were less conclusive. Close contact between infected and healthy rabbits was necessary for disease transmission, but the infected animals showed a tendency to squat in the open by themselves when they were at their most infectious, so he concluded that the virus was not a viable means of rabbit control.

Nonetheless, in 1950, Australia unleashed myxomatosis on its out-of-control rabbit population. In less than three months, myxomatosis had spread 2,000 km and killed 99 per cent of infected animals. In June 1952, the virus was illegally introduced to France by the bacteriologist Paul Armand Delille, to rid his private estate of rabbits. By 1954, it had killed 90 per cent of the wild rabbits in France. The first outbreak in England was at Bough Beech near Edenbridge, Kent, in October 1953. By the end of 1954, myxomatosis had been reported in four counties in southeast England. By 1957, it had been recorded in 34 counties, and six years later there was only one county in England and one county in Wales without any known cases of the disease. The disease decimated the national rabbit population, with knock-on effects for many other species.

Long-term data on rabbits from the GWCT's National Gamebag Census shown in Figure 57 illustrate that there was a relatively stable cyclic trend until the start of the Second World War, then a marked decline until the 1950s. A crash in population numbers can be seen following the advent of myxomatosis, in 1953. This was followed by only a partial recovery, to bag sizes roughly half of what they were at the beginning of the twentieth century. Rabbits are an important prey for both European Polecats and Stoats but, while Stoat populations declined along with rabbits, ironically myxomatosis proved to be something of a mixed blessing for polecats. Although they lost what had previously been an abundant food source, they also benefited from the collapse of the rabbit industry and the associated reduction in trapping pressure. In the absence of rabbits, polecats, along with foxes, were able to switch to alternative prey such as mice and voles. Rabbit numbers began to recover by the 1960s but, by then, it seems that people no longer found wild rabbit meat an appetising proposition. NGC data show

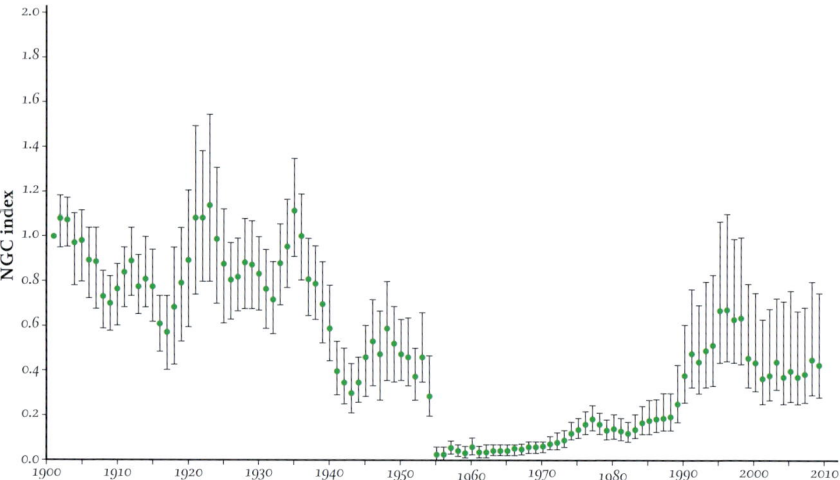

FIG 57. Index of bag density for European Rabbit *Oryctolagus cuniculus* from 1901 to 2009. Redrawn from the Game and Wildlife Conservation Trust's National Gamebag Census.

an increase in the gamebag index up until 2009, with the most rapid rate being between 1989 and 1995, so that by the 2000s, rabbit numbers were thought to be approaching pre-myxomatosis levels (Aebischer *et al.* 2011). More recently, rabbits have experienced significant declines thought to be linked to the spread of new viral diseases such as rabbit haemorrhagic disease.

FIG 58. European Polecat *Mustela putorius* in grassland habitat.

NATIONAL SURVEYS MONITORING THE NATURAL RECOVERY OF POLECATS

Reports suggest that polecats were already expanding their range by the 1960s, but national polecat surveys were then carried out between the 1980s and 2010s, with the most recent being from 2014 to 2015 (Fig. 59). In the 1970s, polecats were still restricted to Wales and the counties of Shropshire and Worcestershire (Tapper 1992). By the 1990s, polecats were recorded in all of the English counties along the border with Wales (Birks & Kitchener 1999). In the 2000s, the polecat was also present in Derbyshire, Buckinghamshire, Berkshire, Wiltshire, Dorset and Hampshire (Birks & Kitchener 1999), as well as appearing in Cumbria, Argyll and Perthshire as a result of unofficial releases. The most recent national survey showed that, by 2015, polecats were widespread across Wales and the West Midlands and are also now established across most of central and southern England (Croose 2016).

Between 1975 and 2015, the polecat's range increased at a rate of approximately 5 km per year. As can be seen from the most recent distribution map, there are currently still gaps in northern England and Scotland. The latest population estimate for polecats is 83,300 (Mathews *et al.* 2018).

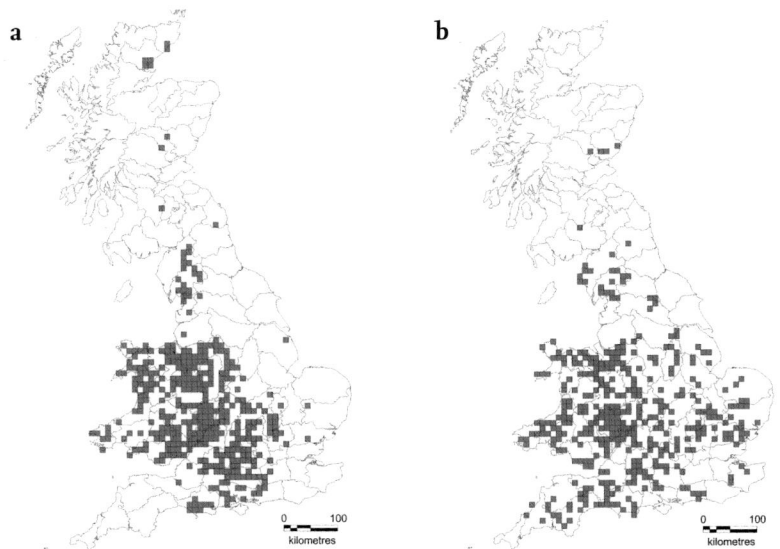

FIG 59. European Polecat *Mustela putorius* changing distribution maps from Vincent Wildlife Trust surveys in (a) 2004–6 (Birks 2008) and (b) 2014–15 (Croose 2016).

The polecat's range expansion and recolonisation has followed the pattern of a diffusion model, whereby population density gradually increases in core areas, which leads to animals dispersing outwards with a 'wave-front' of expansion at the edge of the range. One of the consequences of this is that animals are at low density at the colonisation front, and there is a higher risk, than in the core range, of hybridisation with domesticated Ferrets.

The Ferret is a domesticated form of the European Polecat. Ferrets are kept in captivity as pets or as working animals and have been used to help humans catch rabbits and rats by bolting them from their burrows since as early as the first century BCE. The Greek writer Strabo (64 BCE–24 CE) described Ferrets bred for the purpose being muzzled and sent down rabbit burrows (Thomson 1951). It is thought that polecats were probably first domesticated somewhere around the Mediterranean within the native range of the rabbit. There was some debate about whether domestic Ferrets originate from European Polecats or the Steppe Polecat *Mustela eversmanii*, but recent genetic analyses confirmed that the modern Ferret is derived from the European (or Western) Polecat. Johnny Birks, in his book *The Polecat*, explains that this is unsurprising, given that hunting rabbits 2,000 years ago was the main driver for domestication of Ferrets, but the range of the rabbit and the Steppe Polecat did not overlap at the time (Birks 2015). Because of the previous uncertainty of their origin, Ferrets were considered to be a separate species, originally *Mustela furo* but more recently *Mustela putorius furo*, in acknowledgement of their European ancestry. Ferrets were probably first introduced to Britain by the Normans in the eleventh century, although the word 'ferret' and illustrations of the animal being used for rabbiting do not appear in Britain until the Middle Ages (Fig. 60). King Richard II granted a licence in 1384,

FIG 60. *Women Hunting Rabbits with a Ferret*, from an illuminated manuscript c.1300 (from the *Oxford Junior Encyclopaedia*, 1950, Vol. IX).

FIG 61. Breeders have produced a variety of colours and markings in domestic Ferrets *Mustela putorius furo*. These include sandy, silver, black-eyed white and many more.

allowing one of his clerks to hunt rabbits with Ferrets, and a few years later issued a statute which prohibited the use of 'fyrets' on Sundays. From the sixteenth century onwards, references to Ferrets began to increase in both the scientific and non-scientific literature, and Linnaeus, in 1758, named it *Mustela furo*.

In terms of appearance, Ferrets are very similar to European Polecats in size and shape, but Ferrets have been bred in a variety of different colours and markings. The original colour of the Ferret would have been the same as that of the polecat (fitch), but albinism arose, probably as a result of inbreeding. This was a much-favoured coloration in Ferrets, both for its novelty and for conferring visibility in dense undergrowth, and was selected for by Ferret breeders. Behaviourally, selective breeding has resulted in a social and docile animal with low levels of alertness and fear, so periodically, Ferrets have been crossbred back with European Polecats to increase their hunting instinct. In the 1700s, all Ferrets were described as 'white' (albino) or 'fitch' (polecat). But by the late 1960s, Ferrets started being kept as pets (Fig. 61). In recent years, while Ferrets have increased in popularity as pets, they continue to be used for pest control and in sport, with substantial numbers of working Ferrets in different parts of Britain. The results of a national survey in 2009, carried out by pet food manufacturer James Wellbeloved in collaboration with the Ferret Education and Research Trust (Wellbeloved/FERT 2009), suggested that nearly 20 per cent of Ferrets in Britain were working animals, used to help catch rabbits or kill rodents and, sometimes, other species. In some parts of the country, such as Norfolk, the figure was nearer 40 per cent.

Being so closely related, European Polecats and Ferrets can mate with each other and produce fertile hybrid offspring. This raises the question about how much the recovery of the polecat in the second half of the twentieth century was a result of increases in hybridisation.

WHEN IS A POLECAT NOT A POLECAT?
WHEN IT IS A POLECAT-FERRET

Before the development of molecular genetic methods for identifying species, European Polecats, Ferrets and their hybrids could only be distinguished from each other on the basis of morphological features or appearance, specifically pelage (fur) coloration and markings. Guidelines were developed in the 1990s by Andrew Kitchener and Vincent Wildlife Trust, based on a large study of European Polecat and Ferret skins from the collection held at the National Museum of Scotland in Edinburgh. Animals were categorised according to appearance (phenotype). Those with dark pelage, dark fur on the face extending to the nose with a well-defined white facial 'mask' and classic 'polecat' appearance, were assumed to have a corresponding polecat genotype, or genetic makeup, whereas animals with paler fur, pale paws and cheeks and less distinct markings were classified as Ferrets or Polecat-Ferret hybrids depending on the presence and extent of polecat features.

A study in 1999 by Andrew Davison was one of the first to look at European Polecat genetics. Mitochondrial DNA was extracted from the same museum specimens that had been categorised by phenotype in the earlier study. DNA analysis was used to assess whether British polecats and Ferrets were genetically distinct and to ascertain the extent of hybridisation. The results suggested that all of the polecat and Ferret specimens belonged to one of two distinct cytochrome *b* haplotypes (genetic variants), which differed only by a single substitution. The WP (or Welsh polecat) haplotype was thought to represent the native lineage of the British polecat and was mainly restricted to specimens from Wales and the English border counties. The other haplotype (F, or Ferret) was found elsewhere in England, in Scotland and in some parts of Wales where feral Ferrets (from which it was presumed to originate) occurred. This meant that all of the specimens from outside the core historical range that had been categorised as true polecats based on phenotype were of the Ferret haplotype, highlighting the difficulty of separating polecats and Ferrets by appearance and mitochondrial DNA alone (Davison *et al.* 1999).

Subsequent research by Mafalda Costa at Cardiff University used microsatellite markers to analyse genetic differences from nuclear DNA (Costa *et al.* 2012). Microsatellites are segments of DNA where a sequence repeats. Microsatellites are highly variable and, in combination with mitochondrial DNA, are a powerful means of detecting hybridisation between closely related wild and domesticated species. Costa *et al.* (2013) analysed a much larger sample of polecats and Ferrets than the 1999 study by Davison. They aimed to investigate how genetically distinct polecats and Ferrets are in Britain and to look at whether the nineteenth-century population decline of British polecats resulted in a genetic bottleneck (a loss of genetic diversity when populations are severely reduced). They also assessed how closely the genetic profile, or genotype, of an animal matched with its phenotypic classification. What they found was that some animals that look like true polecats actually have a high proportion of Ferret genes and vice versa. Their study also found that 31 per cent of the wild polecats that they analysed had 'admixed ancestry', meaning that they had both polecat and Ferret genes. The most genetically 'pure' polecats were found well within the polecat's core range in Wales and the west of England, whereas those with a higher incidence of Ferret genes were found along the eastern edge of the polecat's range expansion. No direct offspring from matings between a polecat and a Ferret (first-generation, or F_1, hybrids) were detected. This implies that breeding between the two is probably rare now, and the hybrids seen today are likely a legacy of extensive hybridisation in the past. The pattern of hybridisation and introgression suggests that most of these events occurred as polecats dispersed into new areas outside of the core range where they were more likely to encounter and mate with feral Ferrets than with other polecats. From her analysis of mitochondrial DNA, Costa concluded that hybridisation events at the recolonisation 'front' were most likely between dispersing male polecats and female Ferrets, whose hybrid offspring then mated with polecats. This is consistent with the mating system of polecats, in which males are more likely to disperse while females are more philopatric, settling near to where they are born.

Polecats in Britain have lower genetic diversity compared with other polecat populations in Europe, but this is likely a result of the long-term separation of this population from its counterparts on the continent. The study by Costa (2014) found no clear evidence of a genetic bottleneck in the wild British polecat population that might be attributed to the steep decline in the nineteenth century. This may be because the loss of genetic diversity that would be expected from such a decline has been compensated for to some extent by a long history of crossbreeding with feral Ferrets. In Costa's (2014) study, she also found that 28 per cent of Ferrets from captive stock were admixed as a result of intentional crosses between the two species carried out by Ferret breeders.

As the Ferret is now accepted to be simply a domesticated form of our native European Polecat, it is something of a chicken-and-egg debate regarding which genes belong to which, though there has obviously been some divergence. During the course of domestication, we have favoured docile and sociable Ferrets with coat colours that are highly visible in the undergrowth. This means that as a

FIG 62. Polecats *Mustela putorius* visiting a garden for food.

result of mutations that humans have selected for, Ferrets now have some genes which probably confer a disadvantage in the wild. It is likely that hybrids with the polecat phenotype have higher fitness in the wild, and this is supported by evidence from Vincent Wildlife Trust's periodic surveys which show that as the polecat expands its range, the polecat phenotype is becoming more widespread. However, genetic studies have shown that while polecats can be distinguished from Ferrets in Britain based on appearance, phenotypic features alone cannot be used to identify hybrids of the two species. This is something of a challenge for naturalists and biological recorders, who need to be confident of their identification of polecats and Polecat-Ferrets, especially in areas where polecats are newly recolonising and are relatively rare.

From a conservation perspective, the introgression of domestic genes into the genome of a wild species can be detrimental for two main reasons. Firstly, it may have a negative effect on the fitness and therefore viability of animals in the wild. In the case of the polecat, hybridisation may have been occurring at low levels for hundreds of years, ever since Ferrets were first introduced into Britain, but this probably increased in frequency as polecats started to expand their range after the severe nineteenth-century population decline. A second issue arising from hybridisation is that it can hamper conservation policies and legal protection for wild species if there is confusion over what is 'pure' and what is the status of hybrids. Polecats currently have some legal protection on Schedule 6 of the Wildlife and Countryside Act, making it an offence to intentionally set a trap for polecats without a licence. However, Ferrets have no such protection and can be legally trapped and killed. There is no provision in the law setting out the status of Polecat-Ferrets, nor any way of trapping for one and not accidentally catching the other.

The polecat has made a significant recovery over the past 50 years, although DNA analysis has revealed that the presence of hybrids with polecat phenotype has masked the true range of the polecat in Britain. The putative polecat's distributional range continues to increase, even if the edge of the expansion includes animals with admixed genes from several generations of backcrossing with 'pure' polecats and crosses between hybrids. We need to ask ourselves if it really matters if some of our polecats have a bit of Ferret somewhere in their ancestry. Rather than dismissing an unknown proportion of Britain's expanding polecat population as genetically 'impure', perhaps we should take a more pragmatic approach and accept that if an animal looks like a polecat and acts like a polecat, then we can call it a polecat.

It is even possible that some of the genes inherited from their domesticated relatives might be beneficial to modern polecats in our highly modified human-dominated landscapes. Polecats live side by side with humans in many areas. They appear to have adapted to some level of disturbance and live in farm buildings and under sheds.

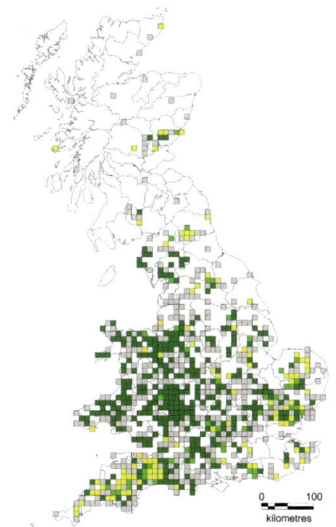

FIG 63. Distribution of 10 km × 10 km squares (hectads) with verifiable records of true European Polecats *Mustela putorius* (dark green), Polecat-Ferrets *Mustela putorius* × *Mustela putorius furo* (yellow) and both true European Polecats and Polecat-Ferrets (lime green) during Vincent Wildlife Trust's 2014–15 survey.

Meanwhile, Wales and the English border counties remain the strongholds of the European Polecat in terms of genetic 'purity'. It is sensible to focus conservation efforts in these areas. By promoting population increase and further natural expansion from the core to surrounding areas, introgression may become

FIG 64. Polecat-Ferret *Mustela putorius* × *Mustela putorius furo* in a garden.

gradually diluted across increasing areas of Britain. This should be monitored periodically, as part of national surveys to assess whether natural expansion of 'pure' individuals from the strongholds is indeed removing the legacy of Ferret introgression from the genetic pool of the British polecat. It is also sensible to try to reduce escapes and releases of fertile Ferrets into the wild, to make sure that the opportunity for any future hybridisation between polecats and Ferrets is minimised. This is especially critical in areas of low polecat population density and high human and rabbit population densities where Ferret escapes are more likely to occur. However, the re-establishment of the genetic identity of the British polecat and the persistence of native extant lineages of the species critically depends on the successful conservation of the Welsh polecat population.

PINE MARTEN RECOVERY IN BRITAIN – THE NORTH–SOUTH DIVIDE

Pine Marten distributional range expansion between 1975 and 2015 has been estimated at approximately 1.7 km per year in Scotland. This is much slower than that of the European Polecat further south over the same period. The difference, however, is unsurprising and relates to differences in the ecology of the two species. Polecats are faster breeding, have larger litters and can occupy a wide range of different habitats. In contrast, Pine Martens are adapted to inhabit old-growth forests which, if undisturbed, are a relatively stable environment. As a result, Pine Martens have steadier, lower productivity, longer lifespans and fairly stable populations with relatively low mortality among the adults. The downside of this is that they are not able to compensate quickly for any sudden increase in mortality. Pine Marten populations are slow to recover from heavy losses, such as those suffered during predator control in the past, and are particularly vulnerable to both local and total extinction.

By the beginning of the First World War, in 1914, Pine Martens had been extirpated from almost all of southern Britain. The only populations that remained were restricted to the northwest Highlands (Ross-shire, Sutherland and Inverness-shire), with some animals possibly persisting in southwest Aberdeenshire, Perthshire and North Argyllshire (Langley & Yalden 1977). At that time, the Scottish Highlands were the Pine Martens' largest remaining stronghold in Britain, and numbers here, although small, were considerably larger than in the few scattered populations that persisted in England and Wales. The population began to show signs of a slow recovery in Scotland in the 1930s, with a reported increase in marten numbers in northwest Sutherland and a southeastward range expansion. As with polecats, this recovery can be attributed to the reduction in trapping pressure from gamekeeping after the First World War (Lockie 1964). Nonetheless, Pine Martens continued to be trapped and killed, albeit to a lesser extent. Pine Martens subsequently reached the north side of Loch Ness by 1946 and became established south of the Caledonian Canal by the early 1960s (Lockie 1964). Records of carnivores from northern Scotland submitted to MacPherson the taxidermist at Inverness between 1912 and 1970 show that 198 Pine Martens ended up there, 40 from Assynt alone. This MacPherson (again no relation to the author) was the largest taxidermist in the area for this period but there were several others. It is reasonable to assume that similar numbers of martens were submitted to them, too. It is also worth noting that in the first national Pine Marten distribution survey, published in 1983, 230 records were of dead Pine Martens, with trapping, shooting and snaring listed as the most frequent cause of death (Velander 1983). This first survey, carried out between 1980 and 1982, indicated that the marten population was continuing to expand its range, although some local extinctions had occurred in the northwest. By the early 1980s, the Highland population had extended east to between the south side of Loch Ness and the Monadhliath Mountains and south to Glencoe and Dalmally, with occasional records in Speyside and Deeside (Velander 1983). However, it is apparent that in the years before Pine Martens became fully protected by law in 1988, when the species was added to Schedule 5 of the Wildlife and Countryside Act (1981), there was only a small increase in reported range. The database of Pine Marten records from the Global Biodiversity Information Facility shows that during the 30-year period from 1929 to 1958, there were Pine Martens recorded in 57 hectads, compared with an increase to 94 hectads in the following 30 years, 1959 to 1988. There can be little doubt that legal protection has been a significant factor in the improving status of martens.

The second wide-scale distribution survey, in 1994, showed that Pine Martens were no longer confined to the Highlands and had recolonised further areas of central and eastern Scotland (Balharry *et al.* 1996). The most recent distribution surveys carried out by Vincent Wildlife Trust in 2012 and 2013 indicated considerable further range expansion, with martens having recolonised much of Sutherland, Caithness, Aberdeenshire, Perthshire, southern Argyll, Stirlingshire, limited areas of western Angus and Fife and part of the industrialised Central Belt (Croose *et al.* 2013, 2014).

The increase in the rate of range expansion observed since Pine Martens were protected suggests that lethal control, which became illegal in 1988, was

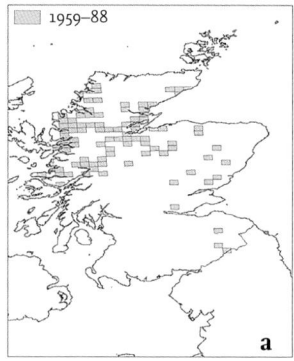
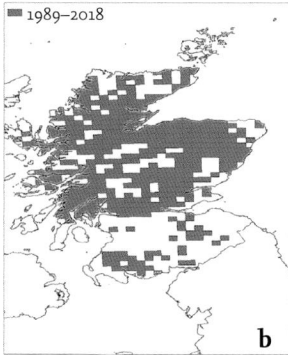

FIG 65. Hectads with Pine Marten *Martes martes* records in (a) the 30-year period prior to legal protection compared with (b) the 30-year period following legal protection. In the 30 years after Pine Martens were added to Schedule 5 in 1988, their range in Scotland increased significantly from 94 hectads to 448 hectads.

FIG 66. Pine Marten *Martes martes* near Oban, a sign of the species' recent range recovery.

having a significant limiting effect on the species' recovery in the years prior to legal protection. Some of the recent range expansion in southern Scotland can be attributed to a reintroduction of a handful of animals which took place in Galloway Forest in the 1980s, and a very small number of recent releases of rehabilitated martens into the Scottish Borders, but these account for a fraction of the newly occupied hectads. In addition to range expansion of the core population, unofficial translocations and releases have led to the establishment of a marten population on the Isle of Mull (Roy *et al.* 2014), as discussed in Chapter 7, and martens are also now established on the Isle of Skye, having colonised from the mainland via the land bridge (Croose *et al.* 2013). By 2018, Pine Martens had an almost contiguous distribution north of the Central Belt, with a distinct but possibly now contiguous population in southwest Scotland and the Scottish Borders. The British Pine Marten population was then estimated to number 3,700 individuals (Mathews *et al.* 2018).

Occasional records of Pine Martens were documented in parts of northern England throughout the twentieth century, suggesting the persistence of a small marten population here. However, there had been no evidence of a population recovery in this area until recently. Pine Martens are now recolonising parts of Northumberland and Cumbria from the Scottish Borders, where martens are becoming more widespread. Since 2018, there have been regular records of Pine Martens in the Kielder Forest block, with occasional records further over the border in northern Cumbria. It is likely that this recolonisation will continue but will be dependent on the availability of suitable habitat, low mortality and minimal conflict with human interests.

PINE MARTENS IN SOUTHERN BRITAIN

It is widely accepted that in the early twentieth century, there were still some very small, isolated remnant populations of Pine Martens in the more remote upland areas of northern England and Wales. But there has been considerable debate about whether and where these 'mythical' martens persisted much beyond that time. In Kathy Velander's national Pine Marten survey of 1980–82, she found no field signs of Pine Martens outside of Scotland; nevertheless, reported sightings by local naturalists suggested that there were still some places south of the border where Pine Martens were hanging on. In 1988, further surveys were carried out in the five core areas of Wales and northern England with the highest likelihood of remnant Pine Marten populations. Field surveys were based on 2 km survey transects to search for Pine Marten scats. Scats that were presumed to be from Pine Martens were found in all five areas, but this identification was based solely on scat appearance and observer judgement. Some of the potential issues with this are discussed in Chapter 8 on survey techniques. Suffice to say, misidentification cannot be ruled out in the days before DNA verification of scats. Even accepting that at least some of the 'positive' Pine Marten records from the 1988 surveys were valid, the conclusion was that all of the populations outside of Scotland were by now at vanishingly low densities and seemed to be contracting. During the twentieth century, there also seemed to be a shift from unequivocal records, such as Pine Marten carcasses, to unverifiable reports of sightings.

Compilation of reported sightings was an accepted way of detecting Pine Martens and assessing distribution before more systematic census techniques were developed. Where other affordable detection techniques are likely to fail, reported sightings can be a source of information on potential Pine Marten presence. However, in the absence of frequent corroborative evidence, caution has to be applied when using sightings from casual observers to determine Pine Marten distribution. This is especially important where the target species is rare and observers are unfamiliar with its appearance, leading to the risk of misidentification and incorrect reports. During the twentieth century, there was a marked decline in the numbers of dead martens recovered that could provide corroboration for sightings. This may be because people were less likely to report Pine Martens they had killed (deliberately or accidentally) once the species was legally protected. The scarcity of supporting evidence added further doubt to the credibility of sightings. Nonetheless, Johnny Birks and John Messenger at Vincent Wildlife Trust made it their mission in the 1990s to develop a robust way of evaluating these sightings records.

UFOS: UNIDENTIFIED FURRY OBJECTS

During the second half of the 1990s, Vincent Wildlife Trust produced and distributed posters and leaflets appealing for recent evidence of Pine Martens

FIG 67. A poster produced by Vincent Wildlife Trust to encourage sightings reports of Pine Martens *Martes martes* in England and Wales.

in England and Wales (Fig. 67). This appeal was promoted through talks, articles in magazines and newspapers and radio and television interviews. To maximise the value of this publicity, the trust attempted to target it both geographically and by interest group. They broadly focussed on Wales and northern England because these included the main areas of previously recorded Pine Marten range. Some people reporting sightings of Pine Martens might be mistaken in their identification, so such reports cannot be accepted uncritically. Therefore, a sightings-based recording system requires a mechanism for discriminating between those reports that probably relate to Pine Martens and those that are more likely to relate to other species.

Birks and Messenger came up with a structured questionnaire that could be completed by an experienced interviewer in approximately five minutes during an informal conversation over the telephone or in person. The main objective of the interview was to gather information to determine the extent to which a reported sighting of a Pine Marten could be considered reliable. The quality of each sighting reported was subjectively scored following the interview. Of the sightings records received between 1996 and 2007, 524 were scored as likely to have been of a Pine Marten (Poulton *et al.* 2006). The distribution of high-scoring sightings showed a highly significant spatial clustering that was independent of the distribution of the original publicity. The highest concentrations corresponded with marten populations identified by earlier studies. When a sighting was reliable and recent, Vincent Wildlife Trust staff went out to search the location for field signs but with very little success. From 1990 until 2010, there were only 16 indisputable records of Pine Martens in England and Wales. These included verified photos of live sightings and carcasses of dead animals. Despite the fact that these records came from the same geographical areas with the highest concentrations of unverified sightings, the sparse nature of unequivocal records gave rise to some doubt as to the origin of these animals. Some asserted that continuous reports and intermittent evidence of Pine

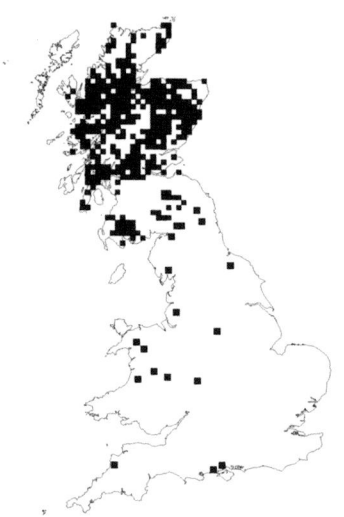

FIG 68. Records of Pine Martens *Martes martes* from Scotland, England and Wales for the period 1990–2014. Data from Vincent Wildlife Trust.

Martens from parts of England and Wales indicated the long-term persistence of 'relict' populations in these areas (Birks & Messenger 2010). Conversely, others argued that these individuals were more likely to be escapes from captivity or illicit translocations and not proof of viable populations in the wild.

In 2012, a study by Neil Jordan et al. (2012) used genetic analyses to investigate the origins of Pine Marten populations in England and Wales. They compared haplotypes, sets of DNA variations that tend to be inherited together, from across Britain, taken from museum specimens (dating from pre-1914 up until 1981) and Pine Marten scats, hairs and carcasses found after 1981. They found no difference between the main haplotype of contemporary populations across Britain. There were two haplotypes, *hap i* and *hap a*, in both the historical and contemporary Pine Marten populations in Scotland, with *hap a* being the predominant one in both periods. However, only *hap i* was found in the historical specimens from England and Wales, whereas all the later samples (after 1924 in England and 1950 in Wales) were of *hap a*. These results suggested that historical Pine Marten populations in England and Wales may have gone extinct in the twentieth century and been replaced by animals that were occasionally released and/or translocated from Scotland. However, the authors noted that the number of samples available for the study was small. As a result, it is possible that *hap a* animals were always present in England and Wales but not represented in the sample of only six specimens from England and seven from Wales.

Prior to this study, a proposal to reintroduce Pine Martens from Scotland to England in the 1990s had been opposed on the grounds that extant populations were still present and could contain relict genotypes of special conservation value. The genetic analyses by Jordan et al. (2012) suggested that this was no longer an issue, as the Welsh/English genetic race of Pine Martens may have already been lost, and therefore reintroductions could be considered.

PINE MARTENS IN IRELAND

There is evidence of Pine Martens on the island of Ireland dating from as early as 2,800 years ago (Montgomery et al. 2014). During this period (the late Bronze Age), woodland cover in Ireland was still high, although starting to decline (Byrnes & Little 2007). It is not known exactly when or indeed how Pine Martens got to Ireland, since it was by then an island, but genetic analysis of the Irish population shows similarities to Pine Marten populations from Iberia and southern Europe (O'Reilly et al. 2021). The fact that there is no evidence of Pine Martens in Ireland before the arrival of people is consistent with the suggestion that martens were one of several mammal species to have been introduced for their potential economic value, either as food or for their pelts (Montgomery *et al.* 2014). Certainly, martens were hunted and highly valued for their fur in Ireland, as they were elsewhere, since at least medieval times. In the early 1500s, an Irish marten skin cost around one shilling, two and a half times that of an otter, which was only fivepence. By the end of the sixteenth century, the cost of a marten skin had increased even further, which is probably a reflection of the increasing scarcity of the animals. Records show huge numbers of skins of other species being exported from Ireland between 1697 and 1819. However, Pine Marten is not among them (Fairley 2001), suggesting that by this time the Irish Pine Marten population had declined to levels where too few furs were available to export.

FIG 69. A member of the Irish population of Pine Martens *Martes martes* with its tree-climbing skills evident.

During the sixteenth and seventeenth centuries, there was also a significant loss of habitat for Pine Martens as Ireland's woodland cover declined to just 2 per cent of the land area. The fortunes of the Pine Marten in Ireland followed a similar pattern to that in Britain: after centuries of being hunted for their fur, martens then had to suffer the label of 'vermin' along with many other carnivorous mammals and birds.

Game shooting began to increase in popularity in Ireland at around the same time as it did in mainland Britain, resulting in widespread intensive stocking of game birds and gamekeepers to manage them. It is almost certain that this resulted in an intensification of trapping and killing Pine Martens along with other predators in order to maximise game bags. James Fairley reports that in the Irish literature of the period, there are frequent mentions of martens being trapped or snared, with comments about how easy they are to trap. They would usually have been caught in gin traps at this time, resulting in horrific leg injuries either from the trap or from the animal trying to bite off its own limb in an attempt to escape. On more than one occasion, a Pine Marten caught (and maimed) in this way was considered enough of a rarity to be sent to Dublin Zoo (Fairley 2001). This rarity was not incentive enough for trapping to stop; however, there were at least two estates (Tomgraney in County Clare and Curraghmore in County Waterford) where the owners protected Pine Martens on their land. (The Pine Martens of Curraghmore later became a focus of study for a group of researchers from Waterford Institute of Technology in the early 2000s, and it was here that I had my first experience of trapping martens with Dr Peter Turner and his team.)

In spite of this level of persecution, Pine Martens were apparently able to persist in every county at least through the first half of the nineteenth century, according to Irish naturalist William Thompson (1856), although we can assume that they were reduced to very small numbers in some regions. Fairley collated records of Pine Martens for each county from 1870 until 1975 (Fig. 70), shortly before the first national survey was carried out, and divided the records into three periods of 35 years each.

Alongside persecution, there was a continuing loss of habitat for Pine Martens. By 1908, it is estimated that forest cover in Ireland had declined to approximately 1.5 per cent of the land area, or 125,000 ha, with a further reduction during the First World War due to fuel and timber shortages. As they had done in Britain, Pine Martens found refuge in some of the sparsely populated, craggy parts of Ireland. In the 1950s and 1960s, there were still Pine Martens on the Inishowen Peninsula, an almost entirely open and windswept area of County Donegal.

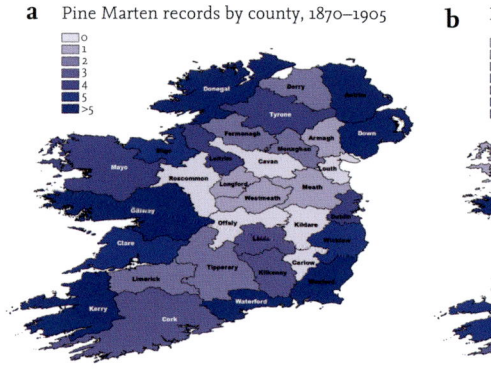
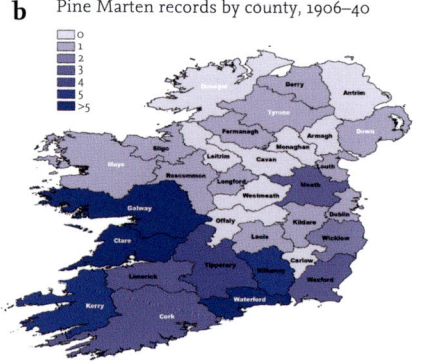
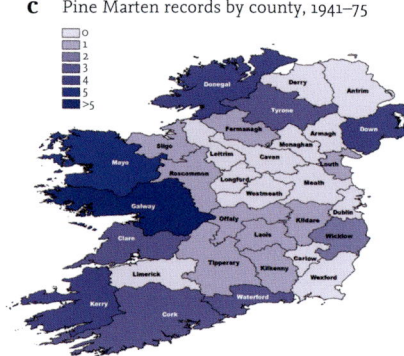

FIG 70. (a) From 1870 to 1905, there were 109 records of Pine Martens *Martes martes* from 26 counties, (b) and from 1906 to 1940, there were 61 records from 23 counties, (c) but the period from 1941 to 1975 yielded just 46 records from only 19 counties. Data from Fairley (2001).

Due to concerns in the 1970s about the status of the Pine Marten in Ireland, the first national distribution survey for the species was carried out between 1978 and 1980. The results of this survey showed that, by then, Pine Martens had declined significantly and undergone a major reduction in range since the 1930s. By the 1980s, the population was mainly confined to woodland and scrub in the midwestern area west of the River Shannon, with smaller populations in the Slieve Bloom Mountains in the midlands and in Counties Meath and Waterford (O'Sullivan 1983). In Waterford, Pine Martens were still found at Portlaw, within the estate at Curraghmore. During the course of his survey, O'Sullivan reports that he met and talked to many families who had, for up to four generations previously, trapped and snared Pine Martens for their pelts. In two counties (Cork and Kerry), this had ended by the late 1950s, due to the scarcity of the animals. However, in County Clare, trapping and snaring of Pine Martens had actually increased during the 1950s and 1960s in response to increased demand for rabbits for food and other mammals for their skins. In spite of this, Pine

FIG 71. Pine Martens *Martes martes* also managed to persist in the suboptimal habitat of sparse Hazel *Corylus avellana* scrub on the limestone pavement of the Burren district in County Clare.

Martens survived in County Clare, while they disappeared from many other parts of their former range. It was suggested that the decline may have been caused by an increased demand for meat and wool in the 1950s, which led to the introduction of government grants to improve sheep farming in the poorer upland areas. To protect their sheep and lambs from predators, farmers laid carcasses laced with strychnine on woodland edges bordering the mountains. Multiple land ownerships meant that some areas were often poisoned repeatedly by different farmers. Pine Martens will readily take carrion as part of their diet and often hunt in edge habitats, so poisoning would have had a disastrous impact on their population. Levels of sheep farming were always much lower in County Clare than elsewhere, which may explain why this area remained a stronghold for Pine Martens.

Following the introduction of legal protection in the Republic of Ireland through the Wildlife Act 1976, Pine Marten numbers built up slowly in its refuges in the west and south. From there, facilitated to some extent by the expansion of forestry, it has spread into the midlands and the northwest. In recent years, small numbers have reached the east coast. Pine Martens were given full legal protection in Northern Ireland slightly later by the Wildlife (NI) Order 1985. In addition, the laying of strychnine baits was banned across Ireland in the 1990s. Legal protection, combined with increases in woodland, particularly since the 1980s, led to a recovery of the Irish Pine Marten population. Forest cover has now increased to around 927,000 ha for the island of Ireland (DAFM 2023, Forest Research 2023). Although this is the highest level of forest cover in Ireland for 350 years and has greatly increased habitat availability and connectivity for martens, at 11 per cent of the land area, it is still relatively low, and only a very small proportion (less than 2 per cent) is native woodland, with the majority being comprised of non-native commercial conifer.

Further island-wide Pine Marten surveys in 2005 (O'Mahony *et al.* 2006), 2012 (O'Mahony *et al.* 2012) and again in 2019 (Lawton *et al.* 2020) have recorded steady range expansion and recolonisation. Pine Martens are now present in every county in Ireland once more, and the population has recovered to such an extent that its conservation status is listed as favourable and of Least Concern, according to the International Union for Conservation of Nature (IUCN) Red List categories (Marnell *et al.* 2009). The fact that Pine Martens have recovered their range in spite of relatively low forest cover shows the resilience of the species and its ability to make use of suboptimal habitat. However, the historical bottlenecks that both the British and Irish Pine Marten populations have been through have left a genetic legacy that may have an impact on the conservation and recovery of the species in future. A recent study found that the Irish Pine Marten population has a low effective population size and low genetic diversity relative to those in continental Europe (O'Reilly *et al.* 2021). It is not known what impact this might have, but the authors of the study reported that values are below the accepted thresholds for long-term population viability. They suggested that further losses in genetic diversity might affect the population's ability to adapt to future environmental change. Therefore, it might be time to consider the possibility of translocating animals from elsewhere in Europe to supplement the genetic diversity of Pine Martens in Ireland.

CHAPTER 6

Pine Marten Reintroductions to Wales and Beyond

Pine Martens undoubtedly survived in some parts of Wales well into the twentieth century. Occasional unequivocal records between the 1980s and early 2010s suggested the ongoing persistence of martens in small numbers in certain areas (Birks & Messenger 2010). However, despite these occasional records, extensive survey efforts from the 1990s to 2010s failed to detect evidence of Pine Martens in viable numbers. It was therefore concluded that any remnant 'population' was probably functionally extinct and highly vulnerable (Jordan 2011). At this point, discussions around translocations began to gather momentum. An external stakeholder group was set up to focus on the objective of restoring Pine Martens to England and Wales. A strategy document was produced, which concluded that intervention, in the form of translocations, would be necessary to prevent complete extinction of the Pine Marten south of the Scottish border and to restore viable populations of martens to their former range in England and Wales (Jordan 2011).

Conservation translocation is defined as 'the managed movement of animals or plants from one location to another to achieve a measurable conservation benefit for the population, species or ecosystem' (IUCN 2013). This covers a range of interventions, from reinforcement, when animals are released into an area with an existing (but often small) population, to reintroduction, when a species is restored to part of its natural range where it has become extinct. In recent years, the spectrum of translocations has expanded beyond the dictates of historical species ranges to also include conservation introductions. These are more risky interventions, where the objective is to establish new populations of a species beyond what has previously been its natural range (Seddon 2010, IUCN 2013).

Nonetheless, it is acknowledged that these will sometimes be required to respond to habitat change from causes including anthropogenic impacts and climate change.

Reintroduction as a management tool is increasingly used to re-establish wildlife populations in situations where natural recovery or recolonisation is unlikely. Nevertheless, all animal translocations are inherently complex, high-risk activities, and reviews of the outcomes have often reported low rates of success (Fischer & Lindenmayer 2000, Berger-Tal *et al.* 2020). The apparent ease with which some species (including small mustelids) have established and become invasive after being introduced outside of their native range (discussed further in Chapter 7) belies the challenges of conservation translocations where species restoration is the primary objective. The science of reintroduction biology has evolved over recent decades as a result of these challenges, along with guidelines for the justification, design and implementation of conservation translocations. A series of these has been published by the Conservation Translocation Specialist Group of the IUCN (2013). The IUCN guidelines are also now incorporated into UK-specific codes of good practice in Scotland (NSRF 2014), England and Wales (DEFRA 2021).

The first question to ask when considering translocations is whether translocation is necessary, or the best option. Habitat management to improve connectivity and the likelihood of natural recolonisation is an alternative conservation action with fewer risks and lower costs. In 2011, this was not a practical option for restoring Pine Martens to southern Britain. Wales and large areas of England are relatively isolated from what was then the main distributional range of Pine Martens in Scotland. The distances involved, as well as intervening areas of unsuitable habitat and urbanisation, made it highly unlikely that Pine Marten populations in Scotland would expand southwards much beyond some counties in the north of England. In 2014, I began work at Vincent Wildlife Trust on a feasibility study to determine if and where, in southern Britain, Pine Marten translocations would be appropriate and most likely to succeed and to identify potential donor populations and priority regions for Pine Marten releases (MacPherson *et al.* 2014).

Habitat suitability and quality are among the most important factors in determining the success of species reintroductions (Griffith *et al.* 1989, Wolf *et al.* 1996). Therefore, it is vital to put a significant amount of effort into properly evaluating areas before considering them for reintroductions. It cannot be assumed that all the historical range of a species will still provide sufficient suitable habitat for a reintroduced population to establish, reproduce and persist. In fact, many things may have changed since a species was last found in an area. The more time that has gone by between local extinction and a planned reintroduction, the greater the likelihood that the habitat and other conditions will no longer be suitable. This means there is a need to evaluate each area regardless of historical occupancy. Detailed knowledge of a species' ecology can provide information on the likely current suitability of any proposed release site, but it is now possible to use modelling methods which enable that knowledge to be set within a landscape context, used to simulate different scenarios and make comparisons between different options. This means that the effectiveness of species reintroduction programmes can be maximised.

Models can be used to predict the suitability of habitat for a species in unoccupied areas based on an assessment of habitat attributes at locations where the species is known to exist. Using these methods at a landscape scale, we identified six regions in Wales and England with large enough areas of high predicted habitat suitability to support viable populations of Pine Martens. Then, for each of these regions, we carried out more detailed analysis to look at factors likely to have an impact on population establishment and spread. These included woodland patch size, connectivity and prey availability. Even a low rate of additional mortality (i.e. in addition to the usual 'natural' mortality) will increase extinction risk and jeopardise the establishment of a population while numbers are still small. Roads can be a significant cause of mortality for Pine Martens. Therefore, the total length of roads in each region and the percentage of those within woodland, combined with the annual volume of traffic, were used to assess the relative likelihood of marten mortality due to road traffic accidents.

Habitat modelling was followed up with ground-truthing of the results. The Pine Marten is a generalist predator and in Britain its diet includes small mammals, fruit and berries, birds, invertebrates and carrion. Pine Martens preferentially den above ground in tree cavities, birds' nests and squirrel dreys, but they will also den in cairns, burrows, tree roots and brash piles. Extensive field surveys were carried out to evaluate both structural and species diversity of woodlands in the proposed release area, and to look at the extent of ground cover and denning sites and the availability of prey and other food.

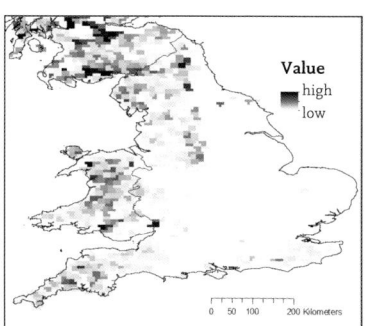

FIG 72. Habitat suitability map for the Pine Marten *Martes martes* in Britain based on MaxEnt model predictions. Darker shades of grey represent higher predicted habitat suitability per hectad. From MacPherson *et al.* (2014).

The results of the feasibility study suggested that the large expanse of well-connected woodland throughout the Cambrian Mountains in central Wales provided a habitat network with the potential to support a viable population of Pine Martens. The total area of woodland combined with low density of roads and traffic in central Wales made it a good candidate for conservation translocations of Pine Martens. This is also a region from which there had been recent DNA-confirmed evidence of Pine Martens (in 2007 and 2012), so releases here constituted a reinforcement rather than a reintroduction. Field surveys were then carried out to inform the final decision on the most optimal release sites.

PEOPLE AND PINE MARTENS

Alongside the biological feasibility of translocations, it was essential to evaluate the social feasibility, and the potential human impacts and consequences of translocating Pine Martens to Wales. All aspects of species restoration projects should involve local communities from the outset, and everyone's concerns should be listened to and addressed if possible before carrying on. It has been shown that translocations can be seriously compromised by opposition from the local community, and this is especially true for predators (Breitenmoser 1998). David Bavin, one of my colleagues at Vincent Wildlife Trust, had carried out a public opinion survey across Wales in 2012, and the results suggested that the majority (almost 90 per cent) of people were in favour of action to prevent the Pine Marten from becoming extinct in Wales. The main reasons for support were the Pine Marten's native status, a perceived positive effect on biodiversity, and its contribution to a balanced environment. Nonetheless, more detailed conversations needed to take place with stakeholders and other land users in the prospective release areas in order to gauge local levels of support (or otherwise) for the project, answer questions and identify any specific issues. All the information gathered during this stage was used to inform the final decision on whether and where releases should take place. Following these discussions, north Ceredigion was agreed upon as the most suitable area for releases.

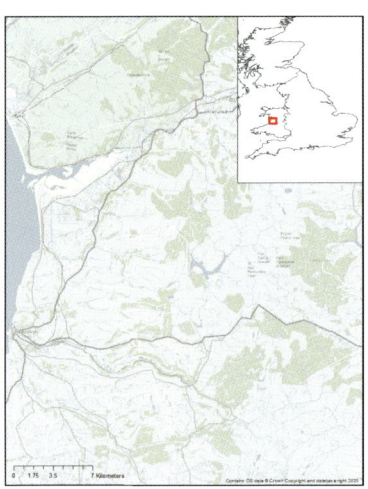

FIG 73. Detail of Pine Marten *Martes martes* release region and (inset) location in Wales.

FIG 74. (a) & (b) The release sites were in a series of well-connected forests in mid-Wales surrounding Devil's Bridge, approximately 16 km southeast of Aberystwyth. All are owned and managed by Natural Resources Wales, the Statutory Nature Conservation Organisation responsible for managing the natural resources of Wales.

TRANSLOCATIONS: A RISKY BUSINESS

While David Bavin was canvassing public opinion about Pine Marten translocations to Wales, one of the most frequently voiced concerns was about predation of other wildlife. There is often a perception that the recovery of a native predator will have a negative effect on native prey species. The possible ecological roles that the translocated species will have in their new environment must be considered and care taken that the conservation interests of other species and habitats are not likely to be jeopardised by the translocation. Introduced non-native predators can have a severe impact on naive prey populations, and examples of this are numerous. Both Stoats and American Mink have devastated populations of some bird species in circumstances where they have been introduced outside of their native range. However, where predators and prey have coevolved over a long period of time, prey species often develop very effective defences. These can be morphological (physical) or behavioural adaptations which reduce their rate of encounters with predators or increase their prospects of escape if detected (Lima & Dill 1990). Across its range, including Scotland and elsewhere in Europe, the Pine Marten coexists with many potentially vulnerable rare bird species. Pine Martens are territorial, have large home ranges and live at low population densities, so their impacts on rare birds are likely to be lower than commoner predators such as Stoats and Red Foxes. However, where a prey species has already suffered significant declines as a result of habitat loss, fragmentation and other factors, its vulnerability to even slight changes in predation rate may be increased. This has been the case for populations of many wild birds in the UK, including woodland specialists and long-distance migrants (Gregory *et al.* 2007, Hewson & Noble 2009). Therefore, in some circumstances, the recovery of a formerly very rare or absent native predator could have a negative impact. A thorough risk assessment, evaluating the ecological roles of translocated animals in their new environment and potential impacts on other species already present in release areas, is an essential aspect of any translocation and was carried out as part of the feasibility studies for the Pine Marten releases in Wales (and later in Gloucestershire).

Another risk of translocations is that of disease. When wild animals are moved from one place to another, there are potential associated disease risks. The animals could carry parasites or pathogens with them that will cause harm to other animals or humans at the new location. There is also a risk that the animals being moved will encounter new diseases in the destination ecosystem and be harmed by these, or that the translocated animals will alter the disease ecology at the release location in some way (such as becoming a new vector, or host species, for an existing parasite or pathogen). An essential element of the planning and preparation for any translocation is a detailed disease risk analysis (DRA). This determines the extent of these risks, and the magnitude of any potential consequences, should they occur. The results can then be incorporated into the final decision of whether or not to proceed with the translocation. If the decision is to proceed, a mitigation strategy that is proportionate to the risks identified in the DRA can then be designed and implemented. The DRA for the translocations to Wales was carried out by Dr Alexandra Tomlinson, a wildlife health and veterinary consultant and specialist in wildlife population health. It was reviewed by a panel of experts in this field and then presented to a wider group of stakeholders in order to reach agreement on what was an acceptable level of risk associated with the translocation and appropriate mitigation measures to put in place.

FINDING A SUITABLE SOURCE: PINE MARTENS IN SCOTLAND

As well as ensuring that the proposed release area was suitable and there were enough resources to support sufficient numbers of Pine Martens to establish a viable population, we needed to find a suitable donor population as a source of animals for translocation. For any wild population, it is important to assess the risk to the donor population of removing individuals for translocation. Results of population viability analyses, supported by experience from translocations of other closely related marten species elsewhere, suggested that to maximise the likelihood of establishment, a minimum of 30–40 Pine Martens would need to be released into an area of woodland sufficient for this number of adjoining marten territories (c.10,000 ha). Wherever possible, wild-caught animals should be used for carnivore reintroductions because they generally show higher survival and better adaptation to new environments than captive-bred animals (Jule *et al.* 2008). The genetic provenance, morphology, physiology and behaviour of source populations should be similar to any remaining wild populations. Ideally, animals should be sourced from areas with similar prey species, competitors, predators and habitats because the animals will then find it easier to adjust to their new home. After a study comparing the haplotype composition of historical and current Pine Marten populations in England, Scotland and Wales had found no differences between the main haplotype of contemporary (post-1950) populations across the UK (Jordan *et al.* 2012), it was concluded that the Scottish Pine Marten population was a suitable source of animals for translocations to Wales. From a practical and logistical point of view, sourcing animals from elsewhere in Europe would have massively increased the costs and complications.

Sites in Scotland that were likely to contain suitable Pine Marten donor populations were identified on the basis of their high woodland cover and length of occupancy by breeding martens. These were all in the Highlands, north of the Great Glen, where martens were present throughout the history of their decline elsewhere in Britain. We then carried out field surveys at all of these sites to confirm that Pine Martens were present in sufficient numbers so the removal of 2–4 individuals at the end of the breeding season was unlikely to have an impact on population viability. Forestry and Land (formerly Forestry Commission) Scotland supported the project, providing help and advice and allowing access to sites for surveys and trapping.

We decided to translocate only a small number of Pine Martens from each of a large number of sites to maximise the likely genetic diversity of the translocated animals and to minimise the potential impact on the donor populations. Pine

FIG 75. (a) Pine Marten *Martes martes* habitat in Scotland, and (b) the author (right) and Huw Denman carrying out a scat survey.

Marten populations can be susceptible to overharvesting when too many animals are removed (Helldin 2000); therefore, the total number of Pine Martens that could be sustainably taken from each donor site was assessed based on the most conservative site-specific estimates of minimum numbers present and population viability models. Following initial discussions early on in the project, formal applications for the relevant licences and permissions to trap and remove specific numbers of Pine Martens were submitted to Scottish Natural Heritage, now NatureScot, and to the landowner, Forestry and Land Scotland.

The aim of the project was to translocate and release Pine Martens in sufficient numbers to restore a viable population in Wales. To ensure long-term success, it was also important to establish community-wide support and community ownership of the Welsh Pine Marten population. If successful, there would be a long-term increase and persistence of the population and an expansion of its distribution in Wales. Success would also provide a blueprint for translocating Pine Martens to other areas where they have become scarce or extinct.

Short-, medium- and long-term success indicators were established at the outset of the project. A long-term programme of monitoring was implemented to enable achievements to be measured against these targets and to identify if and where procedures needed to be adapted. These included:

1. A minimum of 40 adult Pine Martens in approximately equal sex ratio translocated over two years with no loss or injury.
2. Post-release animals remaining within c.10 km of release sites, stable home ranges established, and overall recorded annual adult mortality not exceeding 30 per cent.
3. Evidence of successful breeding by some translocated females in the first year following their release; survival of their young born in Wales.
4. Minimal conflicts with people and widespread community involvement.
5. Persistence, increase and distributional range expansion of the Pine Marten population in Wales over time.

PLANNING, TRAPPING, TRANSLOCATION AND RELEASE

As part of the feasibility study and associated translocation plan, we assessed the number, age class and sex ratio of individuals required to maximise the chances of the translocation achieving its goals. Historically, there have been declines and local extinctions of many populations of *Martes* species, including Pine Marten and the closely related American Marten *Martes americana*, Sable and Fisher. In attempts to reverse these, a number of reintroductions and reinforcements have taken place, not all of which have succeeded in re-establishing viable populations. Roger Powell and others carried out a study in which they modelled reintroductions of martens and Fishers to predict the variables associated with population persistence, and then tested these model predictions with data from actual reintroductions of these species (Powell *et al.* 2012). The model predicted that the probability of establishing a population increased with more adult females released and more release sites used. In the model, the number of males released did not affect reintroduction success, provided that an effective few were released. However, in analyses of data from actual *Martes* reintroductions, the number of males released did have a strong effect on success, in addition to the number of females released and the number of release sites. The model of Powell *et al.* suggested that releasing more than one adult male for every four adult females should not affect reintroduction success. However, their analysis of real data from reintroductions showed that the number of males released actually had a significant effect on reintroduction success. They suggested that this may be because, while the model assumed that all adult males are equally successful breeders, this is unlikely to be true. There can be a large reproductive skew in males of solitary carnivore species. If this is true for male martens (of all *Martes* species), it could explain the contradiction between the model and data from actual reintroductions. Few males might be needed in reintroductions if all those males are effective breeders. However, it may be difficult to identify which will be effective breeders when they are first trapped, and the data from actual reintroductions suggest that plenty of males are needed in order to ensure that enough effective breeders are released. Differences in survival between the sexes have also been found (Hodgman *et al.* 1994, Bull & Heater 2001, Ruette *et al.* 2015). For example, it is likely that male martens have a higher risk of mortality from roads than females because of their larger home range. Therefore, it is probably best to aim for releasing equal numbers of male and female Pine Martens. Population viability analysis suggested that a minimum of 30–40 Pine Martens should be released in an area to have a reasonable likelihood of establishing a self-sustaining new population; this meant we were aiming for at least 15–20 adult animals of each sex.

The whole translocation process was designed to minimise stress to the animals. While short-term (acute) stress is a natural part of life for any animal and provokes the appropriate physiological and behavioural responses, prolonged exposure to a stressor can result in chronic stress, when a series of acute stress

FIG 76. Pine Martens *Martes martes* mating.

responses merge. This can have a number of negative effects, including an impaired immune system and disruption of normal behaviours, such as failure to escape from predation. Translocations have a high potential to cause chronic stress because the steps involved can initiate multiple, consecutive acute stress responses. The stressors during a translocation include capture and handling, captivity or any form of prolonged restraint, transport, and release into an unfamiliar location. The effects of chronic stress in the initial phase of a translocation have the potential to reduce the likelihood of success. For example, if the newly released animals have high rates of post-release mortality, low reproductive success, or disperse away from the release site, it will reduce the capacity for this population to become self-sustaining (Armstrong & Seddon 2008). While the direct causes for translocation failure may be external factors such as predation or disease, the animals' vulnerability to these is increased by chronic stress. The potential negative consequences of stress for our Pine Martens during the translocation programme included deaths during translocation, but also after release, perhaps as a result of disease, starvation or predation, as well as reduced breeding success. The detrimental effects of chronic stress may affect animals for some time after release until they have fully adapted to their new environment (Dickens *et al.* 2010). It was therefore essential for us to examine the translocation pathway to identify all the acute stressors and take care to minimise them, where possible.

Capture methods, holding, transportation and release protocols were all designed to minimise stress for the animals and maintain the highest standards of animal welfare. Trapping and translocations were carried out in early autumn in 2015, 2016 and 2017 from 18 forests across northern Scotland, previously identified as sustainable donor sites. By this time of year, spring-born young are independent, adults have mated in late summer, and food availability is high. Pine Martens were captured under licence from Scottish Natural Heritage (now called NatureScot), the public body responsible for Scotland's natural heritage, using approved Tomahawk 205 cage traps, baited with whole hens' eggs, peanuts, jam and fruit. The traps were also 'flagged' with an attractive scent lure (Blackie's 'Magnum Call'). This was a glutinous concoction derived from skunk anal glands that could be smelt from a mile off. (The trapping team drew straws each day for which of us would be on lure duty and inevitably end up getting some of it on our hands or clothing, however much care was taken to avoid it.) Each trap was fitted with a wooden bite bar inside the door to minimise the risk of injury to captured animals from chewing the wire mesh trap. Traps were covered with a layer of hay that could be pulled in for bedding and also served as a diversionary activity. This was topped with a thick layer of moss and other vegetation for camouflage and insulation.

FIG 77. The author pre-baiting a Pine Marten *Martes martes* trap. After a two- to three-week period of pre-baiting, during which traps were locked open, they were set at dusk and checked just before dawn the next day.

Captured animals were initially assessed in the trap, and those that appeared to be adults in good condition, and therefore candidates for translocation, were taken to a mobile veterinary unit nearby. Here, the martens were lightly anaesthetised by Alex Tomlinson and Alice Bacon, our wildlife veterinarians, and given a full health check. Data were collected on each animal's sex, weight, body condition, measurements and breeding status, and blood samples taken for further tests. Age was approximated (young of the year, juvenile, adult, aged adult) by tooth-wear assessment (Fig. 78). By late summer/early autumn when we were trapping, it is almost impossible to distinguish young of the year based on their size alone, as by this time they are fully grown. However, the canine teeth of a young marten in its first year are still very sharply pointed before the tips become rounded off with use. They are also discernibly whiter than those of adult animals. In addition to the teeth, the breeding apparatus of juvenile martens is a giveaway. The nipples of females are almost impossible to find in their first year, while the baculum (penis bone) of a young male has usually not yet reached its full adult length (of around 50–55 mm).

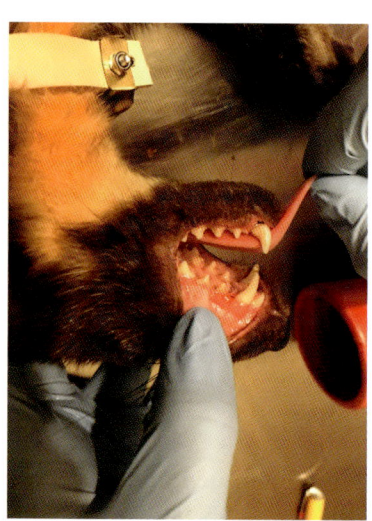

FIG 78. Assessing the age of a Pine Marten *Martes martes* from tooth wear.

FIG 79. The author offering food to a Pine Marten *Martes martes* during one of the rest stops en route to Wales.

Animals not suitable for translocation (not of breeding age, or a surplus of either sex) were released back at their capture site as soon as possible.

Pine Martens selected for translocation were microchipped, in the same way that pet cats and dogs are, so that if they were subsequently trapped in Wales, or found as a roadkill, they could be scanned and identified from the microchip's unique code. The throat patch was photographed for future visual identification, as they all differ individually, and a hair sample was taken to look at how related each of the animals might be and as a record of their genetic fingerprint. They were then fitted with a Biotrack radio collar, incorporating a mortality sensor to monitor post-release survival. Some of the larger (male) animals were fitted with a combined VHF and GPS collar to gather more detailed movement data. Total collar and transmitter weights did not exceed 3 per cent of the body mass of each animal, as recommended (Macdonald 1978, Coughlin & van Heezik 2015, Wilson et al. 2021). Once the animals had fully recovered, they were transported to the release area overnight by road in a specially adapted vehicle. During the journey, stops were made approximately every three hours to check on the animals and offer them fresh food and water.

The release protocol was carefully considered to reduce stress and maximise post-release survival. The options considered were hard release, where animals are released immediately on arrival, or soft release, where they are kept in temporary captivity at the release site. Soft release allows animals to acclimatise to a new site and recover from the translocation process before being released. It can also reduce any homing instinct and allow for the development of social relationships in some species. However, being kept in temporary captivity can induce additional stress and increase the risk of injury if animals try to escape.

For our translocations to Wales, Pine Martens were all soft-released, as it had been shown previously that animals acclimatised to a release site in this way were less likely to move long distances following release (Davis 1983). Large (3.6 × 2.3 × 2 m), temporary, pre-release pens were built by Chester Zoo, one of the project partners, and then flat-packed as panels so they could be transported to the release sites by staff from Vincent Wildlife Trust and the zoo with the help of volunteers (Fig. 80). Pens were sited in large, open canopy areas of woodland near Field Vole–rich habitats, such as rough grassland. This presented some interesting challenges, as the best places to site the pens were well off the beaten track, where everything had to be carried some distance from any vehicle access, and all the groundwork to make the sites as level as possible had to be done by hand. Pine Marten conservation is not for the faint-hearted!

Each pen held only one animal at a time, and where possible, we located the pens so that each male's pen was within 500 m of a female but at least 2 km (the approximate diameter of an average home range) from the nearest male. There was a wooden den box in each pen, mounted approximately 1.5 m off the ground. Whilst in a pre-release pen, each animal was monitored remotely by camera for any visible signs of stress, including abnormal behaviour or loss of appetite. The radio collars were also monitored to make sure they fitted comfortably and worked properly. The pens were visited briefly once a day to provide fresh food and water and collect scats. Blood samples from all translocated Pine Martens were sent to Glasgow University Veterinary School to be tested for canine distemper by virus neutralisation assay. Canine distemper virus (CDV) causes serious infectious disease worldwide in both domestic and wild carnivores, including mustelids. In Britain, the CDV status of wildlife species is unknown, but in domestic dogs, infection is thankfully rare due to widespread vaccination. Nevertheless, the potential for transmission between unvaccinated domestic dogs and wildlife may exist in areas where there are frequent interactions. All of the Pine Martens we tested were seronegative for CDV, but the temporary captivity for soft release meant that we could wait for the results of blood tests for each animal before releasing it. After a maximum of seven days in the pre-release pen, the door of the pen was propped open before dark so the animal could find its way out during the night.

In the autumn of 2015, 20 Pine Martens (10 males and 10 females) were translocated to Wales. After the trapping in Scotland was completed, I returned home and joined the team in Wales to help them in their mammoth task of radio tracking, by this time, all 20 martens each night. These first animals had been released into an area with no resident Pine Marten population and so, understandably, some of them had gone exploring over a fairly large area before

FIG 80. The team from Vincent Wildlife Trust, Natural Resources Wales and Chester Zoo assembling pre-release pens in preparation for the first translocations of Pine Martens *Martes martes* to Wales.

FIG 81. Pine Marten *Martes martes* leaving a pre-release pen. A maximum of two animals were released on any one night to minimise stress on the field team, whose herculean task it was to radio track them all. Supplementary food was provided at each release site for as long as it continued to be taken (up to six weeks).

choosing where to settle. It was both exciting and exhausting to follow them in these early weeks and months. At the end of one of these nights after David Bavin and I had been tracking through the forests and trying to second-guess where each animal might head off to next, we were rewarded with a particularly strong signal off to the side of the forest track that we were driving along. The next minute, the Pine Marten whose transmitter we were listening to suddenly appeared in the headlights, looked at us indignantly for a second and then disappeared again into the trees. In that fleeting instant, Dave and I allowed ourselves the briefest moment of self-congratulation: there were Pine Martens in these woods again – we (and the team) had done that.

The following autumn, a further 19 martens were released into the same areas. In September 2017, the third and final tranche of 12 martens was brought from Scotland to join them, bringing the total number of animals released to 51.

A MATTER OF LIFE AND DEATH – BREEDING AND MORTALITY

After release, each animal was radio tracked intensively until they had established their home ranges, after which efforts were made to locate them at least weekly. Radio tracking was carried out by Vincent Wildlife Trust staff with the help of local volunteers. From the following March–April, when kits are born, there was a further period of intensive radio tracking, with the focus on females to locate possible breeding den sites.

Trapping in Scotland had been carried out immediately after the marten mating season to maximise the chances of translocating females that had already been mated. Pine Martens, as already described, have a period of delayed implantation in their reproductive cycle. Mating takes place in July–August, and then the fertilised egg, or blastocyst, is held dormant until early the following year. Around February, if conditions are suitable, the blastocyst implants and gestation begins. The kits are born usually in late March to early April in Britain. At about this time, in spring 2016, some of the translocated females that we were radio tracking began staying very close to their chosen den site, which suggested that they might have had kits, but we had to wait until the middle of May to find out for certain. By this time, the kits would be quite well developed, and the females would leave them to hunt for longer periods of time, so we were able to check the dens. The first of these checks was a very tense affair as there was a lot riding on it. We hoped that enough time had passed since their release for the Pine Martens to have recovered from the stress of translocation, combined with

FIG 82. One of the first females we released (PM02) had been radio tracked recently staying very close to this one spot and denning in one of the wooden den boxes high up on a tree near the top of the valley.

the challenges of establishing new home ranges in an unfamiliar environment. However, the effects of stress can last for many weeks and can reduce or even prevent reproduction in some species. For example, the Fisher, a relative of the Pine Marten, has been reintroduced to parts of its former range in North America. Fishers and martens have much in common, including very similar reproductive cycles. During four years of Fisher translocations to California and Washington from 2008 to 2011, the effects of timing of release on female Fisher reproductive rates was studied by Aaron Facka and others (2016). They found that denning rates, as an index of birth rate, for female Fishers that were released in early winter (November–December) were significantly higher than those for females released in late winter (January–March). We hoped that by timing our releases in late autumn, we had given our females the best chance of breeding success. Just before dawn on a chilly morning in May 2016, David Bavin, Josie Bridges and I arrived at a secret location in the steep, wooded Rheidol Valley in Wales.

Josie had been checking PM02's movements every day and assured us that she usually left the den box at around six each morning. We were keen to be there when she left so that we could put a ladder up and peep into the den box. We were hoping she had kits in there. Our vigil started just before first light, and we could hear the signal from her radio collar, so she was still at home. We retreated a few hundred metres so she could leave without seeing us. An hour passed and she was still in there. The morning slipped by and she didn't leave the den box. We ate our packed lunches and took turns quietly approaching the den box to check the radio signal. This went on all day until finally, more than 12 hours after we'd arrived, we heard her leave and zigzag away over the top of the hill to hunt for food. We carefully propped our ladder up against the tree under the den box, gingerly lifted the lid and peeped inside to be rewarded with the sight of two healthy-looking kits curled up asleep inside (Fig. 83). After silently replacing the lid of the box with mimed jubilation, we took ourselves off to the nearest pub to celebrate this milestone event. Happily, further similar (though less protracted) encounters that spring confirmed that at least four of the first translocated females had successfully bred and, in total, we were able to count five hearty-looking, Welsh-born Pine Marten kits.

From March onwards over the next two years, while Pine Martens released the previous autumns were still radio-collared, intensive radio tracking of the females continued so that we could determine if any of them were showing signs of breeding – for example, by remaining close to one particular den site. Each year in mid-May, accessible-likely breeding dens were checked, under licence from Natural Resources Wales. Extensive mature conifer woodland, like much of the plantation forestry in the release area, provides martens with plenty of cover. However, even-aged plantations often do not provide good denning opportunities, generally due to the lack of deadwood habitat and hollow trees. Pine Martens often use tree cavities for breeding dens so they can rear their kits safely above ground away from predators for the first weeks after birth. To ensure that there were enough suitable dens available, a network of artificial den boxes was put up throughout the release area and some of these have certainly been used by martens. However, we also tracked animals to the many natural den sites that they had managed to find, including tree holes, squirrel dreys, rocky ledges and derelict buildings. Where Pine Martens bred in natural den sites, rather than artificial den boxes, they often challenged our ingenuity. We had some particularly memorable moments trying to access a den in a cavity more than three metres up a tree growing on a very steep-sided Welsh valley. A complicated arrangement of extendable drain rods to which an endoscope camera was attached came in handy. Camera traps were also used at potential

FIG 83. The first Pine Marten *Martes martes* kits for many decades, seen in Wales in spring 2016.

natal den sites, along with a combination of other evidence. All this detective work confirmed that at least five females bred in spring 2017, resulting in at least 10 Welsh-born Pine Marten kits, four of which were born to PM16, who had been released in autumn 2015. This meant that hers were the first-known Pine Marten kits conceived and born in Wales for decades, which was another big milestone for the project.

The battery life of the radio collars was somewhere between 12 and 14 months, so we made every effort to re-trap the animals and remove the collars while we could still locate them. This was also a good opportunity to check the martens' weights and general condition before letting them go for the last time. Once the

FIG 84. Female Pine Marten *Martes martes* with almost full-grown kit.

females were no longer radio-collared, it became more difficult to keep track of where they chose to have their kits. Nevertheless, in most years we were able to confirm breeding through a combination of checking den boxes during late spring/summer, camera trap images or video footage, and ad hoc observations from members of the public and volunteers. The annual den box checks were not carried out in 2020 when volunteer activity was seriously curtailed due to COVID-19 restrictions. However, some of the local volunteers were able to document evidence of Pine Marten breeding in 2020 from camera traps that they could still monitor privately. From 2016 to 2019, breeding by 10 of the 23 translocated females continued, with breeding by a further two females suspected but not confirmed (Table 6). Four females bred at least twice, meaning that their kits have resulted from a mating in Wales, as opposed to a Scottish mating before their translocation. At least 38 Welsh-born kits were born up until 2020, although the true number may be even higher than this, as we weren't able to locate every female every year. This has continued since and achieved the project's aim number 3, as set out earlier.

Initially, we planned to translocate and release 30–40 Pine Martens to Wales over two years; however, translocating a third and final tranche of 10–20 martens was also an option in our contingency plans. After the first year of releases and monitoring, the decision was taken to go ahead with a third year. The most

TABLE 6. Summary of known or suspected Pine Marten *Martes martes* breeding in Wales between 2016 and 2020.

Pine Marten ID	No. kits observed	Evidence	Year
PM16	1	Seen in owl box	2016
PM02	2	Seen in den box	2016
PM11	>1	Heard and seen in barn roof	2016
PM13	1	Heard and seen	2016
PM16	3	Seen in den box and on camera trap	2017
PM35	2	Seen on camera trap	2017
PM27	≥1	Heard in den box and seen on camera trap	2017
PM26	3	Kit(s) heard in brash pile, and collar chewed by kit(s)	2017
PM34	1	Seen and photographed by forester	2017
PM21 (presumed)	Unknown	Filmed in den cavity. Inconclusive blur on borescope cam; hairs taken from den entrance for DNA	2017
PM22 (presumed)	Unknown	Loyal to den site but impossible to access to check	2017
PM02	2	Seen near den box	2018
PM26	3	Seen in den box	2018
PM27	1	Seen in den box	2018
PM13	2	Seen on camera trap	2018
PM07	2	Seen on camera trap	2018
PM40 (presumed)	1	Seen in den box	2018
PM07	2	Seen on camera trap	2018
PM16 (presumed)	2	Recorded on camera by professional photographer	2019
Unknown	2	Regular visitors to local resident's garden to take food left out	2019
PM07 (presumed)	1	Presumed female and kit seen on camera trap over the winter	2019
PM35	Suspected but not confirmed	Female seen on camera trap looking like she has probably bred	2019
Unknown	≥1	Female and kit seen on camera trap using a squirrel box at Strata Florida	2020
Unknown	Unknown	Lactating female seen on camera trap near Llanafan	2020
Unknown	2	3 martens seen on camera trap on the Hafod Estate in July	2020
Unknown	Suspected but not confirmed	Possible juvenile (unrecognised) marten seen on camera trap near Forge	2020
Unknown	2	Female and 2 kits photographed and filmed by local resident in garden	2020

optimistic models suggested that with a founder population of 30–40 animals, it would take at least 10 years for numbers to double. Although we had confirmed breeding by 50 per cent of the released females in spring following the first translocations, this was lower than the published fecundity rates (for American Martens) that were used in population viability analyses during the feasibility phase of the project. If this were to continue or if breeding rates in Wales varied greatly between years (possibly linked to Field Vole cycles), it would have an impact on the probability of martens becoming established, as well as their rate of spread. As we were observing the animals settling in well-spaced home ranges across a fairly wide geographical area, the population was likely to be at low density and vulnerable to random and possibly catastrophic events, such as a run of particularly poor breeding years or severe weather events (termed demographic and environmental stochasticity) for some time. Therefore, we decided it would be sensible to increase the numbers immediately, rather than possibly having to do it later. A third year of releases also enabled us to take advantage of some of the rapidly developing technology for tracking and noninvasive monitoring. In the 'top-up' year in 2017, eight males and four females, including three juveniles, were translocated. As recorded post-release mortality was higher for males than females (2:1), the larger number of males trapped and translocated in the third year helped address this imbalance. In total, 28 males and 23 females were translocated over the three years.

Mortality

During the course of Pine Marten translocations to Wales, we continually reviewed what we were doing to see if we could refine or improve the methods and procedures for translocation and release. We saw no mortality during translocation, and post-release survival rates were high, with observed mortality rates within the range reported for undisturbed marten populations in the wild. All deaths were monitored, and carcasses were retrieved immediately where possible and sent for postmortem as part of an ongoing health surveillance programme. During the course of the year following the first releases in 2015, 6 out of 20 Pine Martens died. This was within the range of average annual mortality for wild marten populations, which is reported as being between 30–50 per cent of adult animals. Mortality in the first 12 months was higher for males (40 per cent) than females (20 per cent), as expected (and found in previously published studies as mentioned before). The commonest cause of mortality was natural predation, probably by foxes, but the first two animals that died, in November 2015, shortly after being released, had an acute fungal encephalitis. This had not previously been seen by the wildlife pathologist who performed the postmortem examinations. The fungus was confirmed to be a mucoraceous mould, commonly found on vegetative matter such as decaying fruit. This does not usually cause infection but can be associated with disease where the immune system is weakened. It is possible that exposure to a very high infective dose, either on its own or in combination with a compromised immune response following the stress of recent translocation, resulted in these two fatal cases. Those two martens were among the last to be translocated in 2015, in a very wet November, when conditions were optimal for fungal growth. The most likely source for the mould was decaying natural vegetation in pre-release pens, or uneaten food items. In response to this, the translocation period was shortened in 2016 to ensure that all animals were released by the end of September. We also stopped hiding food items in the enclosures for behavioural enrichment so that we could be sure that all uneaten food was removed each day. We had no recurrences of fungal encephalitis among the animals subsequently released.

In the year after the second tranche of translocations, only 1 of the 19 animals that were released is known to have died. She was found on an A-road in the release area, and postmortem examination confirmed that she had probably been run over. This was particularly unlucky as it is not a very busy road, but she was found on a blind bend where the road passes through a forested area. Of the animals released in autumn 2017, only one, an adult male, died while being radio tracked. A second male from the first year of releases was found dead in 2018 after being hit by a vehicle on an A-road near Belper in faraway Derbyshire. The carcass was retrieved and scanned for a microchip, which confirmed the initial identification indicated from his bib pattern. He had been resident in Clocaenog Forest in north Wales for the preceding two years, and it is a mystery how or why he moved from there to Derbyshire almost 200 km away. Two animals were found dead on or next to roads in 2019; and a further one in 2020. Postmortem examination confirmed they were all sub-adult males, most likely hit by vehicles during dispersal and confirming that road traffic poses the greatest threat to Pine Marten survival these days.

MONITORING, RESEARCH AND ADAPTIVE MANAGEMENT

As discussed already, significant amounts of time, money and effort go into planning and conducting translocations, as well as into the ethical and welfare considerations for the released animals. Therefore, it is important to establish clear objectives and criteria for both short- and long-term success, and to monitor and adapt to what happens after the translocation. If a translocation results in establishment, followed by reasonable population growth and long-term persistence, then further intervention may not be necessary, although long-term monitoring is critical to check for unexpected problems. Alternatively, if monitoring reveals low post-release survival and low population growth, then further action may be needed to achieve establishment and long-term persistence. This might include refining the translocation protocols and release site selection for future translocations and further 'top-up' translocations. Finally, if a translocation fails, it is essential to know when and why it failed in order to establish what else could be done to increase the likelihood of success before deciding whether further translocations should be attempted.

Research is key to improving the science of reintroductions and translocations, and it has been suggested that this has been a missed opportunity in many past projects. The Vincent Wildlife Trust, in partnership with the University of Exeter, had a rigorous programme of research associated with the Pine Marten Recovery Project. This focussed not only on the ecology of the translocated animals but also on other species at the release sites, as well as on the socioeconomic impacts of the project. The project provided an opportunity for two postgraduates to study the behaviour of newly translocated Pine Martens in an unoccupied part of their former distributional range. All the Pine Martens were monitored and studied for a prolonged period following release. The data collected have provided a huge amount of information, which can be used to inform subsequent reintroductions elsewhere.

FIG 85. Postgraduate student Cat McNicol doing fieldwork during the Pine Marten Recovery Project.

During her PhD, Catherine (Cat) McNicol used radio-tracking data to investigate the spatial ecology of both Pine Martens and Grey Squirrels. Some of her research, which is summarised in this section, focussed on the martens' ranging behaviour and habitat use in the days and months following release. She looked at how Pine Martens used the landscape in the first couple of years after their release, and how this might impact resident non-native Grey Squirrels. She also studied the diet of the martens before and after translocation to see how this was affected by a sudden change in habitat and prey base. Vincent Wildlife Trust project officer, David Bavin, was interested in variation in the behaviour of individual Pine Martens and whether this was linked to differences in dispersal movements following their release in Wales. Using levels of faecal glucocorticoid metabolites (fGCMs), an indicator of stress, he explored the relationships between behaviour in the wild at source sites, behaviour at the point of release, including levels of stress in captivity, and post-release movement. The second aspect of David's research followed on from the earlier public opinion surveys and focussed on the social acceptability of the translocation. He used Q-methodology, a research method used in social science to study people's viewpoints, to understand the perspectives of residents living in the area where Pine Marten translocations were being proposed. This was extremely helpful with the continuing community engagement that has been an integral part of restoring Pine Martens to mid-Wales.

POST-RELEASE MOVEMENTS OF PINE MARTENS

Pine Martens were radio tracked for up to 10 months from the night of their release. Tracking was carried out from late afternoon and then through the night to collect data on daytime denning and rest sites as well as movements during the active hours after sunset. Each animal was located by triangulation from two bearings taken within 5–10 minutes of each other. All translocations are essentially exercises in forced dispersal, but dispersing animals in many species, even relatively antisocial ones, rely on there being others of their kind (conspecifics) present as an indication that this is a good place to settle. Translocated animals, especially in the first releases, are unlikely to find others of the same species, in which case they may not settle in suitable habitat at the release site, choosing instead to go further afield in search of conspecifics.

What happened with our martens? How did they cope in their new home? Cat McNicol's analysis of the radio-tracking data showed that for our translocated Pine Martens, there were two clear stages of post-release movement within the

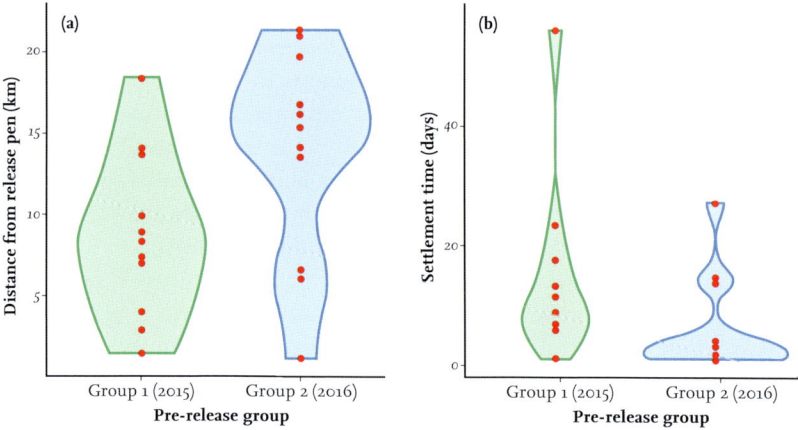

FIG 86. Summaries of post-release movement (red dots) of translocated Pine Martens *Martes martes* away from pre-release pens over 100 days after release: (a) distance from pre-release pen at which martens established stable home ranges and (b) time since release at which martens switched from the exploration phase to settlement into established home ranges. Redrawn from McNicol *et al.* (2020).

first 100 days: a period of exploration, followed by settlement. The first Pine Martens released in 2015, into 'empty' habitat, initially made some relatively long-distance exploratory movements, often using wooded river valleys to travel around, before tracking back to the release area. The exploratory phase lasted for an average of just over two weeks before the martens settled. Those released in the second year, when there were other martens already present, settled much more quickly and established territories after an average of about seven days of being released (Fig. 86b). The Pine Martens released in 2015 settled closer to where they were released than those that were translocated a year later, with animals in the first year settling at a mean of 8.7 km from the release site, whereas animals in the second year travelled a mean of 14.0 km (Fig. 86a).

The longest time that it took for an individual to settle down was 56 days. The furthest at which a tracked marten settled within 100 days was 21.5 km from its pre-release pen and the closest was 1.1 km. While there were differences between years, there were no differences between males and females in either settlement times or settlement distances. The average home range size for Pine Martens once they settled was 9.5 km². This is within the 5.6–23.6 km² reported for martens in northern Scotland (Caryl 2008). Home ranges appeared to be more defined in the second-year animals, which may be a result of stronger territorial distinction once the cycle of mating, breeding and dispersal is under way. Most of the Pine Martens translocated in the first year eventually established home ranges in and around the release area, whereas those released in the second year dispersed further away, probably because the best patches were already occupied and being defended by then. However, those animals in the second year spent less time searching for suitable habitat near to their release sites and quickly settled in forest on the periphery of the core population. The initial retention of translocated animals close to their release sites is integral to the establishment and long-term viability of a new population. One of the performance indicators for the Pine Marten translocations to Wales was that the majority of released individuals settled within approximately 10 km of the release sites. In the first year, the average distance of Pine Marten dispersal prior to settlement was 8 km but this was slightly greater, at 14 km, in the second year.

The release sites were chosen for the diversity of denning and foraging habitats that they provide (Fig. 87). Cat McNicol assessed habitat use by the radio-tracked martens in years one and two to see if they used the habitats in proportion to their availability and found that preferences for habitat and forest type differed between years. Whereas the animals released in the first year used forest, agricultural and grassland habitats in equal proportion to their availability, those released a year later showed preferential use of forested habitat but lower use of agricultural and grassland habitats than expected, in relation to their availability. Conversely, the first releases showed some finer-scale preferences within forest habitat, showing a strong preference for felled areas (compared with the categories open, young, broadleaf or conifer), while martens released in the second year showed no preference by forest type.

FIG 87. The release sites consisted of large areas of commercial (conifer) forest surrounded by a mosaic of rough pasture, farmland and moorland.

These results suggest that the first Pine Martens to arrive had little competition for the best home ranges in the heavily forested, core release area, leaving the second wave of released animals to establish their territories in the remaining vacant habitat. These territories tended to be in smaller, non-commercial woodlands in farmland on the edge of the core population. This might explain the year-two animals' broad-scale preferential use of forested habitat regardless of forest type. Pine Martens released in the first year showed an apparent preference for felled woodland, which was quite surprising. That being said, thinned and felled compartments within large commercially managed forests usually contain the kind of woody debris that offers shelter and den sites for martens. This includes trees blown down by the wind (windthrow), the exposed roots upended when a tree is blown down (root plates) and the piles of brash or dead branches left when trees are trimmed down during felling. Field Vole densities have been shown to increase for up to 15 years after clear-felling, as have those for Bank Voles, Wood Mice and shrews (Bogdziewicz & Zwolak 2014). Therefore, the preferential use of felled areas by Pine Martens in the first year of releases may reflect high prey abundance in close proximity to den and rest sites.

THE LONG-DISTANCE DISPERSERS

Post-release movements of some animals were unpredictable and, in each year, a few individuals (six in the first two years) were lost shortly after release. Four of these were subsequently found again and some had travelled exceptionally long distances, including the second male that we translocated (PM03, nicknamed 'Sid Vicious' to reflect his temperament at first capture). He was discovered 172 days after being released, when a photo of him was taken by a camera used to monitor a Red Squirrel *Sciurus vulgaris* feeder 103 km away in north Wales. Slightly higher proportions of long-distance dispersal have been recorded in studies of other translocations of marten species (Davis 1983, Slough 1989), which may be attributed to a degree of local overcrowding. It is not clear what caused our animals to disperse (or what stopped them in one place rather than another), but David Bavin's data suggest that individual personality and levels of stress may play a role.

Anyone with experience of keeping pets will tell you that different animals can show quite marked personalities. The importance of animal personality in conservation is now being studied and acknowledged more widely (de Azevedo & Young 2021). We had some experience of how different individual Pine Martens responded to being trapped and translocated, resulting in the various nicknames

FIG 88. Pine Marten *Martes martes* at the door of a pre-release pen.

we had given to particular animals. Sid Vicious expressed his displeasure at being trapped very clearly, whereas other animals appeared much more stoic. Some silently accepted the Blueberries that were offered to them on recovery from anaesthesia, some took them more grudgingly with accompanying huffs and growls, and a few ate them with audible relish. PM07, a young female, was quickly christened 'Miss Piggy' for devouring everything she was given before, during and after her journey to Wales. We provided supplementary food at the open pens for some time after each animal was released. The majority ignored it, choosing instead to forage for themselves. However, Miss Piggy was radio tracked not only returning to her own pen to take advantage of the free meal but also visiting other pens to take what was on offer.

This, however, was all anecdotal evidence of personality. In the second year of our Pine Marten translocations to Wales, David Bavin studied 14 animals through the entire process. Data were collected from footage recorded by trail cameras prior to trapping at source sites in Scotland and videos of each animal's behaviour in pre-release pens up until the point at which it left the pen. Scats were collected from the pens each day to measure levels of faecal glucocorticoid metabolites (fGCMs) (Fig. 89) as an index of stress. David recorded consistent variation between individuals in a standardised set of behaviours that could be referred to as 'personality'. He also looked at whether stress responses to the translocation process varied between individuals and if they were related to post-release outcomes recorded from radio tracking (dispersal distances, time to settlement, etc.). He found that Pine Martens that exhibited more vigilance behaviour seemed to have a lower stress response in captivity (lower levels of fGCMs), left the pre-release pens sooner, travelled further from the release site

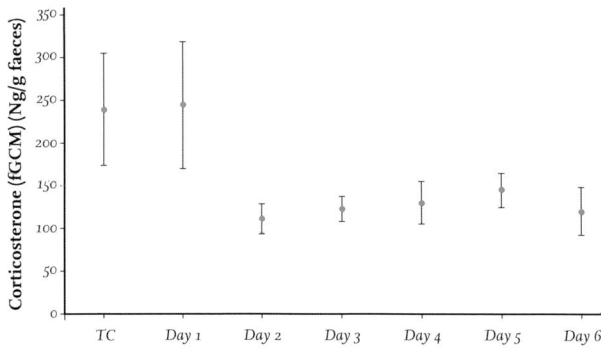

FIG 89. fGCM levels as an index of stress, from Pine Marten *Martes martes* scats in translocation crates (TC) and each day in pre-release pens. David Bavin, unpublished data.

and took longer to settle. His results suggest that Pine Martens in his study could be characterised as proactive and reactive. The proactive type showed lower levels of stress in temporary captivity and rapid, extensive dispersal, in contrast to the reactive type, which experienced greater stress in captivity and slower, less extensive dispersal, and they also ended up with smaller home ranges (Bavin 2021). High mortality and/or dispersal away from the release site in the early stages are two of the main risks to population establishment following translocation. David's study suggests that behavioural assessment of Pine Martens before and during translocation might give an indication of which individuals should be prioritised for monitoring as they are most likely to travel and settle furthest from the release site. Monitoring fGCMs while Pine Martens were in pre-release pens also gave us a useful indication of how supportive the soft release actually was. We found that levels of these stress hormones were, predictably, high during translocation and in the first day following trapping and translocation to their new home. However, there was a significant decrease by the third day (day two in the pen). By contrast, immediate (or hard) release while martens were still showing high levels of stress would very probably have had an impact on their ability to orientate in their new environment, and to find food and safe shelter. Our small dataset suggests that a short period of temporary captivity prior to release may be beneficial in future projects like this.

RADIO TELEMETRY AND GPS EQUIPMENT

The Biotrack VHF radio transmitters that we used on Pine Martens were reliable and worked extremely well in challenging terrain. In 2015, we attached small GPS loggers (from Ecotone, another manufacturer) to 10 of the marten radio collars to test their efficacy. GPS loggers search for a signal at pre-programmed intervals, calculate and then store the coordinates of the unit (and therefore the animal) at that time. The data can then be retrieved either remotely via a receiver if it is within range or by retrieving the logger and downloading the data. Pine Martens are a tricky animal when it comes to using GPS tracking. Because martens are relatively small, the GPS logger and its battery must also be small and light. Also, in contrast to many of the species that are tracked with GPS collars, martens spend most of their time under dense forest cover or in rocky crags, where communication between satellites and the GPS logger is often interrupted or prevented altogether. There were some issues with the GPS loggers that we trialled in the first year. The batteries drained more quickly than anticipated, probably as a result of the extra time needed to find satellites under thick forest cover, so remote download failed when this was attempted a few weeks after release. The data are stored onboard, but the few loggers that were retrieved yielded very little. However, the technology in this area of wildlife tracking is moving forward all the time, and so, by the second year, a new type of combined GPS and VHF collar, the Lotek Litetrack-30, was available that was small enough for larger, male Pine Martens. The costs were four times that of a standard VHF radio collar, so we tried out a small number on five of the martens released in 2016 and waited with bated breath to see if they would be able to cope with the challenges of pinpointing a location (known as a 'fix') in the craggy habitats and thick forest cover that martens occupy. Miniature (less than 80 g) GPS collars still have severe battery limitations, so fix interval must be traded off against battery life when programming schedules. We programmed the Litetrack-30 collars to collect a GPS fix every four hours through the night (6pm–6am) to maximise the length of time animals could be tracked with the VHF transmitter on the collars as well as collect GPS data. In early 2017, we were able to get close enough (within 50 m) while they were denning and remotely download the data from four of them and were thrilled to find that they had worked. The success rate of the GPS units (percentage of scheduled attempts that found enough satellites to obtain a fix) was relatively low (mean of 18.75 per cent). However, these data added to the VHF fixes obtained from radio tracking to provide more detailed information on the movements of these animals since release. Combined radio and GPS data showed that martens were using steep-sided, wooded river valleys to move through the landscape and travel between forest blocks. As expected from previous studies, our Pine Martens tended to avoid open habitats such as moorland. In some cases, the GPS data revealed outlying locations where an animal had not been radio tracked. For other animals the GPS locations provided information on where they were during periods when we had not been able to locate them by radio tracking. The results were promising enough for us to invest in more of these collars,

TABLE 7. Fix success rate (FSR) of GPS transmitters on Pine Martens *Martes martes* in Wales.

Pine Marten ID	Period	No. of successful GPS fixes	No. of attempts	Overall FSR %	FSR 'cold starts'	FSR 'hot starts'
PM36	28/9/16–4/12/16	65	269	24.16	24.16	N/A
PM37	5/10/16–4/12/16	28	241	11.62	11.62	N/A
PM31	21/9/16–9/11/16	43	197	21.00	21.00	N/A
PM23	7/9/16–7/2/17	108	612	17.65	17.7	N/A
PM40	28/7/17–21/2/18	61	755	8.1	5.03	15.6
PM42	10/9/17–29/1/18	105	644	16.3	15.5	17.9
PM47	3/10/17–21/2/18	176	670	26.3	22.3	34.6
PM48	9/10/17–18/10/18	13	280	4.6	0	5.8

which we used on nine martens the following year. Recent studies had shown that miniature GPS collars are more efficient at collecting data from a 'hot start', with short fix intervals (two hours or less), rather than the four-hour interval (or 'cold start') that we had used in 2015. In 2017, we reduced the fix interval to two hours for the first month after release to see if this resulted in a higher fix success rate. However, as the results in Table 7 show, it did not. Data were retrieved from only four out of the nine GPS collars, and the results were disappointing.

Our conclusion was that the Litetrack-30 collars used in 2016 and 2017 were relatively effective for Pine Martens, given the difficult terrain. The battery life was less than that of VHF transmitters alone, but we were satisfied that by using a combination of VHF and GPS tracking, we collected sufficient data from each tranche of released animals to be assured that they were behaving normally, establishing home ranges within the release region and that there were no welfare issues or lack of natural resources for them.

RECAPTURES

We aimed to re-trap the released martens from each year to check them over and remove radio collars while there was still plenty of life in the batteries. Trap sites were targeted in the vicinity of known den sites for each individual and this resulted in good trapping success. The martens that were recaptured were all in good condition and weighed the same as or a bit more than when they were first trapped in Scotland. This was a very good sign that they were finding enough food in their newly established territories in Wales. Almost all (90 per cent) of the animals were successfully recaptured to remove their collars. They were weighed, checked, a whisker sample taken to be used for stable isotope (dietary) analysis and the radio collars removed. As all of the transmitters fitted to martens released in the third year were mounted on leather collars, recapture of these animals was not absolutely necessary to remove them because the leather will stretch or perish over time and the collar will drop off. This was confirmed by a combination of incidental recaptures (in which case the collar was removed), camera trap footage of animals once they had shed their collars and retrieving collars that were shed and still transmitting. None of the collared martens showed any signs of abrasions or adverse effects from the collars, either when they were checked at recapture or during postmortem examinations of those that died. Some of the recaptures were memorable for different reasons: one female, PM13, proved to have an uncanny knack of knowing when a trap was locked open and pre-baited, in which case she would take full advantage of the free meal it provided. We acquired hours of camera trap footage of her going into our carefully placed traps, but the minute we actually set the traps, she seemed to sense that something was different and avoided them like the plague. It took several months of attempts before we finally managed to catch her and take off her radio collar, which by then had long since ceased transmitting.

FIG 90. Pine Marten *Martes martes* in a pre-release pen, showing combined GPS/VHF collar.

SAVANNAH MARTEN

One recapture and collar removal was particularly notable. In April 2018, the Georgia Port Authority in Savannah (USA) was alerted to the fact that there was a Pine Marten running round the port that had presumably got off a container ship there. It was remarkable for two reasons: firstly, it was a European Pine Marten, and therefore non-native in North America; and secondly, it was wearing a radio collar. It was trapped by a biologist from the US Fish and Wildlife Service, who removed the radio collar, and the marten telegraph (mostly through social media) started buzzing with speculation about where the mystery marten could have come from. I was sent some photos and noticed that the radio collar was identical to those that I had been fitting onto most of our translocated Pine Martens. But I also thought that the manufacturer in Dorset must have provided those collars to a number of Pine Marten projects right across Europe. However, when I contacted them, I was told, 'No, just you'! At that point I began frantically trying to track down the marten and the collar and eventually found that the animal was alive and well and enjoying the hospitality of Chehaw Zoo in Albany, Alabama. Kara Day, a biologist with the Georgia Department of Natural Resources, was extremely helpful and sent the collar to me in the post. Meanwhile, I got in touch with the director of the zoo, and he told me that the vet and trapper who had dealt with the animal initially thought that it was a neutered male due to a lack of scent, and that it was very calm around people. American Martens are reportedly very aggressive, which this Pine Marten was not. The conclusion was that it was most definitely accustomed to humans and probably captive-bred. I responded that in my experience of having trapped, microchipped, collared and transported a large number of them, European Pine Martens are very calm when being handled and don't have a strong smell. The radio collar, when it arrived, had the handwritten number .330 on the transmitter, which meant a radio frequency around 173.330 MHz. I checked my records, and we had one Pine Marten, PM38, whose collar frequency matched and who had last been heard of about 12 months earlier near the Welsh coast, but she was a female. The mystery deepened.

On 1 May, the Savannah marten was anaesthetised by the zoo vet for vaccination and a full health examination. This confirmed it was, in fact, a female. The vet confirmed that she was in ideal body condition with good teeth, believed to be a young adult, weighing 3.3 lb (1.5 kg). She was vaccinated for canine distemper and rabies and scanned for a microchip but none was found. I had hoped that the microchip would be conclusive evidence that this was our PM38, as I had microchipped her myself. However, it is not unheard of for these tiny tags to

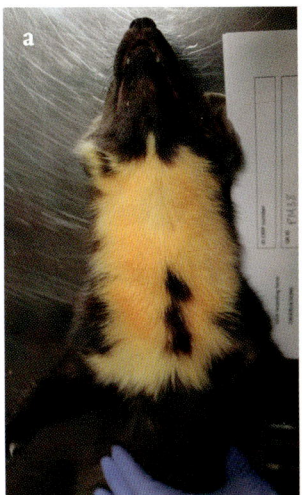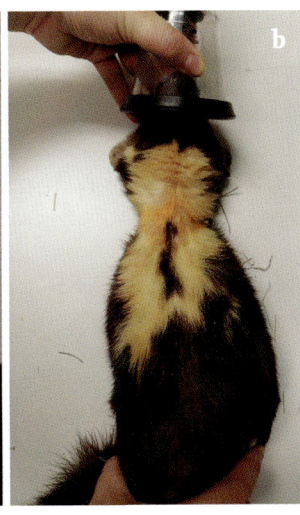

FIG 91. (a) Pine Marten *Martes martes* PM38 and her bib pattern at capture in Scotland, and (b) the bib pattern of the marten captured in Savannah.

be lost, so it could not be ruled out. The radio collar was compelling, but I really needed to see some photos showing the details of her bib pattern for comparison.

Three weeks later she was anaesthetised again for a blood test, and one of the zoo curators used the opportunity to photograph her bib and throat patch. She sent me the pictures, commenting that 'her fur is shorter and less dense now than when she first arrived. I suspect that's due to the warm weather'. (I suspected that was an understatement. If this Pine Marten had grown up in Scotland, then overwintered in Wales, she would have found May temperatures in Georgia roughly double those that she was used to!) The keeper also commented that the patchiness in the marten's neck fur was from when the vet had shaved her neck for a blood sample three weeks previously. I looked at the photos side by side with those that we had taken of PM38 back in Scotland, and I was convinced it was her.

The next question was, what would be the most humane course of action for the animal? I contacted the zoo in Albany to ask if repatriation was a possibility if that was decided to be in her best interests. The zoo curator admitted that the local climate meant that their zoo was unsuitable to house a Pine Marten outdoors, but they had recently been contacted by an AZA (American Zoological Association) accredited zoo in New Jersey, which was interested in taking her for exhibit. Having taken her from the wild for translocation to Wales in the first place, I felt a huge responsibility to this animal, and so I discussed the options with a number of people. Unfortunately, by now she had been in captivity for some months and her keepers had been spending time in her enclosure in an effort to tame her. They had named her Nicki and become very fond of her, as zookeepers often do of their charges. However, they thought that this, and the stress of transportation, would compromise her chances of success if she was repatriated and returned to the wild. The New Jersey zoo was located in a climate that was conducive to her being comfortable and they had experience of housing American Martens, so in the end it was decided that was the best option for her – quite an adventure and almost certainly the longest post-release dispersal distance recorded. We can only speculate as to how she ended up on another continent, but close to her last recorded location in Wales was a company that exports mineral water all over the world, including to the USA. Perhaps she stowed away in a container and that was how her journey began. We will never know. But we would never have heard about her emigration at all had it not been for the fact that she was wearing a radio collar.

LONG-TERM MONITORING IN WALES

Once it was no longer possible to check on the Pine Martens by radio tracking, noninvasive methods, including camera trapping and scat surveys, were used to monitor the population. A systematic distribution survey was initiated in 2018, once translocations were completed and territories established, to collect distribution data, act as a baseline for future surveys, enable population change to be assessed, and provide a better understanding of the rate of range expansion of translocated populations. The first method for collecting distribution data was by using trail camera 'traps'. These were deployed principally as part of Vincent Wildlife Trust's Camera Trap Loan Scheme, whereby a large number of cameras were loaned to volunteers to monitor Pine Martens, as well as other wildlife in their area. Cameras were deployed both in and around the original release sites and in areas that martens were thought to be moving into, covering 26 hectads. The cameras were not placed in a pre-defined survey area or pattern but were deployed ad hoc to confirm the presence of Pine Martens in an area where they had not yet been recorded; to monitor individual martens, identified by their unique chest patterns; or to determine breeding success of particular individuals. Cameras were active for varying periods of time during 2018 and 2019, ranging from two weeks to a few months. Cameras were usually deployed with bait (typically raw hens' eggs, sardines, peanuts, raisins, peanut butter or jam). Volunteers monitoring the cameras were given guidance on how to operate and check them and how to select sites to optimise the likelihood of recording Pine Martens. The resulting videos and images were screened to determine if Pine Martens had been recorded.

FIG 92. An example of camera trap footage confirming the presence of Pine Marten *Martes martes*.

The second method comprised scat surveys, which have been used in national and wide-scale Pine Marten distribution surveys in Scotland (Velander 1983, Balharry *et al.* 1996, Croose *et al.* 2013) and Ireland (O'Mahony *et al.* 2012). Scat surveys were undertaken in a targeted area, defined by mapping hectads where Pine Martens were recorded during 2018 and delineating a boundary of 30 km from these squares. Hectads in which Pine Martens were recorded in 2018 and where camera traps were already deployed were excluded from the scat survey, with the exception of those on the edge of a marten's range where it was only recorded on one occasion. Four 1.5 km transects per hectad, distributed as evenly as possible, were pre-selected for scat surveys. In hectads without sufficient woodland to accommodate four transects, as many transects as possible were installed. A total of 210 transects were surveyed for scats, covering 72 hectads. Surveys were completed between May and November 2019, by a combination of volunteers and Vincent Wildlife Trust staff. Transects were walked slowly whilst searching for scats, and all potential Pine Marten scats were collected in a plastic zip-locked sample bag with a unique reference number. All possible Pine Marten scats were sent to the Waterford Institute of Technology, in Ireland, for DNA extraction and analysis.

DISTRIBUTIONAL RANGE EXPANSION

During 2018 and 2019, using records collected from both camera trapping and scat surveys, Pine Martens were recorded in 22 hectads in Wales (Fig. 93), but

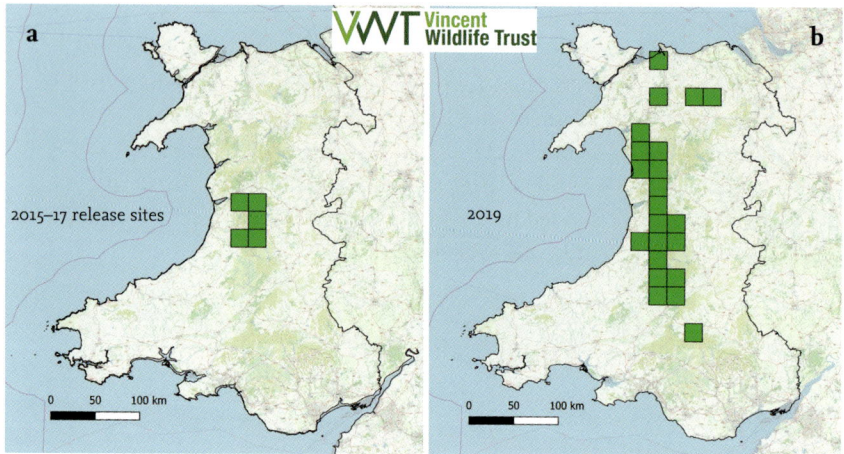

FIG 93. (a) Hectads where Pine Martens *Martes martes* were released in 2015–17 and (b) those with confirmed evidence of Pine Marten occupancy by 2019 from the volunteer survey data.

there are a number of challenges in surveying mustelids, discussed further in Chapter 8. One of the issues is that Pine Martens can be difficult to detect when they are at low densities; therefore, it is likely that what we found was only the minimum extent of the Pine Marten's distributional range in Wales at the time. Pine Martens may have been present outside of this area but were not detected nor records received during the survey period.

By 2021, our surveys and ongoing monitoring showed that a core marten population had become well established in the original release area, centred around the horseshoe of forestry that follows the Ystwyth and Rheidol Rivers (around the villages of Devil's Bridge, Ponterwyd, Cwmystwyth and Llanafan). Beyond this, Pine Martens appear to have a relatively continuous range along the spine of the Cambrian Mountains, from Coed y Brenin in the north to the Tywi Forest in the south. North of the release area, Pine Martens quickly expanded their range through southern Snowdonia (Eryri), from Machynlleth and Dolgellau into Coed y Brenin up to Trawsfynydd. As already mentioned, Sid Vicious, one of the first Pine Martens released in 2015, made his way up to Denbighshire and quicky settled in Clocaenog Forest, where he was spotted on camera taking advantage of the peanuts in a Red Squirrel feeder. It soon became clear that he was not alone. Soon afterwards, there were Pine Martens recorded elsewhere in north Wales, including Gwydir. South of the release area, they are now well distributed in the Tywi Forest south to Llanwrtyd Wells and Rhandirmwyn and have spread down into Brechfa. There is also recent evidence of dispersal into

Bannau Brycheiniog (the Brecon Beacons) and Radnorshire. Further records have since been collected, but so far, there appears to have been limited expansion of the population eastwards. This is not unexpected due to the large expanse of open moorland habitat east of the Cambrian Mountains. It seems that the core population is still mainly to the west of Llanidloes, Rhayader and the Elan Valley, although there are undoubtedly some individuals east of there, and the distribution has now expanded to the point where it is has almost joined up with the population subsequently re-established into the Forest of Dean.

THE IMPORTANCE OF COMMUNITY

All aspects of species restoration projects should involve members of the local community and include meaningful opportunities to engage and contribute. It is increasingly acknowledged that local participation and connection should be a fundamental objective of translocation projects, not just part of the planning process. Establishing support for the project from across the community at the earliest stages was critical. The local communities in and around north

FIG 94. The Pine Marten Den visitor information centre at the Vale of Rheidol Railway in Devil's Bridge.

Ceredigion were very open to being involved in the project and have been absolutely crucial to its ongoing success. We also had help and support from key landowners and forest managers, including Natural Resources Wales, as well as several local businesses in mid-Wales. As well as attending events and meetings, many local volunteers were directly involved in the project by radio tracking the martens, monitoring camera traps, collecting data and providing other assistance in the field. Continuous community involvement throughout the process has been key. Over the course of the operational stage of the project between 2015 and 2020, a total of 10,519 volunteer hours were recorded and, on average, 75 volunteers contributed time to the project each year, with 42 contributing 470 hours to scat surveys alone. There have been 24 community meetings and Pine Marten events in which more than 300 people have participated. Ultimately, this local support is as important to the long-term survival of Pine Martens in the Welsh Cambrian Mountains as the science that underpinned the translocations. Several local businesses and community hubs have provided space and/or resources for Pine Marten interpretation panels. In 2018, the Pine Marten Den information and visitor centre was opened in the heart of the release area, in partnership with the Vale of Rheidol Railway, which celebrates the natural and cultural heritage of the area, including the native Pine Marten.

Before proceeding with translocation and then throughout the releases, there was a continuous and constructive dialogue with local shoots and estates as well as farmers in the release area. Since then, Vincent Wildlife Trust staff have maintained regular communication, rapid response and a sympathetic approach to any potential issues. Willingness to listen and take concerns seriously have been key components of building trust. It is important that communities have confidence in the longevity of conservationists' commitment to them and to the species that they have restored. This is a factor that has been identified as unique to wildlife reintroductions or translocations (Auster *et al.* 2020). To date, there have been very few potential conflicts between Pine Martens and householders, farming or game interests, but Vincent Wildlife Trust staff continue to respond quickly to any potential issues as and when they are reported. In fact, several people living in Wales now enjoy having Pine Martens visiting their gardens, where they put food out for them. Some of these animals have names and provide a lot of pleasure for those householders.

MEASURING 'SUCCESS'

There is no single definition of what constitutes a successful species restoration or reintroduction (Robert *et al.* 2015, Chauvenet *et al.* 2016). One frequently stated objective is the establishment of a viable population in the release area. This is often assessed as having been achieved if there is recruitment to the adult population and requires that the population is monitored long enough for individuals to reach reproductive maturity (Bubac *et al.* 2019). It can take a long time for released individuals to populate an area. Another common goal is long-term persistence of the re-established population. The specific timescale for evaluating the achievement of objectives such as these is of great importance. Therefore, an appropriate post-release monitoring program is vital. Bubac *et al.* (2019) found that the majority of translocation failures occurred within four years, suggesting that this is a critical window of time for post-release monitoring. However, species differ in their life history strategies and population structure; therefore, translocations of long-lived species with low reproductive output may require decades of monitoring for evaluations of success to be meaningful. Consequently, it is recommended that there should be a sequence of short- to long-term measures which may include survival of the released animals, breeding by released animals and their offspring and persistence of the re-established population. These need to be set within appropriate time frames for evaluation, dependent on the life history traits of the focal species and the length of time of the release project.

In common with a large proportion of published conservation translocation case studies, success in achieving the fundamental long-term objectives of the Pine Marten restoration to Wales is still unknown at the time of publication. For a species like the Pine Marten, which naturally lives at low density and is slow-breeding, re-establishing populations are likely to remain vulnerable to the impact of chance events for many years. Therefore, an appropriate monitoring and adaptive management plan is essential. Almost 10 years after it began, we can say that the Pine Marten restoration project in Wales has been successful in the short to medium term. Released martens have established territories in the release area (McNicol *et al.* 2020), and both breeding and dispersal of juveniles have been recorded, as has some distributional range expansion. The Pine Marten population in Wales is still in the relatively early stages of re-establishment, but so far it has expanded its range in line with expectations since translocations were completed.

A considerable amount of effort and resource was put into the selection of both release and donor sites, with a significant post-release monitoring programme factored in as an essential element of the translocation plan. Translocations, particularly of carnivores, are complex, high-risk and high-cost endeavours. The overall budget for the full seven years of the Pine Marten project was £1.3 million. This included two years of feasibility studies and initial community and stakeholder engagement; three years of translocations and release, with intensive post-release monitoring; and a further two years of monitoring beyond the final releases. Research, adaptive management and ongoing engagement continued throughout.

Longer-term survey efforts are an important component of monitoring the persistence and changing distribution of the Welsh marten population, whilst it is still in the relatively early stages of recovery. Local residents as citizen scientists can have a key role in assisting with long-term monitoring. This, and other opportunities for community involvement, enables greater efficiency and longevity of the project, as well as increasing local ownership of it. The next distribution survey is planned for approximately five years after the initial survey, with subsequent surveys being conducted at five- to seven-year intervals. It may take decades before we can confidently say that the project has succeeded in its overall fundamental objectives. However, having met the shorter-term objectives defined as a means to achieve the bigger aims of the project gives confidence that the recovery of the Pine Marten in Wales is well on track.

BEYOND WALES: PINE MARTENS BACK IN THE FOREST OF DEAN

Having had the experience, over a number of years, of first carrying out all the detailed background work that is essential for any reintroduction, and then with all the planning and practicalities involved in translocating Pine Martens to Wales, Vincent Wildlife Trust shared that knowledge when a similar project began in summer 2016. Gloucestershire Wildlife Trust and the Forestry Commission, with the help of Vincent Wildlife Trust, started to investigate the feasibility of reintroducing Pine Martens to the Forest of Dean and Wye Valley. This area was one of several that were identified as being potentially suitable for martens in the Trust's original feasibility study for England and Wales, but with some caveats, particularly around roads and traffic in the region, as well as the presence of other protected species. These issues were looked at in more detail, with recommendations for appropriate mitigation (Stringer *et al.* 2018), before the decision was taken to proceed with pilot releases in autumn 2019. Vincent Wildlife Trust provided the expertise to survey and select source sites and to carry out trapping and radio collaring, by the same methods that we had used in Wales.

In late summer 2019, having carried out initial scat surveys and with help from both Forestry and Land Scotland and NatureScot (who granted permits and licences), a team of us set up base in north Moray and began trapping Pine Martens. In addition to the animals' welfare, a high priority was to minimise any impact on the Scottish marten population. This meant using forests where we hadn't trapped before and only taking a small number of animals from each site. Based on our experience of the releases in Wales, we also wanted to get the number of martens in the Forest of Dean to the maximum as quickly as possible. As already discussed, we had observed from radio tracking martens after previous releases that early on in the first translocations, when numbers were very low, martens were more likely to make exploratory movements over a relatively long distance before returning. In mid-Wales, this was more of a problem for those of us radio tracking the martens than it was for the animals themselves! However, in an area like the Forest of Dean, with more roads and traffic, roaming widely would increase the risk of martens becoming roadkill. Therefore, we decided that we would trap and translocate in a much shorter space of time. This was hard work, but it meant that during an intense three weeks of trapping, we were able to get 18 Pine Martens radio-collared, transported to Gloucestershire and released in their new home.

This small core population settled and began breeding in the Dean and Wye Valley, but a planned second tranche of translocations had to be postponed during the height of the COVID-19 pandemic in 2020. However, we were able to resume trapping in Scotland in late summer 2021, and a further batch of Pine Martens was translocated, bringing the total number released up to a healthy 35. As in the Welsh project, all the released martens were radio tracked in their first year by the Gloucestershire Wildlife Trust project team, which included Josie Bridges and Catherine McNicol. Both had a wealth of experience with radio tracking Pine Martens in Wales, where there were far fewer roads and tracks than in the Forest of Dean. However, some of the challenges of locating martens in the much busier woodlands around the Wye Valley included inadvertently tracking adders belonging to another research project, which had similar transmitter frequencies to some of the martens.

At the beginning of 2021, Vincent Wildlife Trust produced a long-term strategic recovery plan for Pine Martens in Britain, with support from Natural England and NatureScot (MacPherson & Wright 2021). With limited resources and suitable donor populations for actions such as reintroductions, it is important to identify how to achieve the maximum conservation benefit. We used modelling methods based on data and knowledge of Pine Marten ecology and distribution to develop a framework for optimising conservation measures for this species across England, Scotland and Wales. The results were used to inform a series of recommendations for long-term Pine Marten conservation in Britain.

FIG 95. Distribution in green of Pine Martens *Martes martes* in 2022.

While reintroductions can be very valuable, they should always be a last rather than a first resort. If reintroduction is the only option, then it is an acknowledgement that we have failed to properly protect and preserve a species. Ultimately, we should be aiming to prevent local extinctions in the first place. Looking forward, our focus is now on facilitating natural recolonisation and recovery while developing and implementing effective monitoring schemes for existing and expanding Pine Marten populations, as well as other species.

CHAPTER 7

Where Weasels Aren't Wanted: Small Mustelids as Invasive Non-native Species

Humans have been moving animals and plants around for as long as they themselves have been mobile, with the result that some have successfully invaded distant lands at the expense of local, often unique, species that had evolved in their absence. These invasive non-native species (INNS) arrive in new environments where they themselves are free from the predators, pathogens and competitors that they are usually constrained by and consequently flourish. This can create severe problems for native wildlife and is why INNs currently represent one of the greatest threats to global biodiversity, after habitat loss. In Britain alone, there are more than 3,000 non-native species. One example is the Grey Squirrel, introduced to Britain in the nineteenth century from North America. Grey Squirrels rapidly spread and colonised most of Britain to the detriment of the native Red Squirrel. Grey Squirrels carry a disease, squirrelpox virus, which is harmless to greys but almost invariably fatal to reds.

Human activity, particularly in the last hundred years, has increased both the rate and extent of species introductions. Once humans started to keep domesticated animals, these were taken with them across the globe; however, non-domesticated animals were also translocated. It is likely that sometimes this was accidental, particularly in the case of small, inconspicuous species, but economically and culturally favoured species, including deer and wild-caught songbirds, were commonly and intentionally moved around with humans (Heinsohn 2001). There is archaeological evidence of a wide variety of human-mediated introductions from as far back as the Pleistocene Epoch (Grayson 2001). However, it was during the Holocene (from c.11,500 years ago) that these have had some of the most significant impacts on native animal populations, accelerating in what is known as the Anthropocene. This is acknowledged to have begun at the start of the nineteenth century, when the Industrial Revolution in Britain created the world's first fossil fuel economy, and Victorian Britons were asserting their misconceived assumption that it was their right to exert control over nature at home and across much of the world.

Because of the value of small mustelids as furbearers and as highly efficient predators, they have been moved around by humans to suit our purposes, with varying success and a range of consequences. Across the world, there are many examples of mustelids being moved and establishing very successfully outside of their native range. This has included some of the mustelid species native to or found in Britain today. Mustelids are highly adaptable and efficient predators, so when they arrive or are released in an environment where the native fauna has not coevolved and adapted to live alongside them, the consequences for naive prey species can be disastrous. The history of Stoats exported from Britain to New Zealand is one of the best-documented examples of this.

Stoats and Weasels were shipped to New Zealand from Britain in huge numbers in the latter part of the nineteenth century. At that time, New Zealand sheep farmers were at risk of ruin from having their pastures overrun with European Rabbits. When the first Europeans settled in New Zealand, they saw it not as a unique environment to be treasured but rather as an unproductive, empty landscape from which it was difficult to earn a living. There were no native land mammals in New Zealand, and settlers had no farm animals from which to make a living until their introduction in the late eighteenth century. Many colonists from Britain zealously set about 'acclimatising' New Zealand to a version of the country that they had left behind. The land that could be was converted to productive pasture for sheep and cattle farming, and game species were released into the forests and rivers so that the colonists could enjoy the same field sports as they had 'back home'. Trout and salmon soon swam in unfamiliar rivers, while deer, hares, pheasants, quail and partridges were among the imported quarry that could be hunted on land.

Live rabbits were often carried on ships as a source of fresh meat for sailors, and probably first arrived in the south of New Zealand with whaling ships. From the 1820s, they were shipped from Sydney for trading, along with other supplies, but sightings of rabbits in the wild in southern New Zealand were relatively few up until the 1850s. However, as colonisation proceeded, rabbits were transported to other parts of New Zealand and released by settlers for sport shooting and for their meat. The rabbits thrived in their hospitable new environment to the extent that, by the early 1870s, they had become what was described at the time as a 'plague'. Rabbit burrowing, grazing and damage to valuable sheep pasture (Fig. 96) affected lambing by malnourished ewes. This led to serious reductions in sheep numbers and wool 'clip' (productivity) and a decline in profits and, consequently, the value of pastoral land. Large blocks of grazing land were leased from the Crown for sheep and, by 1875, these Crown lessees were reporting rapidly increasing pasture damage in parts of Southland, the southernmost region of New Zealand. In response, an inquiry was launched by the Provincial Council to investigate the extent of the 'rabbit nuisance' and what could be done about it. After consulting a number of authorities, it was confirmed that within the short period of two years, the rabbits had spread across the whole of Southland. The council's report, in 1876, included figures which showed that the Southland wool clip had been reduced by almost 800 bales compared to previous years, while lamb losses had increased by an average of up to 20 per cent.

Some rabbits were killed by native predators such as the Harrier Hawk *Circus approximans* and the Weka *Gallirallus australis*, a flightless member of the rail family, but they could not keep pace with the invaders. It was suggested that importing 'natural enemies' of the rabbit could solve the serious threat to pastoral agriculture posed by rabbits. The importation of foxes had been

FIG 96. An example of pasture in New Zealand damaged by European Rabbits *Oryctolagus cuniculus*.

prohibited almost 10 years earlier, because they were known to predate lambs, but, throughout 1876, the idea of exporting Stoats, European Polecats and Weasels to New Zealand was debated in the British press, as well as in New Zealand. Those who voiced opposition to the proposal included leading ornithologists, who were concerned about the risks to New Zealand's endemic and flightless native birds. Also opposed to the idea were those landowners who had been investing heavily in acclimatising game birds for sport. One was quoted as saying that 'there is no gamekeeper [in Britain] who does not wage incessant war against [stoats and weasels]'.

In 1876, Sir George Grey proposed a Noxious Animals Introduction Prevention Bill that would have made illegal the importation of Polecat-Ferrets, Stoats and Weasels to New Zealand. However, the Legislative Council, which included many wealthy runholders suffering losses from rabbit damage, refused to consider it, and that same year, the Rabbit Nuisance Committee of the New Zealand House of Representatives recommended that money should be granted for introducing 'weasels'. When Frank Buckland, an English naturalist and leading proponent of acclimatisation was asked for his advice, his response was that shipping such highly strung and aggressive animals all the way to New Zealand would be impossible. Instead, Buckland recommended Ferrets as a more amenable option, being semi-domesticated and relatively easy to handle.

THE TROUBLE WITH FERRETS

Contrary to expectations, it proved to be extremely difficult to transport Ferrets round the world, and shipments were often disastrous failures. Gamekeepers at the time, with long experience of keeping Ferrets, described them as 'shivery creatures that need warm dry hutches, and always thirsty, so they must be given plenty of water'. The Ferrets' high susceptibility to canine distemper, combined with inappropriate feeding and housing onboard ship, led to high mortality in transit, with some shipments of as many as 600 animals being lost entirely. In 1878, a consignment of Ferrets was dispatched from England on a sailing ship, the *Rialto*. Distemper broke out among them on the voyage, and by the time the ship arrived, only three Ferrets were still alive. One landowner imported 600 Ferrets in a single consignment, which all died in transit, and then another lot of 700, of which two survived. In the period between March 1882 and June 1883 alone, out of 25 shipments and a total of 1,580 Ferrets (including 122 listed as 'natural increase' during the voyage), just 376 were landed alive (King 2017a). Official importation of Ferrets came to an end in 1884, by which time it had become apparent that it was not an economic option to import Ferrets for release and that local breeding was a more feasible and attractive proposition.

From relatively recent research on invasive species and the science of reintroduction biology, we know that the successful establishment of a population is dependent on the number of animals that escape or are released (Blackburn *et al.* 2015). The early releases of a few Ferrets on some sheep stations were unlikely to have resulted in viable feral populations, but in 1882, three new Ferret breeding depots were established by the government and many others soon followed. This new policy proved to be very effective, and by the mid-1880s, extraordinary numbers of Ferrets were being released each year. Ferret breeding was a long-term and risky enterprise, so the government offered private breeders a guaranteed price for their Ferrets as an incentive for them to invest in it. Enterprising operators did very well out of the Ferret-breeding business, for as long as demand continued. For example, one contractor, Messrs. Allen and Riggs at Wairuna Hills, had a lucrative agreement from September 1888 to supply the colonial government with 10,000 Ferrets a year for three years, at a price of seven shillings and sixpence each (the equivalent of almost £40 per Ferret in today's money). One unforeseen problem of the Ferret trade at the time was that it provided an attractive opportunity for dishonest profiteering. In districts where large numbers of Ferrets had already been released, the high prices that were paid for live Ferrets was an incentive for rabbit trappers to poach wild Ferrets instead and offer them for sale. In some cases, sheep inspectors were inadvertently buying back their own previously released Ferrets. In order to stop this, the government changed the rules so that only purchases from known breeders were authorised, until all government contracts were ended in 1889. However, private purchasers continued to buy from private breeders, and Ferret production continued in New Zealand for many years. It is thought that overall, more than 75,000 Ferrets were bred and released on South Island during this time (King 2017a). They survived well and quickly spread throughout both main islands of New Zealand but failed to have any sustained or widespread impact on the country's rabbit population.

IMPORTING STOATS AND WEASELS

As was evident from the high losses of Ferrets in transit, the successful transport and introduction of Stoats and Weasels to New Zealand would need skilful management by those with knowledge and understanding of these species. In the late nineteenth century, gamekeepers on the many thriving sporting estates in Britain made it their business to know all about the fierce little predators that they saw as a threat to their livelihoods and were experts in trapping them. Henry Allbones and his son, Walter, both former gamekeepers, were professional vermin trappers near Brigg, in Lincolnshire. In 1882, they were hired to deliver consignments of Stoats and Weasels to New Zealand because of their knowledge of both species. (Ironically, this knowledge was gained while protecting economically valuable rabbits around Brigg, at the time a local centre of rabbit warrening and fur dressing.) Over the next 10 years, Henry and Walter collected and accompanied shipments of at least 7,838 Stoats and Weasels from England to New Zealand. The pair enlisted their network of local gamekeeping contacts to supply them with live Stoats and Weasels, for which they paid six shillings and fourpence each, respectively, provided the animals were in good condition. As demand grew, they were able to advertise more widely and increase the price paid for Stoats to reflect their greater usefulness as predators of rabbits. The Allbones devised a system for collecting and managing the animals in transit based on a sound understanding of their requirements. They took only the healthiest, most vigorous individuals and provided each one with a dark, zinc-lined nest box filled with dry bedding. Throughout the journey, the animals were kept scrupulously clean and dry and given plenty of food and water. For a shipment of 300 Stoats and Weasels, 4,000 live pigeons were carried, to provide each little carnivore with a generous ration of fresh meat. The care and attention to detail was rewarded with consistently low mortality rates of less than 10 per cent, with rare exceptions. Their methods were expensive, but the animals arrived in good condition and were highly sought after. In January 1885, one newspaper (the *Auckland Star*) reported that £3 (equal to £310 today) was paid for Weasels at auction in Christchurch, while Stoats had fetched £5 5s each (around £550 today). This was reported with astonishment back in England, where these species were regarded as vermin. Over the next year, 1,222 Stoats and Weasels were sent by the Allbones on steamships to New Zealand. Complaints from fellow passengers brought trade to a temporary halt in early 1886, but by May of the following year, a solution was found, and Henry was able to resume, using cargo ships to transport his animals. By the start of 1888, a further 1,886 Stoats and Weasels had been delivered (King 2017b). Unfortunately, as is often the way of these well-intentioned 'experiments' in biological control, the animals did not behave as they were expected to. By 1889, Stoats and Weasels had already started to spread into native forests, where they killed birds rather than rabbits. Objections were raised by settlers until the government was persuaded to end all official imports of both Stoats and Weasels, but the little predators continued to spread throughout the country. The Stoats certainly did eat some rabbits, but the effect they had on the rabbit population was negligible. Conversely, their impact on the native fauna was catastrophic, coming as it did on top of the damage already inflicted over the previous 600 years by rats and cats that the Polynesians had brought with them since they first began arriving in New Zealand.

The financial impact that rabbits were having was disastrous, and importing mustelids as a solution to the rabbit problem must have seemed very straightforward at the time. By 1881, the government had seen its rental income from half a million acres (202,342 ha) severely affected by rabbits decrease from £2,288 a year to just £619, and the financial cost to the whole colony in lost exports was estimated to be in excess of £500,000 a year (the equivalent of about £49 million today). Earnscleugh, one of the largest sheep stations in Otago, was abandoned by the leaseholder in 1895, by which time it could support only 12,000 sheep, half the number it had in 1879 (King 2017a). The use of 'natural enemies' of rabbits must have seemed an attractive option, particularly on inaccessible and unoccupied Crown lands in the more remote areas. In such places, labour-intensive methods of rabbit control would have been well-nigh impossible.

BONUS ON STOATS AND WEASELS.

(UNITED PRESS ASSOCIATION.)
WELLINGTON, Oct. 26.

The Cabinet have decided to give bonuses on stoats and weasels landed in the Colony for distribution on Crown lands on stoats 20s and on weasels 10s.

STOATS AND WEASELS.

TO THE EDITOR.

SIR,—In a letter to the Awatere Rabbit Board read at their meeting yesterday, Messrs Holmes and Bell state that the cost—£3 per head—of the shipment of stoats and weasels imported by them compared very favourably with the cost of those imported through other firms. Allow me to state that the last order completed by the Company I re-present, numbering 483 animals, averaged £2 14s per each stoat and weasel delivered to the Rabbit Board, and I do not antici-pate that the order now being filled will exceed that figure. I may state that our last five shipments, comprising 833 animals, averaged £2 19s per head.—I am, &c.,
T. H. HANNA.

TO THE EDITOR.

SIR,—The warning re introduction of noxious animals was too late! I see by the Rangitikei Advocate of November 27th that " five hundred stoats and weasels which arrived at Wellington by the last direct steamer were distributed over the largest station in North Wairarapa on Saturday. The cost was about £5 each." Five hundred stoats and weasels! Farmers and mothers look out for your poultry and children! Now they are introduced the only thing for the settlers to do is to shoot or destroy them wherever they see them. Remember the rabbits and sparrows and minahs! They were protected! Don't let any law protect stoats and weasels, there are five hundred now, in a few years with protection they will be millions! Where will your poultry be then? The tables may be turned upon the runholders. The kea was not naturally a carnivorous bird, but circumstances changed its nature, and the stoats and weasels may find it easier to kill the young lambs than to go after the rabbits. Let us hope it may so happen, and then we shall soon get rid of the stoats and weasels.—I am, &c.,

A PARENT.

FIG 97. Some local newspaper comments from 1888 and 1889 at the time that Stoats *Mustela erminea*, Weasels *Mustela nivalis* and Ferrets *Mustela putorius furo* were being imported into New Zealand.

FIG 98. Stoat *Mustela erminea* with European Rabbit *Oryctolagus cuniculus* kill, Cornwall.

With the benefit of hindsight, and our increased understanding of the complex and multifactoral nature of ecological interactions, we can see that there are significant differences between the environment in which these small mustelids evolved and the one in which they found themselves in New Zealand. These differences are key to understanding what happened and why the somewhat naive expectation that they would control rabbit numbers was sadly disappointed.

Where Stoats and Weasels evolved, throughout northern Eurasia and North America, they are part of a guild of similarly shaped small predators. Despite extensive niche overlap, coexistence is possible because of differences in body size, giving each species the competitive advantage in different circumstances, depending on habitat, local and temporal fluctuations in rodent populations and the availability of alternative prey. As previously discussed, Weasels are more efficient predators of small rodents than are Stoats, and female Weasels can respond quickly to a glut of prey by producing second litters and increasing their productivity. However, Stoats have the upper hand in aggressive encounters with Weasels (direct or interference competition). Stoats are also able to switch to alternative, larger prey when small rodents are scarce. This is important in the context of New Zealand.

In the UK, Stoats specialise on rabbits, as evidenced by the severe declines in Stoat numbers on game estates following the population crashes in rabbits from myxomatosis. Stoats are vulnerable to interference competition from larger predators such as foxes and owls, and the reduction in rabbits as a result of myxomatosis led to increased competition among small carnivores for other prey.

By the time Stoats and Weasels were introduced to New Zealand to control rabbits, other (larger) mesocarnivores had already been tried. Huge numbers of cats had been released on farms long before any mustelids, establishing feral cat populations that are now widespread. Then from the mid-1870s, thousands of Ferrets were also released. These were able to successfully compete with cats for rabbit prey and are now the most common mustelid in New Zealand in prime rabbit habitat. Stoats and Weasels, when they arrived in this mix, could not compete with the larger cats and Ferrets for rabbits in grassland. An experimental study carried out in 2012 showed the extent to which Stoats will avoid cats and Ferrets. Each Stoat in the experiment was kept for three nights in an enclosure with a caged cat or Ferret at one end and an empty cage at the other. All the Stoats avoided the area containing the larger predator, adjusted their behaviour both spatially and temporally, increased their vigilance when the other species was present, and took less food at the risky area between their cages (Garvey *et al.* 2015).

In New Zealand, Stoats were quickly able to take advantage of the forests where there were plenty of refuges from larger predators and there was also an abundance of new prey in the form of lizards, ground-nesting birds and wetas,

FIG 99. Stoat *Mustela erminea* with avian prey.

large-bodied insects, that they found in their new home. New Zealand has no voles, and so Weasels, being small-mammal specialists, did not fare so well in the absence of their principal prey. In the southern beech forests, they could switch to mice and do very well in mouse peak years. However, they probably could not survive the mouse-poor years in between. While Stoats and Ferrets have established as invasive pests, Weasels are uncommon and, at the moment, far less significant even though many more Weasels than Stoats were imported originally.

Among their many abilities, Stoats are known to be good swimmers (Fig. 100). They have been reported swimming almost 2 km from land in New Zealand's Lake Te Anau, and trials carried out with captive Stoats have shown that they can swim against a continuous current for up to two hours nonstop (King *et al.* 2014). They have been recorded on at least 90 offshore islands that lie within 3 km of the New Zealand coast, and this colonisation is thought to have been primarily through swimming. In the wild, Stoats have also been seen floating on woody debris, demonstrating that they are resourceful enough to be able to use such assistance to cross water. Therefore, it is entirely plausible that the Stoat which arrived in 2009 on Kapiti Island, some 5 km offshore, did so without any help from humans (Veale *et al.* 2012, Veale 2013). Some Stoats in the Fiordland National Park were recorded as having home ranges completely bisected by the fast-flowing Eglinton River, which they regularly swam across, even in flood conditions (Murphy & Dowding 1994).

FIG 100. Stoat *Mustela erminea* demonstrating its swimming capability.

Ferrets are less willing swimmers and, in New Zealand, viable offshore populations only occur on islands that are connected to the mainland at low tide (King et al. 2017). Weasels were present on two islands (Maud and Matakohe), although it is not known how they got there, and they have since been eradicated (Clout & Russell 2006).

THE CONSEQUENCES OF INVASION AND ATTEMPTS AT ERADICATION

Many of the offshore islands which Stoats have colonised in New Zealand (and elsewhere) are of high conservation value and are inhabited by rare and endangered endemic species.

The most drastic losses of New Zealand native species began before mustelids and other mammals were imported by Europeans. Many extinctions occurred in early Polynesian times and coincided with the introduction of Pacific Rats *Rattus exulans*, dogs, cats and pigs. Nonetheless, mustelids are certainly one of the biggest current threats to some of New Zealand's endemic birds (King & Powell 2007).

FIG 101. The Kakapo *Strigops habroptilus*, one of New Zealand's vulnerable endemic flightless birds.

Analysis of Stoat diet carried out in the late 1970s confirmed that Stoats do eat a lot of New Zealand's native birds but shed no light on what impact that might have on those bird populations (King & Moody 1982a). However, they are unequivocally implicated in some of the historic losses on offshore islands that occurred shortly after Stoats first appeared on them. Resolution Island once held plentiful populations of Piopio *Turnagra tanagra*, Tieke *Philesturnus carunculatus*, South Island Kokako *Callaeas cinerea*, Mohua *Mohoua ochrocephala* and Weka *Gallirallus australis*, as well as Kakapo *Strigops habroptilus* (Fig. 101) and Little Spotted Kiwi *Apteryx owenii* that had been translocated there as a refuge from introduced predators on the mainland. In 1894, Richard Henry was appointed as curator of Resolution Island, which was stocked with species such as Kakapo and kiwi that were threatened by mustelids on the mainland. This early conservation attempt failed when the first Stoats found their way onto the island by 1900, and in the following years all of these bird species disappeared (Edge *et al.* 2011). Mustelids are considered to be one of the biggest contemporary threats to the conservation of many of New Zealand's endemic species, including invertebrates. Predation by Stoats (and cats) is responsible for unsustainably high annual chick mortality in all five species of kiwi. Periodic peaks in mouse and, subsequently, Stoat numbers after mast (seedfall) years in southern beech forests can result in catastrophic levels of predation on Blue Duck *Hymenolaimus malacorhynchos*, Kaka *Nestor meridionalis*, parakeets, Mohua, Rock Wren *Xenicus gilviventris*, Brown Creeper *Mohoua novaeseelandiae* and some native bat species (King *et al.* 2017).

New Zealand has led the way in developing research programmes and conservation strategies for controlling introduced predators and mitigating their impacts. As discussed previously, predator control can only be effective in reducing numbers of mustelids when it adds to, rather than substitutes for, natural mortality, and also when rates of mortality exceed natural rates of replacement. Because of limited budgets, decisions inevitably have to be made as to the best use of resources. This is usually to direct them to areas where the threat is greatest to populations of endangered birds. The most common method of trapping is to use a wooden box, or Stoat tunnel, with a small entrance at one end. A baited, 'humane' kill trap is placed in the tunnel to kill the Stoat when it enters. Traps have to be visited and then manually reset if they have been sprung. Whilst not as labour-intensive as live trapping, this is still a time-consuming operation, so recent trials have focussed on new designs of self-resetting Stoat traps that could be used in more remote areas. Despite many years of refining and improving baiting and trapping methods, it is acknowledged that in any mustelid population, there will be some trap-shy individuals, but to exterminate a population it is necessary to eliminate every single one. Stoats and Ferrets are both difficult species to control since they are bait-shy, trap-wary and have high rates of reproduction. It only requires a small number of breeding females to evade capture to undermine an entire control operation. This has been seen on Secretary Island and Resolution Island, in New Zealand, where attempts to completely eradicate Stoats through meticulously planned and intensive trapping programmes have so far been unsuccessful (Elliott et al. 2010, McMurtrie et al. 2011).

Methods that were effective in removing all the Stoats from Chalky (514 ha) and Anchor (1,130 ha) Islands by 1999 and 2001, respectively, were subsequently scaled up for the much larger Secretary (8,000 ha) and Resolution (21,000 ha) Islands (Edge et al. 2011). Secretary Island was covered with a 108 km network of trap tunnel lines at a density of one tunnel (each containing two kill traps) per 9 ha. Traplines were also established on the adjacent mainland, as well as on 'stepping-stone' islands within Stoat swimming distance, to reduce the likelihood of recolonisation. Trapping commenced in 2005, but Stoats have continued to be caught on the island every year since, indicating that the population is being maintained at a low level with no further decline. Recent estimates of Stoat home ranges on Secretary and Resolution Islands indicate that a few Stoats may have smaller home ranges than previously thought and therefore have not encountered a trap. Research by Andrew Veale, based on genetic analysis and age of the Stoats captured, showed that at least one male Stoat had survived on Secretary Island even after several years of intensive trapping in the initial 'knockdown' phase of the programme and that subsequent captures were a mixture of resident Stoats and some immigrants (Veale et al. 2013).

The Resolution Island Stoat Eradication Programme began in 2008 and, 12 years later, there were still Stoats present. Both of these examples illustrate the challenges of attempting to eradicate Stoats from large islands in close proximity to the mainland. Secretary Island is 950 m from the mainland at its narrowest point and Resolution Island is just 520 m away, meaning that both are within the swimming range of Stoats from the mainland. It has been estimated that if a single, undetected pregnant female Stoat reaches an island, this will initially result in fewer than 10 Stoats for up to 31 months, but then numbers would rapidly rise (Choquenot et al. 2001). However, rates of migration of Stoats to islands are much higher when Stoat numbers are high. This is related to beech mast and rodent abundance in beech forests. Unusually high immigration was detected on Secretary Island after a beech masting year in 2006, followed by a rodent and Stoat 'plague' on the mainland in Fiordland. As recently as 2020, there has been a significant increase in Stoats on Resolution, following a peak in mouse numbers due to the beech mast. Department of Conservation biodiversity ranger Peter McMurtrie reported that all the female Stoats had been breeding at capacity, having 10–12 young, and there were 200 Stoats caught that year, compared to the usual 30 or 40.

Although Stoats have not been completely eradicated from Resolution and Secretary Islands, in 2022 the New Zealand Department of Conservation was confident that they had been reduced to levels low enough to allow the recovery of endangered wildlife. Nonetheless, continued, ongoing control will be needed on these large islands to protect the native species present and prevent Stoats from re-establishing.

Closer to home, Stoats have also become an invasive species on some of the UK's offshore islands. While Stoats are native to mainland Britain, they were never present on the Orkney Islands, 10 km off the northeast coast of Scotland. Stoats were first confirmed on mainland Orkney and the linked island of South Ronaldsay in 2010, but it is not known when or how they got there. Possible explanations include accidental import with hay or straw, or deliberate release (perhaps to control rabbits!). However, regardless of how they first arrived, their numbers increased rapidly, and Stoats have now become established and widely distributed throughout mainland Orkney, Burray and South Ronaldsay (Auld *et al.* 2019). In 2014, a report was commissioned by NatureScot to assess the risks that a Stoat population presented to Orkney's native wildlife. The Orkney Islands have no native ground predators, and so the introduction of one, such as the Stoat, poses a serious threat to many native species. These include the Orkney Vole *Microtus arvalis orcadensis*, Hen Harrier *Circus cyaneus*, Short-eared Owl *Asio flammeus* and many other species of ground-nesting birds, such as Corncrake

FIG 102. Corncrake *Crex crex* on Orkney are one of many ground-nesting birds potentially impacted by predation from non-native Stoats *Mustela erminea*.

Crex crex (Fig. 102), Curlew *Numenius arquata*, Red-throated Diver *Gavia stellata* and Arctic Tern *Sterna paradisaea*. In recognition of their importance, there are a number of nationally and internationally designated sites across the islands. These are likely to be negatively affected if the Stoat population cannot be controlled. Orkney's nature and landscapes provide significant socioeconomic benefits, so a decline in native wildlife could damage tourism, which makes a significant contribution to the local economy. Free-range poultry farming is also common practice in Orkney and was economically viable due to the absence of mammalian predators. It is expected that Stoats would have an impact on this industry, too, if successful predator control and mitigation measures are not implemented.

Initial attempts to remove Stoats through volunteer effort and small contracts proved unsuccessful so, in 2017, a study was carried out to assess the feasibility of a full-scale professional project to eradicate Stoats from the Orkney Islands. This concluded that it would be possible to eliminate Stoats, given their range at the time, but that any delays and spread to other islands could threaten the feasibility of eradication. To achieve eradication, a high number and density of humane lethal traps (16 per km^2) was recommended at a cost of around £4.5 million (Auld *et al.* 2019). In late 2019, the Orkney Native Wildlife Project began. The project is led in partnership by NatureScot, RSPB Scotland and Orkney Islands Council, with the aim of eradicating Stoats to protect the unique native wildlife. Some of the challenges include trapping in a landscape that includes both rural and urban habitats. Lessons have been learned from Stoat eradication programmes in New Zealand and also from the experience of removing another non-native mustelid, the American Mink, from offshore islands in the UK.

AMERICAN MINK IN THE UK

The American Mink, as already described, is a semiaquatic mustelid native to North America. After a significant decline in demand for fur at the end of the Middle Ages, it became much sought after once again during the nineteenth century. Demand from the fashion industry for the mink's luxuriant thick fur led to the development of fur farms, which began as far back as the 1870s in the USA. Today, the majority (80 per cent) of pelts used in the fur industry come from fur farms, and the furbearing animal most often farmed is the mink (50 million per year). Since the early twentieth century, American Mink have been introduced to fur farms in many parts of Europe (Dunstone 1993) and, as a result of deliberate releases or accidental escapes, feral mink are now recognised as an invasive species in Iceland, Scandinavia, the former Soviet Union, Germany,

France, Spain, Patagonia, Argentina and the UK (Wildhagen 1965, Gerell 1967, Westman 1968, Aliev & Sanderson 1970, Cuthbert 1973, Smal 1988, Birks 1990, Previtali *et al.* 1998).

The EU Invasive Alien Species (IAS) Regulation provides for a set of measures to be taken across the EU in relation to invasive alien species included on the EU list. These are prevention of the intentional or unintentional introduction of IAS of concern into the EU; early detection and rapid eradication; and concerted management action to prevent them from spreading any further and to minimise the harm they cause. However, the IAS Regulation recognises that some species may provide economic benefits in certain member states. In spite of the fact that

FIG 103. An example of a mink farm in England in 1954.

FIG 104. An example of colour variation in farmed American Mink *Neovison vison*, some of which are now living and breeding in the wild.

the American Mink significantly affects native mammals and birds, it is a hugely profitable species in some countries (e.g. Denmark, the world's largest producer of mink skins where, during a five-day fur auction at Kopenhagen Fur, skins worth more than €130 million can be sold). The UK is no longer a member of the EU but is likely to continue following the principles set out in the IAS Regulation.

The first British mink farm opened in England in 1929 and the first escapes must have begun shortly after, as mink have been killed in the wild since the 1930s. However, it was not until July 1956, when a female with kits was seen on the River Teign in southwest England, that mink were recorded breeding in the wild. Mink were seen as an attractive 'get rich quick' option for farmers and landowners at the time. By the peak of the UK fur trade in the early 1960s, there were at least 600 mink farms in Britain, and feral mink were living wild in more than half the counties of England and Wales and in much of lowland Scotland.

Mink are highly adaptable generalist predators that can exploit both aquatic and terrestrial prey. This includes small- and medium-sized mammals, fish, crustaceans, birds, eggs, amphibians and insects. Within their native range, the mammal species predated by mink are predominantly Muskrats *Ondatra zibethicus* and hares; however, the proportion of mammals in the diet and indeed of individual prey species varies significantly depending on local availability and abundance. Studies have shown that mink diet may differ between individuals, between sexes and between seasons (Dunstone 1993). Both in their native North America as well as in Europe, mink occupy a wide variety of wetland habitats, including streams, rivers, lakes, freshwater and coastal habitats. Their territories are usually based on the water's edge, such as along riverbanks or stretches of coast. At the time when mink were establishing in Britain, the European Polecat and the Eurasian Otter, two native mustelids whose ecological niches have some overlap with that of the mink, had both undergone significant population declines and were absent from much of the country, creating space for mink to occupy.

Although polecats are generally more terrestrial than mink, both species are often found in wetlands and riparian habitats, where they exploit many of the same prey resources. At around 1 kg, both polecats and mink are similarly sized opportunistic predators. The main difference in their diets is that the mink is a food generalist and will commonly hunt aquatic prey (fish and crayfish), whereas the polecat is more specialised on prey such as small mammals, rabbits, rats and amphibians (Hammershøj et al. 2004, Zalewski et al. 2021). Polecats and mink will also make use of the same type of sites for denning and resting, such as rabbit burrows. Otters exploit the same suite of aquatic prey as mink, resulting in some competition between the species. However, otters (at 7–10 kg) are the larger of the two and expected to be the dominant competitor in direct interactions.

FIG 105. A Eurasian Otter *Lutra lutra* hunt in Buckinghamshire, 1937. The typical hounds, which have probably cornered an Otter in this backwater, are shown. The men each hold a stave that helped to block the Otter from escaping by doubling back past the hounds. Otter hunting was banned in England and Wales in 1978.

Interspecific aggression between the two species has been recorded (Simpson 2006), and local, short-term declines in mink abundance have been observed as otters have recovered in numbers. However, mink are able to exploit a wide range of prey and can adjust their dietary niche by consuming less fish in the presence of otters. When mink were colonising much of Britain in the mid-twentieth century, polecats were still largely confined to Wales, for the reasons discussed previously, while otter populations had declined almost to extinction in England and to a lesser extent in Scotland. This was attributed to hunting, habitat loss and aquatic pollution from widespread use of organochlorine pesticides in the 1950s.

Wherever the American Mink has been introduced, there have been negative impacts on native species. These include a correlation with the decline of European Mink *Mustela lutreola* in Eastern Europe, harmful effects on waterfowl in Iceland and Finland and reduced breeding success of ground-nesting seabirds in Finland (Bjornsson & Hersteinsson 1991, Kilpi 1995, Nordstrom *et al.* 2002) and Scotland (Craik 1998, Rae 1999, Craik 2000, Sidorovich & Macdonald 2001, Macdonald *et al.* 2002). As a result, several European countries have made attempts to control or eradicate mink but with varying success (Bonesi & Palazon 2007). The continued presence of active mink farms in many countries presents a persistent risk of reinvasion (Pertoldi *et al.* 2013). In Britain, this is no longer the case. Fur farming had dwindled by the early 2000s to fewer than 20 mink farms in England and Wales, and the last of these was closed in 2002 under the Fur Farming (Prohibition) Act 2000. The last mink farm in Scotland closed in

1993, and fur farming was subsequently banned by the Fur Farming (Prohibition) (Scotland) Act 2002. However, by this time, feral mink were widespread across all regions in England, much of Wales and in all but the far north of Scotland. Mink continued to spread into the northern Scottish Highlands and also to offshore islands within swimming distance of the mainland.

HEBRIDEAN MINK PROJECT

The Outer Hebrides is a complex chain of more than 100 islands located about 70 km off the west coast of mainland Scotland. There are 15 inhabited islands in this archipelago. The main islands include Harris, Lewis, North Uist, South Uist, Barra, Benbecula, Berneray and St Kilda. American Mink arrived in the Outer Hebrides when two fur farms were set up on the Isle of Lewis in the 1950s. They both closed in the 1960s but, by this time, a feral mink population had established, with feral animals recorded on Lewis by 1969. Attempts to prevent this population spreading were unsuccessful, and by 1999, there were breeding populations on North Uist and Benbecula. Mink had also reached South Uist by 2001. The suitable habitat of the islands' coastlines and over 7,500 freshwater lochs helped the invasive species achieve population densities rarely reached in their native North America.

The islands support some of the most important breeding populations of wading birds in Europe. Many nesting areas are listed as Sites of Special Scientific Interest under the Nature Conservation (Scotland) Act 2004 and classified as Special Protection Areas under the EU Birds Directive. These sites support

FIG 106. A floating raft used to detect and then capture and remove American Mink *Neovison vison* during the Hebridean Mink Project.

FIG 107. Important populations of species including Curlew *Numenius arquata* occur on the islands of the Hebrides.

internationally important populations of species, including Red-throated Diver *Gavia stellata*, Black-throated Diver *Gavia arctica*, Great Northern Diver *Gavia immer*, Hen Harrier *Circus cyaneus*, Merlin *Falco columbarius*, Short-eared Owl *Asio flammeus*, Greylag Goose *Anser anser*, Mallard *Anas platyrhynchos*, Teal *Anas crecca*, Wigeon *Anas penelope*, Gadwall *Anas strepera*, Shoveler *Anas clypeata*, Tufted Duck *Aythya fuligula*, Eider *Somateria mollissima*, Shelduck *Tadorna tadorna*, Red-breasted Merganser *Mergus serrator*, Golden Plover *Pluvialis apricaria*, Common Sandpiper *Actitis hypoleucos*, Curlew *Numenius arquata* (Fig. 107), Corncrake *Crex crex* and Common Tern *Sterna hirundo*. Ground-nesting seabirds such as Little Tern *Sternula albifrons*, Arctic Tern *Sterna paradisaea* and Arctic Skua *Stercorarius parasiticus* also occur in significant numbers. Introduced predators such as the mink could cause lasting damage, and it was thought that in just a few decades they would destroy these bird populations that had established over centuries.

The Hebridean Mink Project was set up in 2001 to prevent further significant disturbance and losses to ground-nesting birds and migratory species found on the islands. The aim was to eradicate American Mink from the Outer Hebrides and then to continue to monitor North and South Uist for mink activity. With major funding from the EU Life programme, at the height of the project, 12 core staff worked as teams of trappers to remove mink. Throughout the operational stages of the project, data were collected on the control methods used in order to assess their effectiveness. Work began in 2001 to trap mink in North Uist, Benbecula and then South Uist. The project also included South Harris as a buffer zone against migration from further north. In the second phase of the project, which began in 2006, active management of mink was carried out across

Lewis and Harris. A focussed trapping effort was begun on South Harris, to establish a buffer zone for the Uists. This included targeted trapping around known tern colonies. Trapping then continued all the way to the Butt of Lewis. In 2013, the project moved to a monitoring phase. Footprint tunnels were used to establish the presence of mink and then to trap them. As mink can move great distances in very short timescales, this strategy had to be abandoned when very low densities of mink were reached. At that point, kill traps were used to detect and remove the final few animals. All traps were carefully set to exclude non-target species.

In addition to conservation, the project claimed to have provided benefits for anglers and fish farms, as mink had also caused considerable damage to farmed fish and young wild fish stocks. Removing mink also meant that crofters were able to keep chickens and ducks without the risk of severe predation on domestic poultry. Nearly 2,200 mink were caught up until 2018, when it was announced the scheme had accomplished its aims, with only two non-breeding females and associated males caught on Lewis and Harris in the previous 18 months. However, since then, a number of mink were seen again on the islands, so a trapper was recruited. Only one mink was caught, in 2019, but a network of traps was then monitored across the island to maximise opportunities for complete eradication. However unpalatable it may be for some, there is no doubt that the wide-scale and systematic killing of mink on the islands has had a number of positive results, with seabirds and wading birds returning to breed in the Hebrides, providing a boost for nature tourism (Macleod *et al.* 2019).

MINK ON MAINLAND BRITAIN

Successful eradication of invasive species in mainland systems is rare, generally because the scale of a project is too large and there is continued risk of reinvasion. Whilst there are no longer any fur farms in Britain to act as a source of animals for reinvasion, American Mink are now so widespread that no area subject to local control is completely safe from recolonisation. For this reason, a mink eradication project on the British mainland, where they are almost ubiquitous, has been considered to be unfeasible or prohibitively expensive. Nonetheless, local efforts to trap and control mink in Britain have continued for decades. These have varied in scale, from individuals working on private land to multimillion-pound landscape projects. But, if it occurs at all, trapping is usually temporally or spatially discontinuous. With very few exceptions, mink return as soon as the work stops.

One example of this comes from an opportunistic study carried out over three years at a nationally important site for Water Voles in the lowlands of southwest England. There was a need for mink control at the site, but the methods used were constrained by circumstance. Systematic mink trapping was carried out between 2003 and 2005 at Shapwick Heath and Ham Wall nature reserves (Fig. 109). Prior to that, trapping was carried out on an ad hoc basis, and some mink were removed annually. Traps were placed no further than 350 m apart, preferentially located at points with obvious signs of mink (runs, scats, dens). Cage traps were used so that non-target species could be released unharmed. They were set away from known otter holts and away from open areas and public paths. When traps were set near the water's edge, care was taken to place them well above any possible rise in water level and they were secured to prevent a trapped animal rolling them into the water. Each trap was covered with a purpose-made tunnel

FIG 108. Typical colour morph of feral American Mink *Neovison vison*.

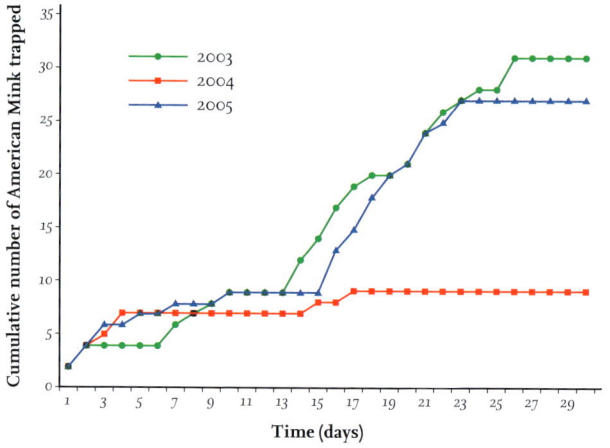

FIG 109. The cumulative number of American Mink *Neovison vison* trapped over time each year 2003–5. The rates of capture in all three years are initially similar but diverge after approximately two weeks. J. MacPherson, unpublished data.

of dark-coloured corrugated plastic and left in the same place until it had been in position for five consecutive nights with no captures. Then it was moved to the next site. Traps were checked twice daily, at first light and just before dusk, to minimise the length of time an animal could be in a trap. (Animal welfare is still a priority, even when trapping a 'pest' species.) Captured animals were sexed and assessed as either adult or juvenile and their approximate age was recorded. Males were aged by palpation of the baculum (penis bone) and females by the presence or absence of healed or recent mating bite wounds on the back of the neck.

All captured mink were humanely dispatched in accordance with published guidelines. In 2004, capture-mark-release-recapture was used to look at the effect on mink numbers of the previous year's control. All captured mink were microchipped for identification in order that they could be individually recognised if and when they were subsequently recaptured. They were then released under licence from DEFRA. In both years when captured animals were removed, the total number of mink trapped was high, but in 2004, when captured animals were marked and released, the total number of animals caught at the site during the trapping period was much lower. This is summarised in Table 8.

Studies suggest that all trappable resident mink are probably encountered within one or two weeks (Dunstone 1993). If this is true, then the number of residents was not significantly different in all three years, but the number of mink being drawn in from the surrounding landscape and removed, once the territory-holding animals are no longer present, may impact on a much larger area than just where the traps are set. This 'vacuum effect' was also reported before the start of the Hebridean Mink Project by gamekeepers trapping over relatively small areas (Roy *et al.* 2009a). A similar mechanism has been observed where localised culling of Red Foxes had the effect of creating a sink (Reynolds *et al.* 1993). Where this occurs in species like American Mink, capable of dispersing over long distances, it has been suggested that the potential source area providing replacement animals may be huge (Heydon & Reynolds 2000). If this is the case, then intensive mink control at 'honeypot' sites with high

TABLE 8. Summary of American Mink *Neovison vison* captured at Shapwick Heath and Ham Wall in March each year 2003–5.

Year	Treatment	Adult males	Adult females	Juvenile males	Juvenile females	Total
2003	Removal	20	9	0	1	30
2004	Capture-mark-release-recapture	6	2	0	1	9
2005	Removal	11	1	3	12	27

habitat suitability and prey availability may be a very effective way of reducing the number of mink in the wider landscape.

The male to female ratio was 2:1 in 2003 and 2004, which is to be expected, taking into account the higher 'trappability' of males and transient males seeking females to mate, but after low numbers in 2004, the sex ratio overall was approximately equal in 2005. One hypothesis here is that females produced more female offspring in 2004, which is a known effect of low population density in mustelids. However, if it is assumed that all trappable residents have been encountered within the first two weeks, then the sex ratio was not significantly different in any year (see Table 9).

A problem with this study is that, due to circumstances beyond our control, removal had to be carried out in alternate years, so although the study provided some useful insights, the results were far from conclusive. It may be that numbers were high in years when trapped animals were removed, as a result of mink being drawn in from adjacent territories; however, the increase may simply represent one year's recruitment. If so, it appears that the sex ratio may be altered in favour of females, as a density-dependent response. This is supported to some extent by the relatively high proportion of sub-adult females trapped in 2005. In order to explore this further, the study would have needed to be extended. Conservation managers are understandably reluctant to licence or allow marked animals to be released for extended periods because of the potential impact on many native species vulnerable to mink predation. Regardless of whether the intention is to kill or not, the traditional method of cage-trapping mink along watercourses requires a considerable number of traps (approximately one per 300 m) and resources. Live traps must be checked *at least* once every 24 hours but preferably more frequently for animal welfare reasons, which is time-consuming and often fruitless. Fortunately, in recent years, this issue has been addressed by the ingenious development of the 'mink raft' (Figs. 111 & 112). Each floating raft consists of a tunnel over a clay pad

TABLE 9. Summary of American Mink *Neovison vison* captured in the first 14 days of trapping each year 2003–5.

Year	Treatment	Adult males	Adult females	Juvenile males	Juvenile females	Total
2003	Removal	8	1	0	1	10
2004	Capture-mark-release-recapture	4	2	0	1	7
2005	Removal	5	1	1	2	9

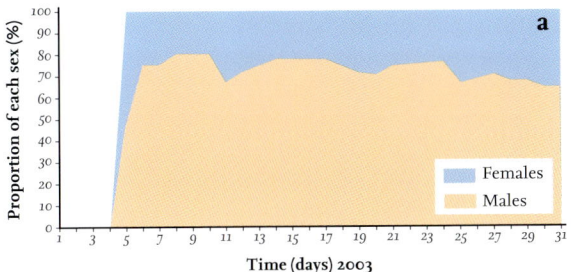

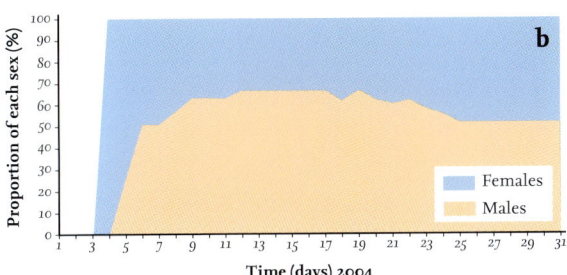

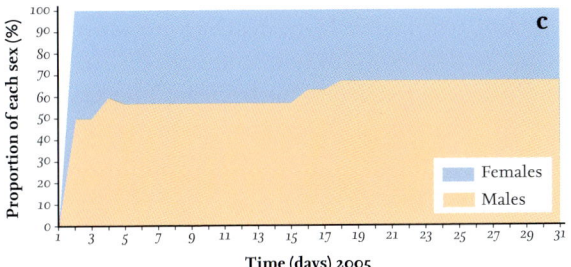

FIG 110. (a), (b) and (c) The cumulative sex ratios of trapped American Mink *Neovison vison* in years when removal (2003 and 2005) or capture-mark-release-recapture (2004) were carried out.

that records footprints of any animal passing through it. Mink will invariably investigate these as they encounter them when travelling along a watercourse. Once footprints have been recorded, a trap can be set on the raft under the same tunnel. This method is considerably more efficient than traditional trapping. Rafts used at relatively low density (one per 1–2 km) are effective even at catching female mink, which are often more difficult to trap than males. Rafts are also successful where mink numbers are low (Harrington *et al.* 2008). The tracking plates can be checked once a week rather than once a day, and traps only need to be set on or near rafts on which footprints have been detected.

FIG 111. American Mink *Neovison vison* footprints on a mink raft tracking plate.

The rafts were originally designed to be constructed from sheets of exterior grade 6 mm plywood, sandwiching a sheet of expanded polystyrene for buoyancy. These are very durable but also difficult to carry any distance. The rafts could be made more portable by replacing the plywood components with corrugated plastic (Correx). This made the raft cheaper and had the advantage of being able to fold the tunnel flat. But a disadvantage is that the Correx version is not as sturdy, and the plastic will not biodegrade if the raft is not retrieved. Once footprints have been detected on a mink raft in mainland Britain, a suitable live-capture trap must then be used, as there is a risk of catching native mustelids such as polecats, whose footprints can be indistinguishable from those of mink. Lethal spring traps can be used for mink on offshore islands of the UK, where there are no native mustelids of similar size to the mink. Otters, being larger, can be safely excluded from mink traps by using otter excluder bars to reduce the size of the trap tunnel entrance.

On the British mainland, live-capture traps must be used and any captured mink dispatched using the recommended most humane method of a shot through the front of the skull with a sufficiently high-powered air rifle. The efficiency of mink trapping can be improved by fitting a trap alarm system that sends a mobile phone text alert to the trapper as soon as a trap has been sprung. This allows for non-target animals to be released and target animals to be humanely despatched as quickly as possible after being trapped.

Mink rafts have proven to be highly successful in aiding local efforts to control mink and have also enabled the largest mainland mink eradication programme in Britain. This was a result of four successive, joined-up projects

FIG 112. An American Mink *Neovison vison* raft with otter excluder bars and trap in place.

each with short-term funding but which, over time, have succeeded in pushing back the invasion of American Mink over a large area. The project began in northeast Scotland in 2004 and started at the relatively small scale of around 30 km² to conserve a remnant lowland Water Vole population. Following on from this, in 2006, mink control was then expanded across 6,000 km² of mink habitat as part of the Cairngorms Water Vole Conservation Project. The project was led by the University of Aberdeen and involved the Cairngorms National Park Authority and local fisheries trusts. A rolling wave approach to trapping was deployed based on the use of mink rafts operated by a large number of volunteers to detect and remove mink. After this project ended in 2009, further funding was then secured for its successor, the Scottish Mink Initiative (2010–14). As a result of this approach, mink were effectively removed from a vast area of approximately 29,000 km², stopping their spread in coastal areas of northwest Scotland.

After a three-year hiatus, with only minimal funding, the Scottish Invasive Species Initiative was able to resume coordinated mink control efforts in the region. Volunteers have been invaluable in implementing this, initially recruited to monitor rafts near to where they lived or worked. They included land managers on sporting estates who had an interest in protecting native wildlife. Subsequently, more volunteers were recruited, and since spring 2018, mink have been controlled across the northern third of Scotland with a network that extends from the Firth of Tay in the south, north to Durness in Sutherland, covering 43 river catchments. By the end of 2020, about 650 rafts and traps were monitored by around 350 volunteers. In spite of the challenges of short-term funding, continuous and coordinated mink control has been achieved across a large area. Mink remain absent or very scarce throughout the region, with significant recovery of Water Voles and other tangible benefits (Lambin *et al.* 2019).

Based on the development of recent technologies such as trap alarms and mink rafts, and the experience gained from landscape-scale control in northern Scotland, it has been suggested that it might be plausible to eradicate mink from Britain (Martin & Lea 2020). This would doubtless cost tens of millions of pounds, take decades and be a mammoth logistical task. However, it is argued that this would be cheaper and more effective in the long term than the current cycle of control and reinvasion that we are left with as a legacy from the fur trade.

MINK IN IRELAND

The first American Mink farms opened in Ireland in the early 1950s, and by the 1960s there were at least 40 in operation, with an estimated 125,000 mink in captivity, mostly in small establishments (Deane & O'Gorman 1969). The licensing of farms became mandatory in 1965, by which time there had already been some documented escapes. The first was in 1961, when approximately 15 mink escaped from a fur farm in Omagh, County Tyrone. Within five years, there were more escapes from the same farm, and by 1969 feral mink were present on the Strule, Mourne and Glenelly River systems. Since then, there were more escapes and, in some cases, intentional illegal releases with the result that populations of mink were thought to be self-sustaining in the wild by the late 1980s (Smal 1991). While early sightings were in close proximity to the locations of operational fur farms which were mainly in the east, the species has also moved west, and mink are now widespread on the island of Ireland.

Mink are present in or near to many important conservation areas on the east coast of Ireland, as well as inland, and the species is now also encroaching on some of the designated sites in the west. Here, as elsewhere, they pose a serious risk of predation to ground-nesting birds and island colonies of seabirds. There has been some small-scale mink control at specific sites across Ireland in an effort to safeguard important populations of species such as terns and Corncrakes. These include several major lochs and rivers, such as Black Islands and Carrownure Bay, Lough Ree; Frans Callow, Tower Callow and Borranangh Callow, River Shannon; Inch and Crolly Loughs, County Donegal; Carrowmore, Cross, Conn and Cullin Loughs, County Mayo; Wexford Wildfowl Reserve and Lady's Island Lake, County Wexford; Baltray, Meath; and Kilcoole beach, County Wicklow (Roy *et al.* 2009b).

In 2008, using baited live-capture traps, 49 mink were caught over 2,944 trap nights. The average number caught per trap night was 0.044, with the maximum

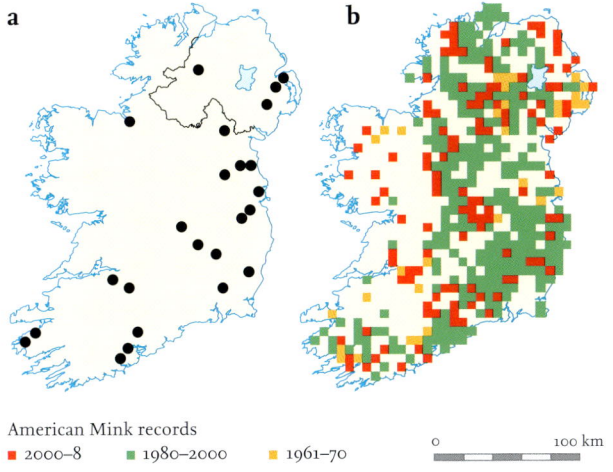

FIG 113. (a) Location of Irish American Mink *Neovison vison* fur farms from 1900 to 1960 and (b) mink sightings from 1961 to 2008. Redrawn from Roy *et al.* (2009b).

being 0.26 at Black Islands, Lough Ree and Carrownure Bay, and the lowest being 0.0028 at Crolly Lough (Roy *et al.* 2009b). These capture rates are comparable with those recorded at the outset of the Hebridean Mink Project, although capture rates in the Hebrides declined predictably as the project progressed. Roy *et al.* (2009a) reported that in both projects, most mink were caught within the first 2–3 days of trapping. However, in contrast to the Hebrides, mink were found to be active throughout the summer in Ireland. They suggested that this difference might be because trapping in Ireland was focussed in and around colonial ground-nesting bird sites and that these provide a rich and seasonally abundant food source for mink in a relatively small area. It is likely that denning female mink actively seek out these areas during the summer, as has been seen elsewhere (Harrington *et al.* 2009). Once these animals are caught, they are quickly replaced by new colonists. In effect, the mink population is continuously harvested in a small, food-rich area, with new animals then coming in to colonise the empty territories. This 'vacuum effect' often occurs in species removal programmes (Efford 2000). Through this effect, as discussed earlier, large numbers of mink can be trapped with few traps in relatively small areas. This can be a cheap and effective way to locally protect species of conservation concern, but the strategy may not have any lasting impact on the mink population as a whole. While large numbers can be caught year upon year for a long time, this is due to the fact that removed animals are quickly replaced by immigrants from the surrounding area. During the Hebridean Mink Project, trapping was carried out over the whole of Harris and the Uists, including all of the bird breeding sites and surrounding areas, as well as everywhere else, so

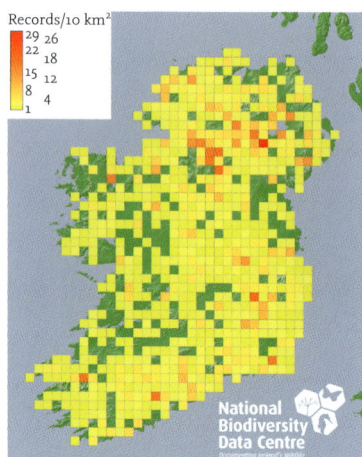

FIG 114. Current American Mink *Neovison vison* distribution in Ireland. National Biodiversity Data Centre, Ireland (accessed December 2021).

there were no animals left to come in and recolonise vacant territories, regardless of how food-rich they were. Nonetheless, trapping results in Ireland and elsewhere demonstrate that strategies focussed on protecting species of conservation concern, rather than on invasive species eradication, can be successful with limited resources.

No mink-specific surveys have been conducted in Ireland, but the species' presence has been recorded during several national Eurasian Otter surveys, a national European Badger survey, numerous smaller surveys and as part of Ireland's annual BioBlitz. The current distribution map from Ireland's National Biodiversity Data Centre includes 2,057 records of American Mink presence in 677 hectads.

Roy *et al.* (2009b) constructed a mink habitat favourability model for Ireland using environmental and landscape variables. The model output gave a relative value of habitat favourability for each hectad which suggests that the landscape of the west of Ireland, with its numerous rivers, water bodies, wetland habitats and offshore islands, provides significantly more favourable habitat for mink than the east. They concluded that mink populations in Ireland would probably continue to increase into the west of the island, where important sites would be at a high risk from mink. They suggested a method for targeting limited resources in high-priority areas, so that those of the greatest conservation importance have adequate resources allocated to them. On the basis of this scoring system used, priority should be given to coastal areas in the southwest such as the Dingle Peninsula, West Donegal coast and Tramore Back Strand, while other areas such as Roaninish, the Wicklow Mountains and Lough Ree were at lower risk from mink. This prioritisation exercise is just one example of how this could be done, but the results would be very different if the focus of mink control was to protect a particular species, such as Corncrakes or terns. It was estimated that a five-year control programme aimed at reducing a mink population by 75 per cent annually would cost more than €1,000 per km^2 and it would be difficult to resource an intensive, year-round control project across the whole of Ireland. However, where feasible, mink should be controlled or eradicated where they pose the greatest threat to globally important populations of vulnerable species. In these focus areas, intensive, year-round management programmes need to be set up, and in areas with the lowest risk of reinvasion, such as offshore islands or a peninsula, mink could be eradicated if sufficient resources were available. For mink control over larger areas, low-level management is the only realistic option, and needs to be carried out and coordinated by dedicated project staff with the support of volunteers and landowners. In these areas, the objective should be to intensively manage the population until virtual eradication is achieved followed by monitoring and low-level reactive management when individuals are reported. This can be achieved through the use of rafts or remote cameras.

For the control of any invasive species to be successful, especially if the desired end result is complete eradication, efforts need to be coordinated at the highest geographical scale. This is essential to prevent recolonisation from unmanaged areas. The Republic of Ireland and Northern Ireland must be considered as a single geographical unit if mink are to be eradicated or managed effectively over large areas of the island. The accidental escape or deliberate release of animals from fur farms continues to be a risk. While fur farming was banned in Northern Ireland in 2002 under the Fur Farming (Prohibition) (Northern Ireland) Order 2002, it continued to be legal in the Republic of Ireland, and in 2003, animal rights activists released mink in County Laois in the Irish midlands. In 2019, legislation was approved to phase out fur farming in the republic, but by 2021, there were around 120,000 mink on three fur farms still legally operating. In addition to the one in County Laois, the other two were in Kerry, in the southwest, and Donegal, on the border with Northern Ireland. However, in June 2021, cabinet approval was granted to the Animal Health and Welfare (Amendment) Bill 2021, which abolished fur farming in the Republic of Ireland from early 2022, allowing farmers to see out the 2021 season.

In Britain, fur farming was banned in England and Wales by the Fur Farming (Prohibition) Act 2000, two years after a public consultation in 1998 found that there was overwhelming public support to end the practice. Immediately prior to the ban, 11 fur farms were still in operation, producing up to 100,000 mink pelts each year. Two years later, fur farming was legally prohibited in Scotland by the Fur Farming (Prohibition) (Scotland) Act 2002, although the last fur farm in Scotland had already closed much earlier, in 1993. Nevertheless, it was deemed necessary to enshrine this in law to ensure that a situation did not arise where England and Wales imposed a legal ban to close down fur farms only for them to relocate to another part of the UK.

FERAL FERRETS

The feral Ferret is another mustelid which has been added relatively recently to the fauna of Ireland. Domesticated Ferrets are frequently lost while hunting but they can also escape, or be released or abandoned, and while many of these animals do not survive long in the wild, some do, resulting in the establishment of feral populations. A number of these can be found in the British Isles, although there are few truly self-sustaining feral populations on the mainland (Kitchener & Birks 2008). Small mustelids must compete with, and are frequently killed by, other larger members of the carnivore guild (Korpimäki & Norrdahl 1989, Donadio & Buskirk 2006). Feral domesticated Ferrets are often less wary and aggressive than truly wild animals such as polecats (Kitchener & Birks 2008). This may prevent released or escaped animals from surviving long enough to establish viable feral populations in the wild in environments with many other predators and competitors. In 1949, two Ferrets were deliberately introduced to Great Saltee Island off County Wexford to control the rabbit population, but neither survived through the first winter (Roche & Merne 1977). However, feral populations do occur on a number of offshore islands elsewhere. The largest and best documented is in New Zealand. However, feral Ferrets have also succeeded in establishing viable populations on the Canary Islands, the Azores, the Isle of Man and Jersey. In the UK, there have been free-living Ferret populations on Shetland, the Outer Hebrides, Islay, Bute and Arran.

At the northernmost point of Northern Ireland, on Rathlin Island, County Antrim, there has been a population of feral Ferrets since the mid-1980s (Bodey et al. 2011a). On the Irish mainland, a localised survey carried out from 1997 to 2001 confirmed that there was also a self-sustaining population, largely in County

FIG 115. A Ferret *Mustela furo* being set to work at a rabbit warren.

Monaghan but with additional records from Counties Fermanagh, Tyrone and Armagh (Buckley et al. 2007). Several years later, during 2006–8, an all-Ireland survey for feral Ferrets took place. They were widely distributed geographically, recorded in 19 of the 32 counties of Ireland, but the highest number of records was still concentrated in County Monaghan, followed by Fermanagh and Cork. The survey relied on records submitted by the public, meaning the resulting distribution may be distorted by recording effort; however, the 2006–8 results, combined with previous records, provide compelling evidence that there was a self-sustaining population of feral Ferrets in an area covering at least four counties. The survey results also suggested that Ferrets may have spread into County Louth. Records from Counties Sligo and Donegal also indicated the possibility of a population in the northwest coastal region. The distribution of Ferrets appeared to have increased since the previous survey, but it was not possible to compare the results of both surveys directly because of differences in survey effort and the area covered (Buckley & Lundy 2013).

Using records collated from the 2006–8 survey along with previous records, Buckley and Lundy (2013) constructed a landscape model of habitat suitability for Ferrets to assess the potential extent of further range expansion in Ireland. This showed that Ferret records in Ireland were mainly associated with lowland agricultural landscapes, namely pasture, which is the dominant land use across lowland Ireland. In western Europe and New Zealand, the European Polecat and Ferret are usually associated with lowland habitats. Radio-tracking data from studies of polecats have shown that pasture is a preferred habitat in England, although a higher use of wetlands and riparian forest is reported in continental Europe (Birks 2015). The results of Buckley & Lundy's model imply that a large proportion of the land area of Ireland is potentially suitable for Ferrets if the existing feral population expands or other populations become established as a result of escapes or releases of further animals.

Outside of New Zealand, there are few studies on the impact of introduced Ferrets. It is presumed that the Ferret could have negative impacts in Ireland through unsustainable predation of native species and competition with native predators. The biggest concern is predation on Ireland's ground-nesting birds; however, in Europe, it is small mammals and amphibians that are the predominant prey of polecats, and birds only represent a minor element of their diet (Lodé 1997, Sidorovich 2000). Nonetheless, populations of Corncrake and Grey Partridge occur in lowland agricultural habitat which would be of high suitability to Ferrets, and these species may be especially vulnerable to predation due to their critically low population size in Ireland. However, at present, neither of these species overlaps with any known feral Ferret populations.

The Ferret population on Rathlin Island is still present and, although there have been declines and changes in the distribution of ground- and burrow-nesting birds, despite positive habitat management, this could not be directly attributed to the impact of Ferrets. In 2011, a trial removal of feral Ferrets was carried out on Rathlin Island to assess whether localised control resulted in increased breeding productivity and survival rates in Lapwing *Vanellus vanellus*. The results showed there had been no effect on the hatching success or fledging rate of the birds in the study (Bodey *et al.* 2011c). However, localised removal of Ferrets prior to breeding led to a substantial increase in the post-dispersal Ferret population through the enhanced survival of juveniles (Bodey *et al.* 2011b), illustrating the complex ecological processes and outcomes that can arise from attempts to manage wildlife populations. This example highlights how partial or localised predator control may prove ineffective at best and can actually worsen the issues that management was designed to address. In 2023, an ambitious project began with the aim of removing all Ferrets (and rats) from across the whole of Rathlin Island by 2026.

Aside from any impacts on bird populations, another concern in Ireland would be the potential for feral Ferrets to have a negative effect on the near-endemic Irish Stoat, through competition between the two species. Rabbits and small mammals are the main prey of Ferrets, and they are also the principal components of the diet of Stoats in Ireland (Sleeman 1992). In New Zealand where there are fewer species of small mammals than in Ireland, Ferrets and Stoats can occur in the same habitat but tend to avoid each other (Ratz 2000). Feral Ferrets are undoubtedly surviving in some parts of Ireland and may establish further populations there, but their ecological impact on native biodiversity is yet to be fully investigated. Given the high cost of eradicating invasive species, these impacts would need to be determined ahead of proposing any removal programme. The widespread presence of Ferrets as domestic animals creates a constant risk of reinvasion.

PINE MARTENS ON SKYE AND MULL

Regardless of the justifications for it, the control of invasive animal species raises ethical and welfare issues for many people. An understanding of controlling predator populations and generating tangible and significant conservation benefits is essential. The ethical issues can be especially problematic when a native species that is of high conservation concern elsewhere arrives (with or without human assistance) and establishes itself in an area beyond its natural range. An example of this has been the relatively recent arrival of Pine Martens on the Scottish islands of Skye and Mull.

Pine Martens have been recorded on the Isle of Skye since 1995, the same year that the Skye road bridge first opened. By 2012, it was estimated that the population numbered around 50 animals, some of which had spread as far as 40 km west of the bridge, mostly through wooded areas. Pine Martens are regularly seen using the bridge to the island, and some breeding has also been recorded, so it is almost certain that their number and range will continue to increase. However, as yet there is no evidence that they have had any serious impacts on other wildlife or game birds on Skye. There have been some reports of Pine Martens taking poultry, but this has been reduced by the use of suitable fencing and light-activated doors on hen coops. Both of these measures are also effective in protecting poultry from predation by Red Foxes, which are present and well distributed on Skye (Scott 2011).

In contrast, there are no foxes on the island of Mull but there are plenty of feral domestic cats. Non-native American Mink is also relatively common, especially along the coasts, despite ongoing mink control to protect waders, waterfowl and ground-nesting seabirds. An additional predator may have disproportionately damaging effects on some of these vulnerable species. There were unconfirmed sightings of Pine Martens on the island in 1996 and 2000, but the first unequivocal evidence came from scats found in 2008. Subsequently the frequency of sightings increased, including some of a female Pine Marten who raised at least two kits in the roof space of a kayak centre in May 2010. This prompted an official report on the status of the Pine Marten on Mull (Roy *et al.* 2014). The main aims of the report were to assess the risk that Pine Martens could pose to the conservation status of the island's other wildlife and to establish whether Pine Martens were ever historically found on Mull. There is a Mesolithic record of Pine Marten presence on the island of Oronsay, and some from Neolithic sites on Westray, Orkney and South Uist (Fairnell & Barrett 2007), which are all outside the modern range of the species. It has been suggested that Pine Martens may also have been present on Colonsay, Jura and Islay based on some of the place names on these islands, such as Creag an Taghan (crag of the marten). However, there is no evidence from either historical or archaeological records that Pine Martens were ever native to Mull (Roy *et al.* 2014). Pine Martens are notoriously difficult to detect when present at low densities (Birks & Messenger 2010), but it is thought unlikely that they were present on Mull until shortly before the first sightings (Solow *et al.* 2013). Pine Martens can swim across inland waterways 20 m wide (Van Den Berge & Gouwy 2011), but to get from the mainland to Mull would require them to swim at least 1.6 km in open

FIG 116. Pine Marten *Martes martes* at a wildlife hide feeding table in Scotland.

sea, which is improbable. The most likely explanation of how Pine Martens first got to the island is by accidental or deliberate introduction. According to local rumour, there was a clandestine release of newly weaned juveniles in 1986 (Solow *et al.* 2013). This would not have been legally prohibited at that time, as Pine Martens were only given full legal protection in 1988, and the pattern of Pine Marten sightings over time is consistent with the alleged year of introduction, so deliberate introduction cannot be ruled out (Solow *et al.* 2013). Another possible route of entry is as accidental stowaways on one or more of the many boats transporting timber between the west coast of Scotland and Craignure on Mull (Roy *et al.* 2014). Regardless of how the martens got there, they are now thought to be present over a wide area, both north and south of Craignure. The habitat on Mull is very similar to much of the west coast of Scotland, where there are plenty of Pine Martens, and therefore it is assumed that they will continue to increase and spread. There is a wide variety of potential prey species present on the island, such as voles, rabbits and wading birds. In the absence of foxes, which can suppress Pine Marten populations through competition and predation (Lindström *et al.* 1995), it is possible that martens could eventually reach high densities on Mull. There are several Special Protection Areas on Mull and its offshore islands, and there are concerns that some protected ground-nesting birds and reptiles could be threatened by the Pine Marten. Another issue is that of potential predation by martens on game birds, resulting in conflict with estate owners and managers, although it is possible to reduce the risk of Pine Marten predation on pheasants being reared for release by the use of appropriate fencing (Balharry 1998). There may be some negative economic impacts from Pine Martens, but these might be offset to some extent by the positive benefit to the island's tourist industry. Pine Martens are undeniably a tourist attraction as is evident elsewhere in Scotland, where the potential for marten-watching draws visitors to guest houses and hotels. Martens provide an additional source of income to estates such as Aigas in the Highlands.

OUT WITH ALL INNS?

It has been estimated that invasive non-native species (INNS) cost the UK economy between £5.4 and £13.7 billion in reported losses and expenses over the period 1976–2020 (Cuthbert *et al.* 2021). Aside from the economic costs, there are many well-documented examples of introduced species which have had severely detrimental effects on arrival into novel ecosystems. It is undoubtedly true that non-native species present a high risk of becoming invasive. This is because new areas into which they are introduced are unlikely to contain the same coevolved suite of predators, competitors, parasites and pathogens that constrain their populations and prevent them from behaving invasively in their native ranges. Nonetheless, not all non-native species are invasive and not all invasive species are non-native. The two terms are often conflated, but there are many examples of native species exhibiting invasive behaviour under certain circumstances (such as native Southern Cattails *Typha domingensis* that have displaced Sawgrass *Cladium jamaicense* marshes in nutrient-enriched areas of the Florida Everglades). Conversely, some non-native species are harmless, fail to establish or have no negative impacts on the native wildlife in their new environment. In order to become invasive, a species must usually be capable of rapidly reproducing, spreading and displacing other species in its new locality.

In some cases, the success of an INNS is a secondary symptom of other environmental changes – for example, the ease with which American Mink established in Britain as a result of the legacy of intensive predator control, which decimated many of our native small carnivore populations, leaving a vacant niche and naive prey for mink to exploit. This has had catastrophic impacts on Water Voles, which mink have eradicated from many sites, and there is a significant body of evidence showing the effectiveness of mink control in protecting native birds and other wildlife. In the majority of cases where introduced species are implicated as a primary cause of extinction, they are generalist animals transferred to small islands and other 'island-like' habitats. This is exemplified by the Stoats and other terrestrial mammals in New Zealand. Endemic island species are often behaviourally naive, which makes them highly vulnerable to predation by introduced mammals. Many of New Zealand's bird species are ground-nesting, flightless, or with only limited flying ability, having evolved in the absence of terrestrial predators. Consequently, many have undergone extinctions or declines, and the islands now contain a disproportionately high number of threatened species. Around 4,000 native species are considered at risk, and a quarter of those are in real danger of extinction due to introduced predators such as rats, Stoats and Common Brushtail Possums *Trichosurus vulpecula*. Without active and systematic INNS control, New Zealand risks losing much of its unique natural heritage.

As a conservationist, I understand how fragile island ecosystems can be, and swift action to control any INNS is usually the best option to prevent problems with native species. However, in the face of global climate change, many plants and animals are having to shift their 'native' range into new areas to find conditions that they can tolerate. As this continues, and as species that humans have already moved around establish themselves in their new homes, it may become increasingly necessary for us to make choices between the needs of threatened species and the historical continuity of ecosystems. Instead of assuming that all INNS should be aggressively controlled, regardless of context, it has been suggested that a more pragmatic approach could be adopted. Adaptive management offers a way in which INNS control efforts could be designed and evaluated through scientific research and ongoing dialogue about the condition of specific places rather than generalised assumptions about the harmfulness of INNS.

The case of Pine Martens on Mull illustrates some of the ethical dilemmas that may arise when considering removal of species which have moved only just outside of their known distributional range. Pine Martens are currently viewed by many local people as a welcome addition to the species richness and other attractions of the island. They do not meet the criterion of being a rapidly reproducing species and are most unlikely to become invasive. In 2012, it was estimated that there were probably fewer than 20 Pine Martens on Mull (Roy *et al.* 2014) but, assuming that some breeding has occurred and the arrival of new animals cannot be ruled out, this number has almost certainly increased since then. The 2012 report considered a range of lethal and nonlethal options for managing Pine Martens on Mull should it become necessary. These include eradicating them completely from the island, some localised control to reduce damage or limit the increase and spread of the population, or doing nothing. As a new species to Mull, there are anecdotal reports that Pine Martens may now be having some impact on shore-nesting colonies of gulls and terns (M. Wilson, pers. comm.). In order to evaluate the costs and benefits of different management options and inform decision-making, it will be essential to collect more data and monitor the number and distribution of the species in future. However, as we will explore in the next chapter, for many small mustelids, that is often easier said than done.

CHAPTER 8

Mapping Phantoms: The Challenges of Surveying and Monitoring Small Mustelids

Information about the abundance and distribution of species is essential when making decisions about their conservation and/or management. However, managing to obtain that information can be challenging when it comes to elusive animals like the small mustelids. Traditionally, the only way to get reliable estimates of the number of animals in an area was to physically trap them and use capture-mark-release-recapture methods whereby individuals of the target species are trapped, marked and released between each sampling session, creating capture histories of known individuals. This is a surefire way to collect data that can be used to assess population size or density. A major benefit of live trapping is that animals are physically caught so they can be examined, morphometric data (such as weights and measurements) collected, and samples taken. Marked animals can be identified on subsequent occasions to provide information on population dynamics. Live trapping may be necessary if samples (such as blood) are required that cannot be collected noninvasively, or if physical capture is necessary for condition assessment or to fit GPS or radio tags to collect data on movements and ranging behaviour. However, trapping is extremely labour-intensive and is both costly and time-consuming. It is usually not practical to carry out trapping studies over very large areas, so while detailed information can be gathered for a sample of study sites, this must then be extrapolated based on various assumptions. Trapping has a high potential to stress animals and interfere with their daily activity, particularly during sensitive periods such as the

FIG 117. Weasel *Mustela nivalis* (bottom) and Stoat *M. erminea* (top) scats.

breeding season. There is often a bias in the sex, age class or 'personality' of the animals that are the most trappable, thereby limiting or biasing inferences that can be made from the data. Although these effects can be accounted for, to some extent, by methodological choices and ethical considerations, less stressful and more efficient survey and monitoring methods are often desirable.

Field signs such as tracks and scats can be used to confirm the presence of some animals if scats are identifiable and if the animals are cooperative enough to leave them where they might easily be found. Unfortunately, neither Stoats, Weasels nor European Polecats fall into this category. The droppings of all three species conform to the same mustelid pattern, being long, thin and twisted and usually containing fur or feathers, but they are small and found only rarely. They are usually deposited at their dens, which can be difficult to find as they are often the requisitioned underground burrows of their prey. Even when we are fortunate enough to come across them, the scats of most of our small mustelids can only be distinguished from each other in the field, based on the coarse measure of their size, which is not always a reliable diagnostic. A large Weasel scat may look very similar to a small deposit from a Stoat.

Polecat and American Mink scats are pretty impossible to tell apart just from the look (and smell) of them, although mink tend to mark their territories more obviously with scat, which can sometimes, therefore, be found along the edges of watercourses. The late Rob Strachan, one of the best field biologists I have known, had a connoisseur's nose for mustelid scats. He described the subtle differences in aroma between them in his book, *Mammal Detective*. While both are equally unpleasant, the smell of polecat scats reminded Rob of fetid meat, whereas that of mink scat was more complex, being comprised of rotten meat also but with the addition of burnt rubber.

PINE MARTENS

Pine Martens are probably the most obliging of our small mustelids, in that they conveniently (for us) deposit their scats in obvious places along tracks and trails to communicate to other Pine Martens and indicate their territories. They will also often use the same spots repeatedly, building up quite an obvious marker in the form of successive deposits. As a rule (and speaking from personal experience), an additional bonus is that Pine Marten scats are actually not unpleasant to sniff! The scent of a marten scat is often rather sweet and has been likened variously

FIG 118. Pine Marten *Martes martes* scats containing different dietary items: (a) honeycomb, (b) Rowan *Sorbus aucuparia* berries, (c) small-mammal fur and (d) a mixture of fur and invertebrate remains.

to Parma Violets (again, courtesy of Rob), Lapsang Souchong tea (thank you, Huw Denman) or new-mown hay. I can sometimes detect a faint suggestion of aniseed mixed in with this banquet for the nose. However, as you would expect, whatever the marten has recently taken in at the front end will also affect the perfume, as well as the appearance, of what is subsequently expelled from the rear. Delightful, I hear you say. Not usually the topic of polite dinner party conversation, unless you happen to be at a gathering of ecologists who revel in such discussions. Marten diet varies enormously throughout the year, so scats found in the spring are likely to be full of the remains of small mammals. Later on in the summer, they may contain more feathers, whereas in autumn, when Pine Martens can indulge their taste for sweet berries and fruit, their scats will reflect this switch.

All of this suggests that marten scats are easily identified. This was indeed thought to be the case until fairly recently and was the reason that scat searches were confidently used as the main method to survey for Pine Martens. However, the advent of molecular methods of identifying species from DNA extracted from scats revealed that Pine Marten scats could not be reliably distinguished from Red Fox scat based on their size, shape and smell alone. A study carried out at the end of the 1990s by Davison and others showed that in fact even expert naturalists with significant experience in field surveys for the species often misidentified fox scat as being from Pine Marten. The study was published under the genius title of 'On the origin of faeces: morphological versus molecular methods for surveying rare carnivores from their scats' (Davison *et al.* 2002). As a result of Davison's findings, and more recent advances in genetic analysis (O'Reilly *et al.* 2008), species confirmation from DNA analysis is now accepted practice in scat-based Pine Marten surveys. The standardised method for these involves searching transects along tracks or paths in woodland, where Pine Martens usually deposit scats, and testing any likely samples for DNA.

While the issue of misidentifying scats and therefore getting a false positive result can now be largely addressed by using DNA for confirmation of species, there is still the problem that scats may not be found in an area where Pine Martens are present, thereby giving a false negative result. This is a common problem with surveys of many wildlife species. They are rarely detected with perfect accuracy, regardless of which survey method is used. Absence of evidence is not evidence of absence unless the probability of detecting the species is 100 per cent, which leads to a fundamental problem. A species may be recorded as 'absent' at a site if it was present at the site but not detected or if it was indeed truly absent. As a result, occupancy (presence/absence) can be confounded with 'detectability'. To complicate matters even further, detectability of a species can vary between different sites, and it may also be related to features of a survey at a particular time, such as weather conditions. The detectability, deterioration and deposition rates of Pine Marten scats (as well as the likelihood of successful genetic verification) may be affected by a number of environmental and ecological conditions. Poor weather conditions during or in the days before a Pine Marten scat survey may affect the results by reducing marten activity and therefore scat marking. Wet weather can also increase scat deterioration by washing scats away or providing ideal conditions for slugs and other invertebrates that eat marten scats. Scats are easier to see on light-coloured paths with little traffic than on darker 'metalled' forest roads where the tyres of passing vehicles can also remove scats or squash them beyond recognition. Where Pine Martens are newly colonising an area or at low density for other reasons, there may be less incentive for them to mark their territory, making it more difficult for surveyors to find scats. Consequently, it is not enough to simply analyse detection/non-detection data as if they are truly presence/absence data. The proportion of sites where a species is detected will always underestimate the true occupancy level in the study area if detection is imperfect. If detectability cannot be assumed to be perfect, it must be accounted for to avoid bias.

In response to this problem, new classes of models called occupancy models were developed. In occupancy modelling, each site is surveyed more than once during a season to build up a history of detections. A critical assumption for occupancy modelling is that occupancy does not change at a site within the sampling season (no immigration or emigration), although it can change between seasons. A second assumption is that survey sites are independent of each other, so that detection of a species at one site is independent of detecting it at other sites. This can be a problem if survey sites are so closely spaced that animals can move between them, and so the same animal(s) could be detected at more than one site. While detectability may be affected by site characteristics (such as habitat), or by characteristics of a survey (such as weather conditions), the true occupancy state of a site is assumed to be constant between surveys. Therefore, repeated surveys provide multiple opportunities for this to be observed within a given season. Occupancy modelling can be used to examine the effects (if any) of both site-specific and survey-specific covariates. Site-specific covariates are those, such as habitat, which are constant between surveys at each site and can be used to model both detectability and occupancy, whereas survey-specific covariates may vary between surveys and can only be used to model detectability, as each site's true occupancy state is assumed to be constant between surveys. The probability of occupancy is either the same across all sites or, if not, then differences in occupancy can be explained by site characteristics (covariates) that have been quantified and included in the model. If a site is occupied, then the probability of detection is either the same across all surveys and sites, or differences in detectability can be explained by characteristics of the site or the survey, which have been quantified and included in the model. Occupancy models have recently proved to be extremely useful in designing and analysing monitoring surveys for small mustelids, including Pine Martens.

In 2016, a study carried out in northeast Scotland used genetic verification of scats and occupancy modelling techniques to quantify Pine Marten detectability from repeated scat surveys (McHenry et al. 2016). The results showed that for transects of 1 km length, which had been the standard protocol for previous Pine Marten surveys, it was estimated that there was only a 35 per cent chance of detecting martens when they were present. When survey transects were increased to 1.5 km, detection probability increased significantly to 58 per cent – better, but still far from brilliant. Detection probabilities increased with path width and decreased with vegetation cover. Not all scats yielded sufficient DNA to enable genetic confirmation of species, resulting in the issue of 'unclassified' scat samples, which is a common problem. The success rates of DNA extraction and amplification are affected by the length of time that samples are in the field, temperature and other atmospheric conditions, including humidity. Scat samples are easy and relatively cheap to collect, but the DNA starts to degrade if they are not picked up while they are very fresh, which is rarely the case. Scat samples can be collected opportunistically without the need for any preparation or baiting, but if a more systematic approach is preferred, hair-snagging devices can be set out and baited with scent lure or food. The genetic material in hair samples is more robust but still affected by environmental conditions and collection time, as well as being more difficult and time-consuming to obtain, noninvasively. Nonetheless, this method can be used to survey Pine Martens using hair samples collected by hair tubes. These are relatively cheap to make from sections of 110 mm diameter PVC waste pipe, cut into lengths of approximately 37 cm and fixed vertically to a tree trunk.

In the laboratory, a technique is used to amplify, or make, billions of copies of specific sections of DNA (microsatellites), which allows for detection and identification using visual techniques based on size and charge (+ or −). Mustelids appear to have lower microsatellite amplification rates from scats. For instance, for Fishers, success rates ranged from 7–10 per cent (Thompson et al. 2012) compared with, for example, felids (members of the cat family), where studies have reported 87 per cent from Bobcat *Lynx rufus* scats (Ruell & Crooks 2007) and 63 per cent for Mountain Lions *Puma concolor* (Ernest et al. 2002). Hair collected from snares results in greater individual amplification, but in more moist environments, hair snares may require more frequent visits to reduce environmental exposure (Stetz et al. 2015).

THE CHALLENGES OF SURVEYING AND MONITORING SMALL MUSTELIDS · 191

FIG 119. (a) Each tube has a lid covering the top, to ensure that the animal can only enter from the bottom, and bait (typically a raw chicken wing) is wired to the inside near the top end. (b) and (c) Two patches of a strong adhesive tape are fixed to the inside of the tube with the sticky side exposed to collect a hair sample when an animal enters to take the bait.

In 2017, we carried out a study of the Pine Marten population in Galloway Forest in southern Scotland using live trapping, alongside sampling from both scat and hair tube surveys (Croose *et al.* 2019). In this way, we were able to compare the effectiveness of the three survey methods for estimating Pine Marten abundance in a given area, and therefore population density. We also considered the financial cost of each method. Previous scat and hair tube surveys at the same study site in 2014 had found that collection and genotyping of

FIG 120. (a) Alternative designs include spring-based hair traps and wooden feeder boxes where the sticky patches are attached to the underside of the hinged lid. (b) The animal then brushes against the sticky patch with the back of its head or neck when lifting the lid to access the feed (usually peanuts) inside the box.

hair samples detected only 33 per cent of known individuals, therefore undersampling the Pine Marten population, whereas collection and genotyping of scats detected 93 per cent of individuals (Croose *et al.* 2016). Consequently, that study concluded that a combined noninvasive approach using DNA from both scat and hair samples was optimal to detect as many individuals as possible and maximise the precision of population estimates.

In our 2017 study, the minimum number of Pine Martens present was 31 (19 males and 12 females). This was the total of known individuals identified from all three methods. Live trapping detected 77 per cent of known individuals and produced the highest population estimate of 30 with capture-mark-release-recapture analyses. Genotyping from scat samples detected just over half (52 per cent) of the individuals known to be present and gave a population estimate of 26, compared with hair tubes, which detected only 48 per cent of known individuals within the population and resulted in the lowest and least accurate population estimate of 15 animals. However, by combining scat samples with those from hair tubes, 81 per cent of known individuals were detected, giving a much more accurate minimum number present (25) than live trapping alone. Only five animals were detected by all three methods and fourteen were identified by two. Six Pine Martens were live-trapped more than once but were never detected from hair tubes or scat samples, and three were never trapped and were detected from hair samples but not scats. A further three individuals were only known from scat samples. Detections of female Pine Martens were highest from scat sampling, the most 'passive' method, whereas male martens were overrepresented in live trapping and hair tube samples (as shown in Table 10). This highlights the fact that each method on its own has different shortcomings.

A combined sampling approach using hair tubes and scats takes account of the sex bias observed with different methods. Nonetheless, this may not be constant across different study areas. A previous study elsewhere in Scotland found higher detection probabilities for female Pine Martens than for males in hair samples (Sheehy *et al.* 2018). As well as differences in 'trappability' between the sexes, personality or behavioural traits may also affect an individual's

TABLE 10. Results of genetic analysis of Pine Marten *Martes martes* scat samples collected by three methods in Galloway Forest, southwest Scotland, in 2017.

Method	Samples collected	Samples sexed from DNA	Samples genotyped	Ratio of male:female detections	No. of individual genotypes (total and male:female)	Percentage of known individuals detected
Live trapping	24*	24 (100%)	24 (100%)	34:13**	24 (17:7)	77%
Hair tubes	157	89 (57%)	69 (44%)	61:28	15 (10:5)	48%
Scat sampling	103	82 (80%)	33 (32%)	38:44	16 (6:10)	52%
Hair tubes and scats combined	261	171 (66%)	102 (39%)	99:72	25 (13:12)	81%

*Hair samples were taken from trapped animals on the first capture only. Animals were marked so that samples need not be taken at any subsequent captures.
**Higher number of detections than number of samples genotyped as this includes all captures, comprising animals that were identified by genotype during the first capture and identified subsequently in recaptures.
Data from Croose *et al.* (2019).

probability of detection by each method. Studies have shown that bold, active, exploratory individuals are most likely to investigate novel objects such as a trap or hair tube, but these methods may under-sample shy, less exploratory animals. However, these individuals may be detected by scat surveys, as this is a relatively passive method and does not require the animal to deviate from its normal behaviour or take the risk of entering a trap or a hair tube.

When comparing the costs of each method used in our study, hair tubes were the most expensive method, accounting for the most personnel hours, highest vehicle mileage and highest genotyping costs. Hair tubes also detected the fewest individuals and produced the lowest and least accurate population estimate. The relatively high costs of this method are due to sampling taking place once a week over five weeks compared with three weeks for trapping (2.5 weeks of pre-baiting and 4 days and nights of trapping). Scat samples were collected at the same time as hair tubes were checked. There was considerable variation in the average number of captures per individual for each of the three methods, with hair tubes providing the highest (4.6) and live trapping the lowest (2). However, with hair tubes being *in situ* for five weeks, the trapping effort was greater. Although each tube could only yield one capture per week (when the sticky patch would be filled and no longer sticky), this could occur at any time over a seven-day period, whereas live traps were checked twice a day for only four days, so the trapping effort may not be directly comparable.

The relative effectiveness of different survey methods may also vary by locality or other unknown factors. In Italy, Bartolommei *et al.* (2012) carried out a study to evaluate scat surveys, hair tubes and camera traps for detecting and estimating population density of Pine Martens. They found that in their study area, camera trapping was the only effective method. Camera traps (also called trail cameras) have been used since their development in the early 1980s to study population sizes of animals, especially large carnivores with distinctive natural markings (such as Tiger *Panthera tigris*). The advantage of camera traps compared to other methods is that they are noninvasive and relatively low-cost. However, their use is often restricted by the ability to identify individuals in species that do not have distinctive markings. Pine Martens are often identifiable by their markings. Although the fur on their head, body, legs and tail is a uniform rich chestnut brown, this contrasts with a pale creamy yellow to apricot bib. The size, shape and outline of this bib varies between individuals. Some bibs are relatively plain, whereas others have spots, blotches or fingers of brown fur extending into them, which can be used to recognise individuals in camera trap images. This of course requires that a decent 'full frontal' image can be obtained. One way to increase the chances of this is by using a 'jiggler', a device pioneered by Erwin Van Maanen, a

THE CHALLENGES OF SURVEYING AND MONITORING SMALL MUSTELIDS · 195

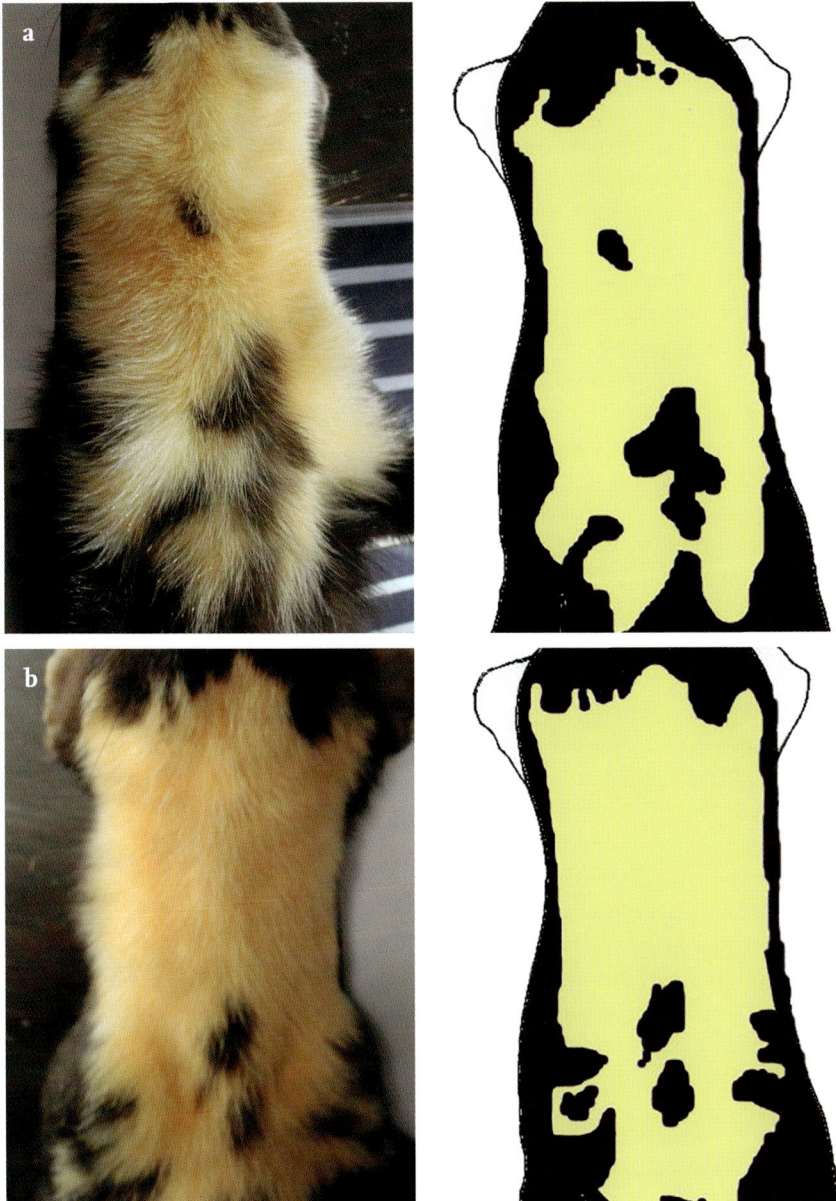

FIG 121. (a) and (b) Photos and bib diagrams of two Pine Marten *Martes martes* individuals (PM31 and PM38) translocated to Wales.

Pine Marten biologist in the Netherlands. The jiggler consists of a metal clamshell tea infuser filled with peanut butter or something equally irresistible to martens, suspended from a length of stiff but springy wire extending about a metre from where it is pushed into the ground. The marten then has to rear up on its hind legs like a Meerkat *Suricata suricatta* to try and grasp the infuser to investigate. If it does this in the right orientation to the camera, the result is a good, clear bib picture. For all the martens we translocated to Wales and Gloucestershire, reference photos of their bibs were taken while they were anaesthetised for health screening and just before fitting each with a radio collar. This resulted in what we called our 'bib-liography' of reference images, from which Josie Bridges, one of the field team, became expert at identifying almost every marten subsequently caught on camera traps. This comprehensive catalogue of bib photos was added to as new individuals were recorded in the wild, sometimes as kits with a known adult female. Some are easier to recognise than others. (One of the females who still regularly visits a garden feeder in mid-Wales is affectionately known as Boobs by Chris, the owner of the property, because of the placement of two round brown patches on her bib.)

In many cases it is not possible to identify individuals with any certainty. Therefore, methods have been developed for estimating abundance without the need for individual identification. These include the random encounter model, a technique for calculating animal densities from camera-trapping rates by modelling the underlying detection process. Other methods which account for this are the space and time to event models and instantaneous sampling estimator. Full details of these methods are given in Rowcliffe *et al.* (2014) and Moeller *et al.* (2018).

Camera traps are a highly effective wildlife survey tool and can detect a wide range of species, but this is also a disadvantage of camera trapping – it is not species specific and can generate large numbers of detections, most of which are non-target species. Reviewing all of the images can be very time-consuming, although there is potential for this to be reduced by the increasing use of web-based platforms that encourage citizen science participation, such as MammalWeb (https://www.mammalweb.org/en/). There have also been promising developments in the use of machine learning, which may be better able to classify camera-trap images in future (Franzen *et al.* 2021, Tabak *et al.* 2022).

STOATS AND WEASELS

In spite of their wide-ranging use, camera traps often fail to detect small and relatively fast-moving species like Stoats and Weasels. The passive infrared sensor used by most cameras is less likely to be triggered or, if it is, the animal may have crossed the camera's field of view too quickly for an image to be captured. Stoat and Weasel footprints are sometimes seen in soft mud or fresh snow, but the ideal conditions for these are rare and ephemeral in Britain and Ireland. Where prolonged periods of lying snow are more predictable, such as in Scandinavia, snow tracking has been used by biologists to estimate Stoat and Weasel home ranges. However, the footprint size of a male Weasel overlaps that of a female Stoat so it is often difficult to distinguish these species from tracks alone (King & Powell 2007). In harsh winter conditions, Stoats and Weasels will sometimes make nests under the snow but above the ground or make use of the nests of their vole prey. After the snowmelt, these nests become visible and can be searched for Stoat and Weasel scat to confirm presence (Zielinski et al. 2020). These nests can also occasionally be found in dilapidated stone walls or under corrugated-iron sheets if these are left in grassland.

Hair trapping with subsequent DNA analysis has been used with varying success for Stoats and Weasels. In New Zealand and Ireland, 20 cm lengths of 45 mm diameter plastic pipe with an elastic band covered in adhesive across the entrance at each end have had some limited success in trials as a means of collecting hair samples from Stoats (McAney 2010, Clayton et al. 2011). In the latter study, which was part of an eradication programme, hair tubes detected about 25 per cent of Stoats that were subsequently captured in kill traps, so they concluded that hair tubes were less effective than kill traps at detecting Stoats, taking account of differences in their deployment. They suggested that the low density estimates from hair tube sampling may have been a result of Stoats being wary of the newly placed hair tubes, in comparison to trap boxes which had been *in situ* and pre-baited for a longer period. Another possible explanation was that the hair tubes needed to be in place for a longer sampling period. Another issue was that of sample degradation, likely due to unfavourably wet or humid environmental conditions for some of the sampling period, which resulted in low genotyping success. This is consistent with other studies showing DNA degradation in hair samples after exposure to environmental moisture (Lindahl 1993, Jeffery et al. 2007). A further problem with hair tubes is that of mixed samples if more than one Stoat goes into a tube. This can be identified in some cases, if three or more alleles (different forms of the same gene) are present at one or more locus (a genomic location where such variation exists). An individual inherits two alleles, one from each parent. However, if there are multiple copies of the same allele in a mixed sample, this would not be detected. Gleeson et al. (2010) proposed that this could be addressed by designing a 'single use' hair tube that closed after the first visit, thereby eliminating the chance of another Stoat

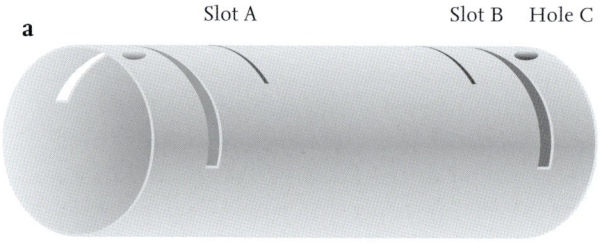

FIG 122. (a) A diagram of the hair tube and (b) the final design for the pilot study in Wales.

contaminating the sample before it was collected. However, this would require some thought and field trials to ensure that any closure mechanism did not deter Stoats from revisiting hair tubes, which is particularly important for such a trap-shy species (Gleeson *et al.* 2010).

We took the hair tube design used in New Zealand and in the Irish survey and modified it to increase the surface area of the hair sampling pad and to accommodate the size range of both Stoats and Weasels (Fig. 122). For a pilot study in west Wales, hair tubes were made from 52 mm diameter PVC pressure irrigation pipe cut into 300 mm lengths. A slot was cut halfway through the tube 65 mm from each end (slot A) and another shallower slot of approximately 4 cm

TABLE 11. The number of sites at which each species was detected from hair tubes in 2011–12.

County	No. of sites (samples)								
	Total	Wood Mouse	Bank Vole	Weasel	Common Shrew	Pygmy Shrew	Hazel Dormouse	Yellow-necked Mouse	Field Vole
Carmarthenshire	23 (96)	20 (43)	18 (41)	13 (16)	10 (12)	5 (5)	6 (7)	3 (4)	0 (0)
Ceredigion	18 (64)	2 (9)	1 (2)	0 (0)	1 (2)	0 (0)	2 (2)	0 (0)	6 (8)
Pembrokeshire	20 (47)	11 (21)	6 (9)	2 (2)	0 (0)	0 (0)	0 (0)	0 (0)	7 (8)
TOTAL	61 (207)	33 (73)	25 (52)	15 (18)	11 (14)	5 (5)	8 (9)	3 (4)	13 (16)

was cut 75 mm from each end (slot B). A hole was drilled 10 mm from each end of the tube (hole C). Size 31 rubber bands were placed into the end slots (B), and a 25 mm × 115 mm strip of plastic secured at the entrance to the tube with a cable tie through hole C. The plastic strip passed over the rubber band, with the end coming out of the tube through the second slot (A) and anchored with the free end of the cable tie.

Double-sided adhesive tape was applied to the plastic strip to collect hairs. Prior to fieldwork, the tubes were tested on captive Stoats and Weasels at the British Wildlife Centre in Surrey. Individuals of both species used the tubes from the first night they encountered them. One study area was subsequently surveyed in each hectad in the counties of Carmarthenshire, Ceredigion and Pembrokeshire in west Wales. Within each study area, 10 tubes were placed at 20 m intervals on the ground along a hedgerow or, alternatively, a stone wall. Tubes were baited with rabbit or chicken meat and anchored in the ground using a cable tie attached to a tent peg. The tubes were left in place for a minimum of seven nights between the end of June and mid-October. At the end of each sampling period, all adhesive pads were replaced and those with one or more hairs present were sealed into individual bags or tubes and labelled with a unique sample number, date and location. Samples were then stored frozen until they were sent for analysis to the Department of Chemical and Life Sciences in Waterford Institute of Technology. Stoat and Weasel DNA were identified using a real-time polymerase chain reaction (PCR) assay (C. O'Reilly, unpublished). Other species of small mammal (Wood Mouse *Apodemus sylvaticus*, Bank Vole *Myodes glareolus*, Field Vole *Microtus agrestis*, Pygmy Shrew *Sorex minutus*, Common Shrew *S. araneus*, Hazel Dormouse *Muscardinus avellanarius* and Yellow-necked Mouse *Apodemus flavicollis*) were also tested for, using a species-specific real-time PCR assay. A total of 61 sites were surveyed across the three counties: 23 in Carmarthenshire (in 2011), 18 in Ceredigion and 20 in Pembrokeshire (in 2012).

Two hundred and seven samples were collected and sent for analysis, of which 145 (70 per cent) yielded sufficient DNA for species testing. Eighteen samples were positive for Weasel DNA, but none were positive for Stoat. Wood Mouse and Bank Vole were the most commonly found species. Weasel was detected at 15 sites across Carmarthenshire and Pembrokeshire, but not at any of the survey sites in Ceredigion.

The location of sites where Weasel was detected in hair tubes, in relation to recent records for the species, is shown in Figure 123. All 15 sites were in hectads with no record of Weasel in the previous 10 years. However, Weasel was not detected in any of the 14 hectads from which the Local Record Centre (LRC) held recent (2003–12) records of sightings.

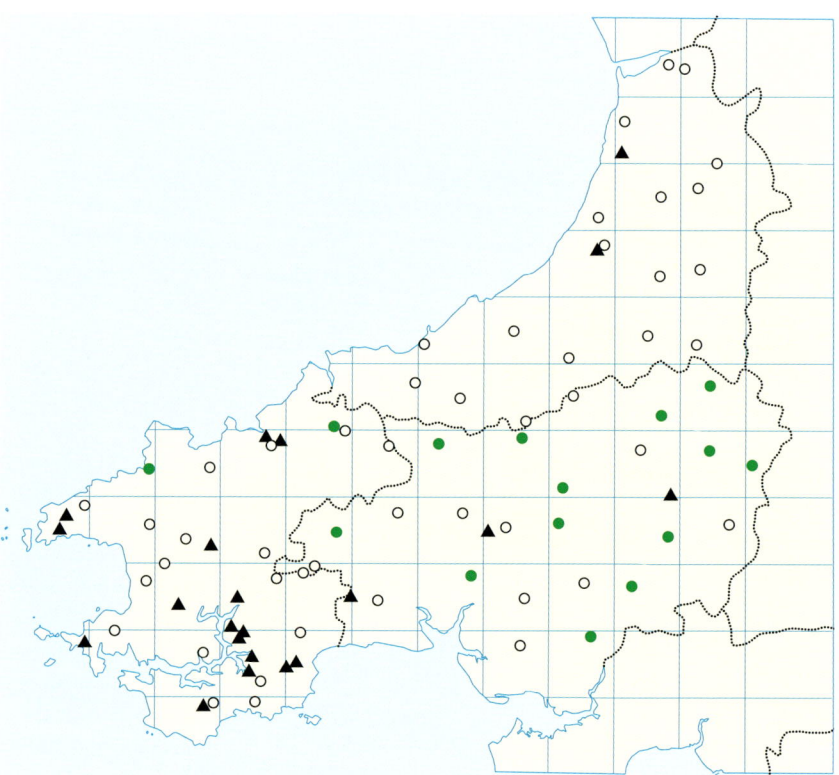

FIG 123. Survey sites where Weasel *Mustela nivalis* was detected (green circles), sites negative for Weasel from current survey (open circles) and locations of Weasel records (2003–12) from LRC (grey triangles).

Stoat was not detected at any of the survey sites. However, the LRC had received only three records of Stoats in the surveyed area during the preceding 10 years, and these were all in the south: two from Pembrokeshire and one in south Carmarthenshire, suggesting that Stoats are very rare in west Wales.

DNA samples can give a lot of information about the animals present, including number of different individuals and their sex, but DNA analysis is still very costly. Small mustelids will readily investigate small entrance holes, and so the Dutch Small Mustelid Foundation developed a monitoring device which takes advantage of this curiosity to direct the animals into a space with a camera trap.

FIG 124. (a) The 'Mostela' (named after Jeroen Mos, who came up with the idea) is comprised of a wooden box (620 mm × 300 mm × 175 mm) with a hole on each side and a plastic tunnel (with one side cut away) running through it, connecting the two entrance holes. A trail camera is fitted at the other end to record any animal that runs through the tunnel and into the box. (b) Weasel *Mustela nivalis* in the Mostela during pilot trials in England.

The diameter of the entrance tunnel (80–100 mm) excludes larger animals from entering, thereby minimising the amount of 'bycatch' (images or videos of non-target species), which can be an issue when using stand-alone camera traps. The Mostela (shown in Figure 124) has been trialled for Stoats and Weasels in the Netherlands and England, with promising results for Weasels.

However, the detection rate for Stoats has so far been very low, and on some occasions the Mostelas have failed to detect Stoats when they were known to be present. During a pilot study in England, Stoats were not detected inside the Mostelas at some sites even though they were seen nearby (Croose & Carter 2019), and in another similar study in the Netherlands, Stoats were not detected at all (Mos & Hofmeester 2020). It has been suggested that this low detection rate may be because Stoats are sometimes reluctant to enter artificial tunnels, as has been found in New Zealand where in at least 20 per cent of visits to tunnel traps, Stoats did not go into them (Dilks & Lawrence 2000), although a number of studies have shown that Stoats will readily enter tubes of 40–50 mm diameter (Brown 2001, Gleeson et al. 2010). Male and female Stoats can squeeze into vole tunnels as small as 22–28 mm diameter (Simms 1979), and dens have been found with entrances of approximately 30 mm diameter (Dowding & Murphy 1996), so it is unlikely that the entrance tube size of 80–100 mm diameter used in Mostelas would deter Stoats from entering. Nonetheless, using Mostelas as a sole method may fail to detect Stoats when they are present, so in order to test this hypothesis, we began trials using a second method (an external camera trap) in combination with the Mostelas to determine the likely rate of false negatives recorded by them (Fig. 125). It may also be the case that Stoats are deterred from entering the Mostela if they have been used by Weasels. In Ireland there are no Weasels, but the Irish Stoat, a subspecies endemic to Ireland and the Isle of Man, is widespread. We carried out a study using Mostelas with an 80 mm diameter entrance tunnel to survey Irish Stoats in Counties Mayo and Galway. Each Mostela was paired with an external camera trap positioned outside of the box. We then used a single-season occupancy model to compare the detection probabilities of both methods and to estimate the probability of occupancy by Irish Stoat at 12 sites.

Over 90 survey days, Irish Stoats were detected a total of 17 times inside the Mostelas at two (17 per cent) of the sites and a total of 25 times on the external cameras at four (33 per cent) of the sites. There were eight occasions when a Stoat was recorded on both the external and internal (Mostela) cameras, but at two sites, Stoats were only detected on the external camera and passed by the Mostela without entering it. Neither method detected Stoats in the first three weeks of deployment (in late April to early May) and detection probabilities were low for both the Mostela (0.09) and the external camera (0.05) (Croose et al. 2022).

FIG 125. A Mostela and external trail camera *in situ* at a study site in Ireland.

A study in the Netherlands found that detection probabilities for Weasels were approximately twice as high for Mostelas with a 100 mm diameter entrance tunnel than when 80 mm diameter tube was used (Mos & Hofmeester 2020). Irish Stoats are generally smaller than British Stoats but larger than Weasels, so the 80 mm diameter entrance of the Mostelas used in our Irish study may have deterred some larger animals from entering.

In 2021, we repeated the study in Ireland using Mostelas with an entrance tunnel of 100 mm diameter and carried out a similar trial on the Eastnor Estate in Herefordshire, England, using Mostelas with both 80 mm and 100 mm diameter tunnels to see if there was a significant difference in detection rates. This is currently being tested in further research by Vincent Wildlife Trust. The 80 mm entrance diameter may have reduced the amount of non-target species being detected; nonetheless, in addition to Irish Stoat, several species of small mammals (rodents and shrews), as well as Pine Martens, were recorded inside the Mostelas in Ireland. In England, as well as Stoats, Weasels and small mammals, the Mostelas have recorded European Polecats (Fig. 126) and American Mink. Overall, both the Mostela and stand-alone camera traps show potential for use in more wide-scale surveys of small mustelids.

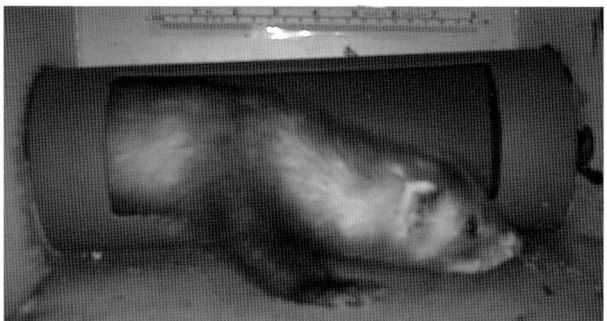

FIG 126. European Polecat *Mustela putorius* in a Mostela, recorded at a study site in Herefordshire in 2021.

The fact that polecats have been recorded inside the Mostelas is interesting. To date, the main method for recording the expanding distribution of polecats as they recover has been the Vincent Wildlife Trust national polecat survey. Records are collected via a national monitoring survey carried out over a 24-month period every 7–10 years, with one running from January 2014 to December 2015 (Croose 2016) and another running from January 2024 until December 2025. The survey uses a citizen science framework for reports of polecat casualties and other sightings from members of the public. Where possible, reports are verified as European Polecats, feral Ferrets or Polecat-Ferret hybrids via photo, video or evaluation of the carcass. These surveys have provided useful information on the expanding distribution of polecats, but the records can only confirm presence, and there is little data on population density currently available.

One of the main advantages of using a Mostela or something similar, when compared to a stand-alone camera trap, is that it reduces the amount of bycatch and therefore the huge amount of data that has to be sifted through and manually recorded. Nonetheless, camera traps are one of the most effective sampling tools when compared to other methods. Camera traps are significantly more effective than other methods at detecting a large number of species and have been shown to perform significantly better than live traps and to be otherwise comparable in performance to other survey methods (Wearn & Glover-Kapfer 2019). Live traps perform poorly compared with noninvasive methods, particularly camera traps, in part because they are typically single-catch traps that require manual resetting after each capture, while camera traps are effectively multicatch traps. Live traps require that an animal is attracted by bait or lure, will then interact with a novel object in the form of the trap and is then successfully physically caught. This may fail at any point in the sequence of events and for a variety of reasons. By contrast, camera trapping only requires that an animal enters the camera detection zone and triggers the sensor, after which the camera is available to 'capture' the next animal that passes without any need for resetting. However, using camera traps alone to estimate abundance can be a challenge if the movement area of animals is not known. This usually requires costly radio or GPS tracking studies to support population size estimates. That being said, these studies have been carried out to some extent for many of the small mustelids and can therefore cautiously be extrapolated to camera trapping studies elsewhere. The images and video footage that camera traps produce also have a range of other benefits. They are a fantastic engagement tool with which to enthuse volunteers and local communities about the wildlife in their area. They can also provide valuable behavioural insights. However, the need to collect genetic samples such as hairs and scats to answer certain questions will mean that camera trapping alone will not be a suitable method for some studies. Moriarty *et al.* (2018) found that for identifying individual Pine Martens and their sex, cameras paired with hair-snagging devices were more cost efficient. Scat detection teams comprised of a human handler and a trained scat detection dog can also help in carrying out area-constrained searches for scats. Scat detection teams are an especially valuable survey tool for small mustelids such as Pine Martens when the objectives include analysing scats to evaluate diet in addition to species distribution.

Paez *et al.* (2021) used a meta-analysis approach looking at a number of different mammal studies to compare the effectiveness of minimally invasive monitoring methods with live trapping. They found that both trapping and less invasive methods, such as camera traps and genetic analyses, produced similar estimates overall, but less invasive methods tended to detect more individuals compared to trapping efforts. As live trapping can be more costly than less invasive methods and can pose more risk to animal welfare, it is generally agreed that minimally invasive methods are preferable for population monitoring if they can be deployed efficiently.

Recent advances in molecular survey methods and the development of environmental DNA (eDNA) analysis mean that it is now possible to detect DNA shed by target species into their environment. This can be used to confirm the presence of single or multiple species (when it is referred to as metabarcoding). Water-based eDNA metabarcoding has revolutionised species monitoring in aquatic environments, but until recently, its application to mammals has largely been focussed on aquatic or semiaquatic species. However, the methods are now being used with some success to detect both semiaquatic and terrestrial mammals in water samples taken from rivers and streams, the rationale being that when a terrestrial animal is near the water's edge, then some of the DNA that it sheds will be washed into the water. eDNA can also be sampled in terrestrial environments by taking soil or sediment rather than water samples. eDNA-based methods are noninvasive, and the use of metabarcoding can provide information at the community level and is therefore relatively cost effective in comparison to some other survey methods. It has been shown that the use of eDNA is comparable to traditional methods such as field signs or camera traps for monitoring *some* mammal species (Sales *et al.* 2020, Fediajevaite *et al.* 2021, Lyet *et al.* 2021), and therefore it has the potential to be an effective addition to existing survey and monitoring methods for some of the mustelids, particularly otters and mink. However, the reliability and accuracy of eDNA-based methods are still being refined. When sampling in aquatic environments like rivers and streams, eDNA detection rates should be higher for aquatic and semiaquatic mammals than for terrestrial mammals due to differences in the amount of time in contact with the water. However, eDNA analysis has performed poorly in the past for some semiaquatic carnivores. This has included the Eurasian Otter, which in some cases has not been detected at all in areas where otters were known to be present (Sales *et al.* 2020a). As with any other method, eDNA surveys are subject to the problem of false negatives, when a species is undetected despite being present. In one recent study, carnivores that were known to be present, including Red Foxes, European Badgers and Pine Martens, were detected only on very few occasions, while others, such as Stoats and American Mink, were not detected at all (Sales *et al.* 2020). Another study, using eDNA from soil samples to detect North American carnivores, had similar results (Leempoel *et al.* 2020). The size and behaviour of an animal will have an effect on how much DNA it leaves in the environment, with the result that some species will rarely be detected while others may be overrepresented. For some species, their ecology or behaviour may result in few or no eDNA records. The probability of collecting target eDNA in any given sample is low for solitary animals with large home ranges that occur at low density or in small, scattered populations. Unfortunately, that describes most of the small mustelids. At the moment, there are still a lot of unanswered questions about the potential of eDNA as a survey and monitoring tool in natural environments, including how frequently or recently an animal needs to pass through a given area for it to be detectable in an eDNA sample.

False negatives can also be due to weather events, such as rainfall or storms, leading to higher flow rates in aquatic environments or washing away eDNA in terrestrial environments which effectively reduces eDNA turnover times (Sales *et al.* 2020). However, low turnover rates can lead to the error of false positive results when slow eDNA degradation, or resuspension of historical eDNA from sediment, suggests the presence of populations that have gone extinct or emigrated away from sampling sites (Goldberg *et al.* 2018). False positives can also result from downstream transportation of eDNA in rivers. It has been shown that eDNA can be detected more than 10 km and potentially up to 100 km downstream from source populations (Burian *et al.* 2021). This is dependent on factors that include eDNA turnover times, shedding rates and the size of source populations. A recent record of Pine Marten was obtained from eDNA in the Peak District National Park, in England, some distance from the known distribution of the species at the time (Sales *et al.* 2020). However, a Pine Marten was found killed on the road just outside the park's boundary at around the same time, so it is not known if the eDNA record was a false positive or not. False positive results can be a result of independent introduction of target eDNA into unoccupied habitats. Such 'contamination' can be from human activities (such as the release of ballast water from ships) or natural, as in the faeces of a predator of the target species. Occupancy models can be used to analyse the results of eDNA surveys but, in addition to evaluating occupancy probabilities (the probability that the target species is present at a given site) and detection probabilities (probability of detection if the species is present), specific eDNA occupancy models also evaluate a third probability, that of eDNA 'capture' (Doi *et al.* 2019). Capture probabilities estimate the chance of collecting target eDNA in a natural replicate, whereas detection probabilities in this case are the probability of detecting captured eDNA with one technical (PCR) replicate. Most occupancy models do not account for false positives, although even low levels of false positives can substantially affect the reliability of model predictions; for example, as few as 2–3 per cent false positives may result in 50 per cent overestimation of occurrence (Ruiz-Gutierrez *et al.* 2016). In order to address this issue, new approaches have been developed that make use of multiple data sources (Chambert *et al.* 2015). One dataset with high credibility (usually for a subset of sampling sites) is needed to establish true positive and true negative results (Chambert *et al.* 2015). These can then be compared with a second dataset (e.g. eDNA survey results) for the subset of sites where both datasets are available. This allows for the estimation of false positive rates with high confidence and therefore enables corrections to be applied, accounting for the occurrence of false positives across the entire eDNA dataset.

The reliability of predictions from occupancy models can be substantially improved if different data sources are combined – even if those datasets vary widely in their reliability (Lahoz-Monfort *et al.* 2016). If we can combine traditional and eDNA survey methods with the use of alternative data sources such as those generated by citizen science surveys, there is huge potential to overcome some of the current challenges we face in monitoring the small mustelids, among other species. Goomber *et al.* (2021) made use of the large source of data held by biological records centres in the UK. This was unstructured and largely collected by citizen scientists. They grouped mammal species into survey assemblages of those that are likely to be surveyed at the same time and applied occupancy models to the resulting derived detection histories. In this way they were able to provide estimates of long-term occupancy trends from 1970 to 2016 for 37 terrestrial mammal species in the UK, including two of the small mustelids. Stoat and Weasel were found to have declined sufficiently across that time period to be classified as decreasing (Stoat) and strongly decreasing (Weasel). The scale of this decline for Weasel is enough to justify changing its IUCN Red List classification from its current status of Least Concern to Vulnerable on the GB Regional Red List. Using occupancy models on unstructured citizen science records in this way has less power to detect trends than repeated systematic surveys; however, it provides a useful way of highlighting where resources for monitoring and conservation action may be most urgently needed. Citizen science surveys and ad hoc records can provide a wealth of data on past and present occurrences for some species but are often biased towards familiar and easily identifiable species such as the European Hedgehog. More elusive or cryptic species such as many of the small mustelids tend to be underrepresented in citizen science datasets or misidentified. For example, the dataset for European Polecat held by the Global Biodiversity Information Facility shows a large number of records in Ireland, where the species does not occur. It is likely that these are sightings of feral Ferrets that have been recorded as polecats by members of the public.

Genetic confirmation of species is more reliable, although not without the issues already discussed. It is clear that eDNA is a promising tool for monitoring semiaquatic and terrestrial mammals, but these methods are relatively new, having only been used for about a decade in conservation management. Nonetheless, the use of eDNA surveys can be a valuable tool to generate an initial, coarse and rapid indication of what species are present in an area and to identify specific areas to investigate further for confirmation of solitary, rare species such as our native small mustelids, the Stoat, Weasel, European Polecat and Pine Marten, as well as the non-native American Mink. At present it is recommended that eDNA metabarcoding is used alongside other noninvasive surveying methods such as camera traps for monitoring invasive species or species of conservation concern to maximise monitoring efforts.

There is no 'one size fits all' method of gathering data on small mustelids. Each species presents its own challenges, as described here. Ultimately the choice of method depends on the goal of the study, the habitat characteristics of the study area, the budget and the personnel involved in the study and the specific traits of the species of small mustelid in question. One thing is certain though, and that is that in order to make informed decisions about their conservation and/or management, we need more and better data.

CHAPTER 9

Mustelid Menus: What Do Weasels Eat?

DNA analysis has not only revolutionised survey methods for some of the small mustelids but is now also contributing to our understanding of what they eat. Dietary studies are essential to our knowledge of carnivore ecology. Information on the diet of small mustelids provides insight into how they hunt, in what habitats and to what extent they compete with each other and with other predators. Three important factors that affect carnivore coexistence are habitat use, diet preferences and time partitioning. Food is a crucial resource for all animals and it is therefore important to know how food or prey is partitioned between them to understand how similar species coexist and what impacts they might have on each other. For this to provide meaningful answers, however, is reliant on accurate detection and identification of what food or prey items have been consumed.

There are a number of different methods that can be used to study carnivore diets: these include direct observation, examination of stomach contents or scats and analysis of stable isotopes. The elusive nature of most small mustelids means that field observation is extremely difficult and unreliable. When working with rare, endangered and/or protected species (such as the Pine Marten), it is usually not possible to examine the stomach contents of carcasses in sufficient numbers as to provide a representative sample. Stomach contents are often available from animals found dead (such as roadkills) and they can provide another source of information about diet.

Scat analysis to look at the undigested parts of prey, such as the hair, feathers and bones in the predator scat by means of morphological identification of hard parts, is a relatively cheap, simple and noninvasive way of determining the

FIG 127. Direct observation is occasionally possible, as in this instance, when the author watched a Pine Marten *Martes martes* selecting from a variety of foods provided by a friendly householder.

diet of those species whose scats can be easily collected. This has been one of the most commonly used methods to date for examining the diets of terrestrial carnivores (Klare *et al.* 2011). Results are usually expressed either as frequency of occurrence (FO) being the percentage of scats containing each food/prey item or relative frequency of occurrence (RFO) which gives the occurrence of each item as a percentage of the total number of occurrences of all food/prey items found. However, sampling error and bias can occur at several steps when using scats to analyse carnivore diet. Firstly, when collecting scats from a given area, there is an assumption that the scats collected represent an independent sample of all available food/prey items. Both the number of scat samples and the area over which they are collected will affect how many individual animals are represented in the sample and the proportional contribution of each to the final analysis. Once samples have been collected, another source of error is that of misidentification of the predator species. As discussed previously, scats of Pine Marten and Red Fox can be difficult to differentiate from their morphology alone, and so DNA analysis should be used to verify predator species before using scats for dietary studies. However, a recent review found that out of 400 published studies of carnivore diet based on scats, only 8 per cent used genetic methods to verify the carnivore species (Monterroso et al. 2019). A third source of error or bias can result from differences in digestibility and therefore detectability of food and prey items. Not all of these will be detected equally; for example, soft and highly digestible items such as fungi and gastropods are not detected at all, and others may be underrepresented or misidentified. Molecular methods which do not depend on the presence of indigestible remains can be used to identify prey DNA within scat samples, but this is more expensive, requires greater expertise and is not without other issues (Morin et al. 2019). Prey DNA may not be distributed evenly within a scat, and it is still not known how the relative size of prey might influence the distribution of prey DNA within a scat. DNA in scats is also subject to degradation, which limits the use of molecular diet analyses to relatively fresh scats. By comparison, those indigestible prey remains that are present in scats can persist for quite some time, meaning that morphological analysis of even old scats is often feasible. Nonetheless, molecular methods are advancing rapidly, although their use may vary depending on the question being asked. DNA metabarcoding can be used to provide an accurate assessment of the range of consumed items, but currently it is only possible to say that an item has been consumed, without any quantitative analysis such as the proportion of each item in the sample. However, this may soon be possible with the development of new approaches which may enable quantitative assessment of consumed prey (Monterroso et al. 2019).

Analysis of stable carbon and nitrogen isotope values in animal tissues can supplement other methods for describing the diet of highly mobile, far-ranging animals. Isotopes are present everywhere, but the balance (or ratios) in which different isotopes of the same elements occur varies between different types of food and ecosystems. As tissues grow and renew themselves, the isotopes that are in the food an animal eats are incorporated into its body tissues. By measuring the ratios of different isotopes in bones, teeth or whiskers and using scientific knowledge to trace them back to the sources that they came from, it is possible to find out many things about an individual, including their diet. Isotopic measurements of different tissues from an individual animal can reveal dietary information on varying timescales because different tissue types have characteristic turnover rates. For example, continuously growing whiskers can produce a sequential record of dietary information over a large timescale, compared with tissues that have much faster turnover rates, such as blood (Beltran et al. 2015). Time series isotope data from whiskers can address important ecological questions that point samples (such as scats) cannot. This can include dietary variation within individuals, seasonal patterns, individual foraging strategies and behavioural responses to perturbations such as being translocated from one region to another. Furthermore, using whiskers in stable isotope analyses for mammals allows for nonlethal, minimally invasive tissue sampling. Whilst all of the mustelids are classified as mesocarnivores, within the mustelid group there is a high degree of variation in dietary strategies. The Stoats, Weasels and European Polecats, in the genus *Mustela*, are the most exclusively carnivorous, whereas some, including the Pine Martens, are more omnivorous, with a diet consisting of plant foods such as berries, as well as invertebrate prey, birds and rodents.

The diets of Stoats and Weasels have been well studied despite the fact that neither leave obvious scats that can be collected and analysed. However, both species are routinely killed as part of predator management in mainland Britain or for invasive non-native species control in New Zealand, providing a ready source of carcasses for research. The stomachs of dead Weasels and Stoats collected in this way have been used for a number of studies of their diets. However, stomach contents only give one sample per individual and represent just a snapshot of what an animal eats.

A Weasel's metabolism is so rapid that its stomach usually only contains what it has eaten very recently and not yet digested. Further down the intestine and in Weasel scats, should you be fortunate enough to find them, the undigestible hair, feathers and bone fragments of their prey remain. These can be identified from a combination of, often microscopic, differences in their structure, and there are some very good published guides to identification, including that of Teerink (2003), which was my bible when sifting through American Mink guts as a student one summer to see what they had eaten. The list of food items that have been identified in Weasel diet is relatively short because they have evolved as specialist predators of small mammals and birds. In their native range they only eat other items such as invertebrates or vegetable matter when they are extremely hungry and if there is no other choice.

The 10–20 g stomach capacity of Weasels and Stoats is about the same as the average weight of a small rodent (15–30 g). This means that they can only eat one small rodent at a sitting, so a single stomach, intestine or scat usually contains the remains of only one prey item. The nutritional value of prey is related to its body size. It is possible to see which are most profitable prey for small mustelids to hunt and eat if their diet is calculated in terms of the weights of the various types of prey eaten rather than their number. This corrects the imbalance between large and small items. Carolyn King (King & Powell 2007) gives the example that seven birds' eggs at 3 g each would equate to two 10 g mice or one meal of 20 g of rabbit meat from a larger carcass. Therefore, if a

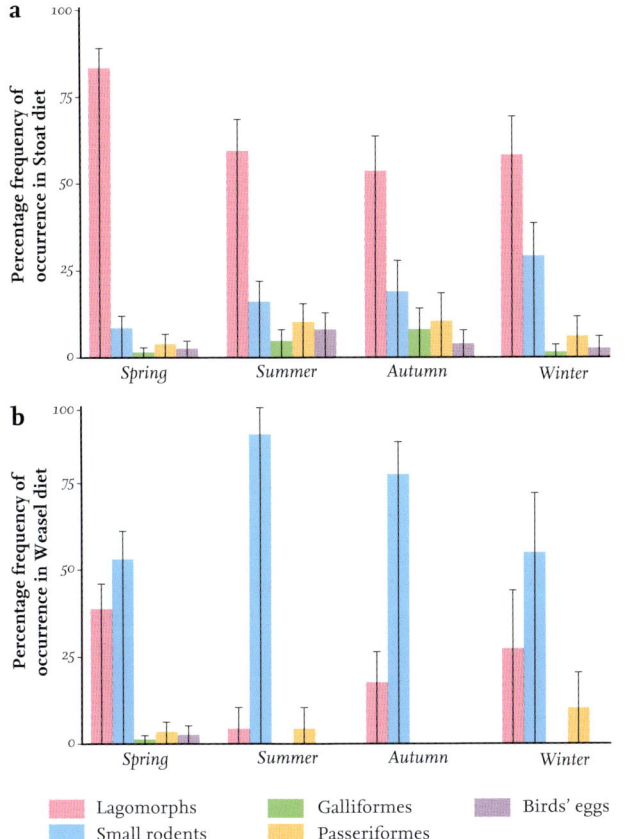

FIG 128. Seasonal variation in the relative importance of the five main prey groups in the diet of (a) Stoats *Mustela erminea* and (b) Weasels *Mustela nivalis* (both sexes combined). Seasons are the three-month periods starting in March (spring), June (summer), September (autumn) and December (winter). Redrawn from McDonald et al. (2000).

sample contained 10 birds' eggs (30 g), 8 mice (80 g) and 6 meals of rabbit (120 g), this would give a total weight of prey eaten of 230 g, of which eggs constitute 13 per cent, mice 35 per cent and rabbits 52 per cent. If, however, the same data were expressed as percentage frequency of occurrence, that would give quite a different picture, with eggs contributing 42 per cent (10/24 × 100), mice 33 per cent (8/24 × 100) and rabbits 25 per cent (6/24 × 100). That being said, the results of weighting prey items by body size must still be treated with caution because the proportion of any one prey item is by definition relative to the total so they are not independent. Therefore, it is not possible to say from such figures if Stoats or Weasels from one area show a preference for one type of prey compared with those from another area. Also, we cannot always tell from undigested remains whether individual prey were adults or juveniles, and these can be very different in both size and catchability. The results are also greatly influenced by the quantity it is assumed is eaten from a single large carcass. Finally, small prey have relatively more bones and hair and so these are not equally represented in prey remains. Weasels (and Stoats) are not able to digest hair, so they will leave skin and hair uneaten if there is plenty of meat, but all the fur that they do eat reappears in their scats (Gamberg & Atkinson 1988). Large bones are also avoided, and small ones can be partially digested; therefore, analyses of bones and hair do not reflect all prey equally. Even so, it is worth calculating the proportion of prey eaten to see that, in general, small prey are much less profitable to eat than large ones, in that they return less energy for effort.

Identifying what the small mustelids eat is relatively easy, but interpreting what the figures mean is a bit trickier. For a start, the prey eaten by each species in different places will be affected by the season, the habitat and the age and sex of the animals sampled. This is true for Weasels, Stoats, European Polecats and Pine Martens. Animals of the same species but different sexes or ages and in different seasons may eat different things. Some foods are highly seasonal, such as birds' eggs, which are most available in spring and early summer. Therefore, it would not be valid to compare the proportion of birds' eggs in samples from two areas if they had not been collected at the same time of year. This is illustrated for Stoats and Weasels in Figure 128.

WEASELS AND STOATS

The large scientific literature on Weasels and Stoats contains many descriptions of what they eat. The first large study of the diet of Weasels and Stoats in Britain was done by Day (1968). For many years, his was the only British study to include animals of both species collected from all over the country, mostly from gamekeepers. He also produced a key to identifying the hairs of small mammals in Britain which has proved invaluable (Day 1966). More recently, a second comparative study of the diet of Weasels and Stoats was carried out by McDonald *et al*. (2000), also using carcasses collected from gamekeepers across Britain. In both studies, small rodents constituted more than half of the diet of Weasels (56 per cent in the 1960s and 68 per cent in the 1990s). Rabbit formed a greater proportion of the diet of both Stoats and Weasels in the 1990s than it had in the 1960s but for different reasons. When Day's study was carried out in the 1960s, rabbit (and Stoat) numbers were low, as a consequence of myxomatosis. Nonetheless, lagomorphs (mostly rabbits) still made up about a third of the prey items eaten by Stoats, with small rodents and birds comprising another third each. By the 1990s, rabbit populations in Britain had largely recovered and this was reflected in the diet of Stoats. In McDonald *et al.*'s (2000) study of 789 Stoats collected from 1995 to 1997, lagomorphs constituted 65 per cent of their diet with only 16 per cent being small rodents. This was presumably much as it had been before the arrival of myxomatosis. Rabbit also formed a greater proportion of the diet of Weasels in the 1990s than 30 years previously (25 per cent in the 1990s compared with 18 per cent in the 1960s). The recovery of the rabbit population meant that Weasels benefited from the greater availability of young rabbits in spring.

In Wytham Woods in Oxfordshire, Carolyn King was able to study Weasel diet using scats she collected from 36 individually marked Weasels caught in wooden box traps (King 1980). She made sure that she obtained a scat sample from each animal by providing a dead white mouse in each trap. The populations of small rodents in the area at the time were low (estimated at 21–39 per ha), and so the Weasels were usually hungry enough to eat the white mouse even in cases where it was not very fresh. Eating this pushed everything else along in the Weasel's gut, resulting in at least one scat sample containing each Weasel's last wild-caught meal. White fur in the scats indicated clearly where the wild prey stopped. King found that the Weasels ate mostly Bank Voles, Wood Mice, Field Voles and birds in the order of their abundance, with Bank Voles being by far the most common small rodent in the wood, followed by Wood Mice. Some Field Voles were eaten by those Weasels whose home ranges extended to a neighbouring young

FIG 129. Stoat *Mustela erminea* with small mammal prey.

plantation. Young plantations are good habitat for Weasels because the ground between young trees quickly grasses over without being mown or grazed. This provides ideal living conditions for Field Voles, which are the Weasels' preferred prey. In young and regenerating woodlands, Weasels are able to prey almost exclusively on voles (King & Powell 2007).

While it is important to know what Weasels eat in different types of woodland, these habitats are in the minority in Britain. In fact, the UK as a whole is one of the least wooded countries in Europe, with woodland making up just 13 per cent of the land area. By nation, Scotland is the most wooded at 18 per cent of total land area, followed by Wales at 15 per cent, England at 10 per cent and Northern Ireland at just 8 per cent. By comparison, farmland comprises just over 70 per cent of the UK land area, meaning that most British Weasels now live on farmland. In open habitats, Field Voles (being a grassland species) are the small rodents most often eaten by Weasels, followed by Wood Mice, which often live in hedgerows and will also forage in fields, especially before harvest. Bank Voles prefer the cover of woodland and scrub, so they are less often caught in open country.

In summary then, Weasels are highly dependent on small rodents and birds and do not do well where these are scarce. The proportion of small rodents in the Weasel's diet is never less than half of the total number of prey items identified and is often nearer to three-quarters (King & Powell 2007). The Stoat's preference is to hunt the largest prey it can catch, so Stoats prefer rabbits if available or the largest prey species in a rodent community, such as the Water Vole. But, in

FIG 130. Weasel *Mustela nivalis* with Brown Rat *Rattus norvegicus*.

FIG 131. Left to right: male and female Stoats *Mustela erminea* and male and female Weasels *Mustela nivalis* show marked size differences between the sexes, as well as between the species.

common with Weasels, Stoats will also take Field Voles, especially when they are at very high density.

All of the small mustelids show marked sexual dimorphism, with males being approximately a third larger than females of the same species. It has been suggested that this size difference reduces niche overlap of the sexes and therefore intraspecific (within species) competition for the same prey, if prey size is related to predator size (Dayan & Simberloff 1994). McDonald *et al.* (2000) found significant differences between the diet composition of male and female Stoats, and this difference was affected by season. The interaction between sex and season significantly affected the proportion of lagomorphs and small birds in the diet of Stoats. Rabbits and hares were always eaten more frequently by male Stoats than by females and this difference was especially marked in winter. Lagomorphs were most commonly eaten by Stoats of both sexes in spring, when perhaps the smaller females can specialise on young rabbits. Small birds were more commonly eaten by male Stoats than by females and this difference was also most pronounced in winter. However, birds were most commonly eaten by Stoats of both sexes in summer. Small rodents were eaten more by female Stoats than by males all year round. Male and female Stoats were equally likely to eat birds' eggs, most commonly in summer when they are most available.

For Weasels, McDonald et al. (2000) did not detect any significant difference in the diet of the two sexes. However, Weasel diet did vary with season. Lagomorphs were most commonly eaten by Weasels in spring and rarely eaten in summer, which again is likely due to the availability of smaller, more catchable young rabbits in spring. Small rodents were eaten most in summer, but there was no detectable seasonal bias in Weasel consumption of birds. Throughout the year, female Stoats had a broader dietary niche than male Stoats. The dietary niche of Stoats of both sexes was narrower in spring than in any other season. This is because both sexes specialised on lagomorphs at this time of year. In Weasels, the dietary niche of females tended to be narrower than that of males, though this between-sex difference was not as pronounced as in Stoats. The dietary niche of Weasels was broader in spring and winter, when predation on lagomorphs was highest, than in summer and autumn, when they specialised on small rodents. Weasels ate eggs mainly during the nesting season, as did Stoats, although the rare occurrence of egg remains in the guts of Stoats killed in winter suggested either that they had cached eggs and then eaten them later in the year or, perhaps more likely, been able to take eggs of domestic poultry from somewhere. The fact that female Stoats have been found to eat a higher proportion of small rodents than males, while male Stoats eat more larger prey such as lagomorphs, supports the theory that prey size in Stoats is related to their body size. There was no detectable difference between the diets of the two sexes of Weasel in McDonald's study, although such differences have been reported previously. In the 1970s, Tapper (1979) found that male Weasels ate more lagomorphs and fewer voles than females but this difference was only significant in spring. Both Tapper (1979) and Dunn (1977) found that when rodent abundance was reduced, Weasels switched to eating small birds. Conversely, in New Zealand, King (1983) found that the overall numbers of birds eaten by Stoats were highest in years of abundant mammalian prey (mice). Similarly, it has been shown that predation on Mohua (or Yellowhead) *Mohoua ochrocephala* was most severe when rodent populations were highest (O'Donnell et al. 1996). This is due to the fact that the non-native Stoat population increases in response to an increase in food supply, in what is known as the numerical response. As a consequence, it is recommended that in order to produce the greatest benefit for bird conservation in New Zealand, the most effort is put into Stoat control in the years when rodent populations are highest.

IRISH STOATS

The Irish Stoat is found only in Ireland and on the Isle of Man, neither of which have Field Voles or Bank Voles as part of the native small mammal fauna. This means that Wood Mice *Apodemus sylvaticus* and House Mice *Mus musculus* are the most widely available small rodents for Irish Stoats. In Ireland there are only two native small mammal species: the Wood Mouse and Pygmy Shrew *Sorex minutus*. The House Mouse and Brown Rat *Rattus norvegicus* were introduced historically, and in recent years two more non-native small mammal species have become

FIG 132. The Pygmy Shrew *Sorex minutus* is one of only two native small mammal prey for the Irish Stoat *Mustela erminea hibernica* in Ireland.

FIG 133. Irish Stoat *Mustela erminea hibernica*.

established in parts of Ireland. The first of these was the Bank Vole, which was discovered in County Kerry in 1964 and has since spread across the southern half of the country (McManus *et al.* 2021). More recently, the Greater White-toothed Shrew *Crocidura russula* was found in owl pellets in 2007 (Tosh *et al.* 2008). Both of these could potentially add to the prey base of small mammals for predators such as the Irish Stoat and Pine Marten.

Irish Stoats are generally smaller than those in Britain, although contrary to what is known as Bergmann's rule, there is a north–south gradient of increasing size. Bergmann's rule argues that species which vary in size over their geographic ranges should increase in size with latitude – the reason being that increased body size results in a decreased surface-to-volume ratio and a lower rate of body heat loss, which is an advantage in colder, northerly climates. However, studies have shown that the European Stoat does not conform to this rule, being smallest in the coldest regions where it occurs, such as northern Sweden, and largest in Britain, which has the mildest winters (Erlinge 1987). Irish Stoats too are a smaller subspecies overall, but they are largest in the milder south of the island where the average body weight of male Irish Stoats is similar to or even larger than that of male Stoats in Britain. Female Irish Stoats are significantly lighter across the island than their British counterparts. Nonetheless, both males and females have to take large prey such as rabbits and rats for want of anything else. They also have to make use of prey and food items that would be unusual in the diet of Stoats elsewhere, such as fish (Fairley 2001, McGreal 2010) and fruit (Buckley *et al.* 2015). Irish Stoats are legally protected, unlike Stoats in Britain, so carcasses for dietary studies can only be collected opportunistically, from road-killed animals or, in rare cases, where localised control of Stoats is licensed for a specific reason, as it was in the Boora area of County Offaly from 1996 to 2009 (Buckley *et al.* 2015). In all studies of Irish Stoat diet to date, lagomorphs are the main prey, followed by shrews, birds, mice and Brown Rats (Fairley 1971, Sleeman 1992, King & Powell 2007). At 18.8 per cent, the high prevalence of Pygmy Shrews in the diet of road-killed Stoats in Ireland is surprising, given that Pygmy Shrews are thought to secrete a foul-tasting substance, which deters many predators, such as cats. However, this may be related to their abundance in the habitats sampled, and shrew predation may be a necessity where there is limited availability of preferred lagomorph prey.

POLECATS

In Britain, European Polecats in common with Stoats also eat predominantly lagomorphs, but they have been recorded as eating a wide variety of food items across Europe and are therefore usually described as generalist predators.

FIG 134. A European Polecat *Mustela putorius* hunting a European Rabbit *Oryctolagus cuniculus*.

Rodents and amphibians are common prey items throughout their European range. However, while polecats show diversity in their diet across all regions, there is some evidence of regional specialisation. Examples of this are seen in the polecat's specialisation on European Rabbits in the Mediterranean (Santos et al. 2009), while elsewhere they have been found to feed predominantly on amphibians (Zalewski et al. 2021). This flexibility in their foraging strategy enables polecats to occupy a diverse range of habitats including lowland, grassland, farmland and wetlands. It may also be an important factor in population persistence when some prey are unavailable, such as when rabbit populations in Britain were devastated by the myxomatosis epizootic. During this period, mammals comprised a much smaller proportion of polecat diet.

Although rabbit populations had recovered to their pre-myxomatosis levels by the 1990s, they have since declined again across Britain (Harris et al. 2020). This is most likely due to the rabbit haemorrhagic disease (RHD) virus which has had a devastating impact on rabbit populations right across mainland Europe. RHD was first confirmed in a domestic rabbit in 1992, and by 1994, it had caused significant declines in wild rabbit populations of the southeast of England. The disease has significantly impacted wild rabbit populations throughout the world; however, within Europe, outbreaks have been most virulent in Spain (Forrester *et al.* 2009). This has resulted in dramatic declines in several of Spain's charismatic species as a result of reduced prey populations following this disease, notably the Iberian Lynx *Lynx pardinus*, which is highly dependent on rabbits. The impact of RHD on the rabbit population in Britain has not been as severe as elsewhere, which is thought to be due to a non-pathogenic RHD-like virus having been present in the region, resulting in a resistance to the disease being developed (Trout *et al.* 1997, Calvete 2006). There is substantial site-to-site variation in rabbit numbers related to habitat and landscape factors. However, year-to-year variation within sites has been shown to be greater, likely as a result of the impacts of disease outbreaks (Trout *et al.* 2000). Analysis of changes in rabbit records from 1996 to 2018 show that numbers have declined by 53 per cent in England, 58 per cent in Wales and a staggering 83 per cent in Scotland (Harris *et al.* 2020).

Dietary differences between the sexes have been seen in Stoats, as already described. In polecats, one study in Britain found that female polecats ate more birds and fewer rabbits than male polecats but the difference was not significant (Blandford 1986). However, as rabbit abundance has declined in recent years, it is possible that increased competition for available rabbits may have led to more pronounced dietary differences between male and female polecats.

FIG 135. European Polecat *Mustela putorius* searching for small prey in grassland.

After the severe population declines in the nineteenth century, polecats have been recolonising Britain from west to east. As they have done so, they have moved from a landscape in Wales that is largely comprised of unimproved and semi-improved pasture and woodlands, to the much more mixed farming and improved grasslands of central England and beyond, to the predominantly arable farmland in the eastern counties. With the polecat's range now extending right across the country, Sainsbury *et al.* (2020) were able to carry out a study to see if any changes in resource use had accompanied the expansion process. They explored dietary variation and niche breadth in polecats during a period in which the species had recovered large parts of its former range at the same time as variation in rabbit populations. They analysed stomach and gut contents from polecat carcasses collected between 2012 and 2016 and compared their findings with earlier analyses of polecat diet in Britain from studies in the 1960s (Walton 1968), the 1980s (Blandford 1986) and the 1990s (Birks & Kitchener 1999). Sainsbury *et al.* (2020) found that lagomorphs dominated polecat diet composition in Britain in all the samples from the 1980s to the 2010s, but there was seasonal variation in the 2010s, reflecting the polecat's opportunistic foraging strategy. They found no evidence that there has been a reduction in rabbit consumption by polecats since the 1990s, despite the declines in rabbit populations recorded over this period. They did find an increase in the frequency of occurrence of mammals in the diet of polecats since the 1960s, as well as an increase in lagomorphs between the 1980s and 1990s. This is consistent with similar increases of lagomorphs in Stoat diet over the same time period (McDonald *et al.* 2000). In Sainsbury *et al.*'s (2020) study, there was no evidence of resource partitioning between male and female polecats, with lagomorphs being equally important prey for both sexes in the 1990s and 2010s.

Unlike the 1950s myxomatosis epizootic, the more recent rabbit declines have not been uniformly spread across the landscape, so polecats have still been able to find rabbits as a major prey item. In the 1980s, polecats were still mainly confined to Wales, where rabbit numbers remained low. At that time, lagomorphs comprised a lower proportion of polecat diet than in the 2010s, and niche breadth was, of necessity, greater than it was in the 1990s or 2010s. There was also some evidence of differences in resource use between male and female polecats in the 1980s. When there was limited availability of rabbits, females consumed fewer lagomorphs than males, but there was no evidence that this resource partitioning was occurring in the 2010s.

The frequency with which birds occurred in polecat diet in Sainsbury's study was consistent with that observed in previous decades. However, it has been shown that bird remains are often underestimated when analysing stomach contents (Reynolds & Aebishcer 1991). Whilst it is known that polecats eat eggs, Sainsbury's study did not

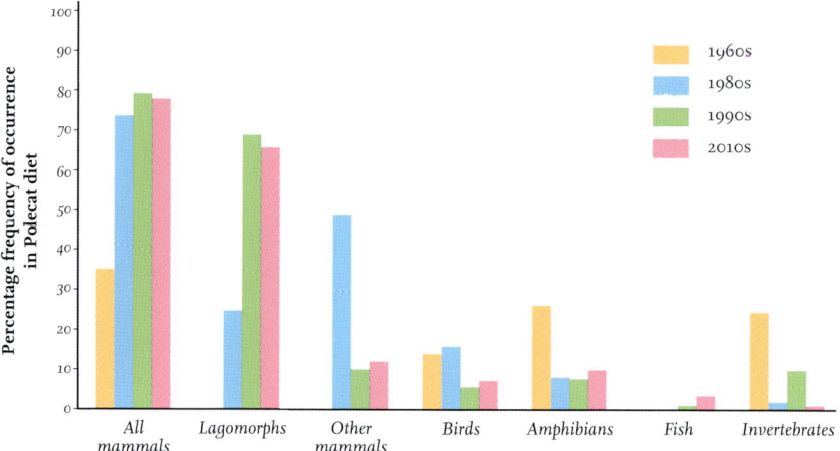

FIG 136. European Polecat *Mustela putorius* diet in Britain. The results show how niche breadth and diet composition have varied over time, illustrating how important long-term studies can be in determining if a species like the polecat is a generalist or more of a specialist. Data from *Walton (1968), Blandford (1986), Birks & Kitchener (1999) and Sainsbury *et al.* (2020). *Lagomorphs not separated from other mammals.

detect any evidence of this, which may be because polecats usually break them open and then eat the contents without consuming the shell (Weber 1989).

Sainsbury *et al.*'s (2020) study highlighted the way in which the proportion of lagomorphs in polecat diet in Britain has increased over the long term, as the polecat population has recovered its range. When rabbits almost disappeared from Britain in the 1960s, the diet of polecats had to be significantly more diverse. The dietary niche of the species narrowed as rabbit populations recovered. Rabbits are undoubtedly an important prey item for polecats in Britain. It is of concern that rabbit populations are once again declining, but there is no evidence from recent studies that this has yet resulted in a reduction of lagomorph consumption in polecat diet. However, given the spatial variation in impacts of current diseases on rabbit populations, it is possible that any effects on predator diets may only be evident from studies carried out at a finer spatial scale.

PINE MARTENS

Pine Martens are the most omnivorous of the small mustelids, eating whatever is locally abundant. This includes small mammals, invertebrates, amphibians, birds,

FIG 137. Pine Martens *Martes martes* eat a variety of small mammals including Wood Mice *Apodemus sylvaticus*.

fungi, plants and carrion. It has been shown that they will preferentially switch to fruit and berries in the autumn, regardless of the availability of other food items (Caryl 2008). Many generalist predators, including Pine Martens, have a preferred prey species that maximises ease of capture and energy intake (body size of prey) (Roth *et al.* 2006). In Britain, the Pine Marten has been shown to exhibit a preference for Field Voles (Lockie 1961, Balharry 1993, Halliwell 1997, Coope 2007, Caryl *et al.* 2012b). Field Voles are thought to be a profitable prey because of their size, comparatively high densities and relative lack of anti-predator behaviours (Balharry 1993). Open areas within woodland and adjacent patches of rough grassland are an important resource for Pine Martens, as this is the preferred habitat of Field Voles (Hansson 1983, Petty 1999, Bogdziewicz & Zwolak 2014). Vole populations can be highly cyclic and increase up to tenfold during peak years compared with years of low density. Field Voles in Britain have been shown to undergo population cycles of between three and five years (Charles 1981, Lambin *et al.* 1998) but these are much less pronounced than at more northerly latitudes (Lambin *et al.* 2000). In Kielder Forest, in the north of England, Field Vole population dynamics have been studied for a number of years, and a clear dampening has been observed in the amplitude of cycles since the late 1990s (Lambin 2017). When Field Voles are at low density in Scotland, alternative foods recorded in the diet have been predominantly invertebrates, but also passerine birds (Balharry 1993, Bright & Smithson 1997).

In Scotland, the breadth of Pine Marten diet was found to increase threefold in the summer as a greater variety of resources became available (Caryl 2008). Small mammals and birds are important throughout the year, with fruit, notably Rowan,

Bilberry and Blackberry, making up a large proportion of the diet in the summer and autumn. Some studies have shown an increased occurrence of birds in the diet during the spring/summer months when nestlings and fledglings are accessible (Balharry 1993, Caryl 2008). The most frequently occurring bird species in the martens' diet are common woodland passerines; Robin *Erithacus rubecula*, Wren *Troglodytes troglodytes*, Blue Tit *Cyanistes caeruleus*, Great Tit *Parus major*, Coal Tit *Periparus ater* and Chaffinch *Fringilla coelebs*, though other woodland species such as Treecreeper *Certhia familiaris*, Mistle Thrush *Turdus viscivorus* and Woodcock *Scolopax rusticola* also occur (Lockie 1961, Halliwell 1997). Woodpigeons *Columba palumbus*, which breed year-round in Britain and Ireland, represented one-third of the birds in Pine Marten diet at one Scottish study site (Halliwell 1997).

Bees and wasps are sometimes eaten, most likely as bycatch when a marten raids a nest for honey and grubs. Carrion from large mammals, such as deer, can supplement marten diet significantly in the winter, while large-bodied invertebrates like ground beetles are only available in the warmer months. The importance of carrion in the diet of Pine Martens in Scotland is probably underestimated when quantified by frequency of occurrence but becomes much more significant when analysed as estimated weight of intake (EWI). Conversely, the relatively high frequency of occurrence of invertebrates and plant material in the diet of martens overestimates their importance when compared to their EWI (Kubasiewicz 2014). Table 12 shows how Pine Marten diet varies across different studies within Scotland.

Other prey species such as squirrels, lagomorphs, amphibians, reptiles and even fish are eaten opportunistically and constitute a small proportion of the overall diet, though there is geographic variation based on the presence/availability of prey species.

Andrzej Zalewski carried out a review of published dietary studies of Pine Martens to look at geographical variation across most of the species' range (Zalewski 2005). He found that while there was some variation in the species consumed, small mammals are the most important prey of Pine Martens throughout Europe. Overall, small mammals constituted 47 per cent of all prey eaten in winter and 42 per cent in summer, followed by plant material (mainly berries) at 16 per cent in winter and 21 per cent in summer, then birds (15 per cent and 13 per cent), medium-sized (150–2,500 g) mammals (10 per cent and 4 per cent) and invertebrates (5 per cent and 15 per cent). The rodent species most represented in the diet of Pine Martens varied among regions, with Wood Mice comprising the largest proportion in the Mediterranean but declining towards the north. Bank Voles were the most prevalent in northern temperate and boreal forests, where Pine Martens also ate Lemmings *Myopus chisticolor* and

TABLE 12. Pine Marten *Martes martes* diet as analysed from scat samples collected at sites in Scotland.

Location	No. of samples	Small mammals %	Birds %	Invertebrates %	Ground beetles %	Fruit %	Carrion %	Other %
[1]Easter Ross, Highlands	2,449	41.4	24	1.4	–	30	3	–
[2]West Highlands, Morvern peninsula	174	41	17	15	51	31	1.7	15
[3]Novar, Highlands	1,938	27	17	23	–	9	8	–
[4]Kinlochewe, West Highlands, Strathglass	1,304	62.8	19.7	35.5	–	9.1	11.9	9.6
[5]Loch Teacuis, West Highlands	546	31.4	16	11	–	17.2	–	11.5
[6]Abernethy, Marr, Inshriach, Darnaway	546	55	19.4	67.4	–	39	12	18

Data from [1]Caryl (2008), [2]Coope (2007), [3]Putman (2000), [4]Halliwell (1997), [5]Balharry (1993) and [6]Kubasiewicz (2014).

Lemmus lemmus. The per cent occurrence of rodents in Pine Marten diet was positively related to rodent abundance in what is termed a functional response to fluctuations in prey numbers. In years when rodent abundance was low, Pine Martens used different alternative food types depending on region, with more birds, amphibians and ungulate (such as deer) carrion consumed in lowland deciduous forests. Further north they ate more large birds, squirrels, birds' eggs and fruits in years when rodents were scarce.

The frequency of plant material in the diet of Pine Martens increased along a north–south gradient as would be expected. Fruits are more available in the south than in the north, especially during winter. In southern Europe, Pine Martens fed on a wide range of plant species, including Rowan berries, Carob *Ceratonia siliqua* fruit, myrtle berries, juniper berries, rose hips, figs and even citrus. Blackberries and Rowan berries were the most reported plant material eaten by Pine Martens in central Europe, whilst further north they ate mostly Bilberries, Lingonberries *Vaccinium vitis-idaea* and Rowan berries. Mushrooms were also part of the diet of northern European Pine Martens. Zalewski found that the frequency of large prey in the diet of Pine Martens increased in both summer and winter further

FIG 138. Pine Martens *Martes martes* preferentially switch to eating fruit such as Rowan *Sorbus aucuparia* berries when they are available.

north but, again contrary to Bergmann's rule, the body size of Pine Martens was inversely related to latitude, so martens are smaller in the north but eat larger prey (up to the size of a 4 kg hare), whereas in the south, the martens are larger but eat smaller prey such as shrews or insects. His suggested explanation for this is that hunting larger prey was a more efficient foraging strategy in the north, reducing the amount of time and energy expended while hunting. Smaller martens in the north have lower energy requirements than their larger southern counterparts and so need to hunt less often. Therefore, in colder temperatures, they can stay longer in insulated dens and rest sites and in that way minimise heat loss.

In the middle and southern latitudes of their continental range, the distribution of Pine Martens overlaps that of the Stone Marten. Where they coexist, the two species reduce competition by selecting different habitats, and though there is some dietary overlap, with both showing a strong preference for small mammalian prey, research from Poland suggests that Pine Martens eat proportionally more small mammals and birds, whilst Stone Martens consume more fruit and insects (Posluszny 2007). Both are more available to martens in southern Europe even in winter. Additionally, research from Peneda-Gerês National Park in Portugal suggests that in the presence of Stone Martens, Pine Martens exploit less profitable prey to reduce the chance of interaction (Monterroso 2016).

Most recent studies in Ireland have shown that fruit, mammals, birds and arthropods occur in Pine Marten diet with equal frequency but with seasonal variations. Fruit was found to be most important in spring, autumn and winter, arthropods were consumed all year round, small mammals and birds were primarily taken in winter, spring and summer, and frogs in low amounts all year except in winter (Lynch & McCann 2007). Reptiles, amphibians and earthworms occur throughout the year, although they were found to be less important in the diet. When the marten population was at its lowest in Ireland, the west remained a stronghold for the species. Some of this region is karst limestone pavement, where small mammals are likely to be less available, which may have contributed to a lower dependence on this type of prey.

Optimal foraging models usually focus on the net rate of energy intake in relation to prey availability and the time costs involved in catching and handling different prey types (Křivan 1996). However, it has been shown that for many species, the aim of foraging is to acquire specific nutrients rather than specific prey (Kohl et al. 2015). A study by Remonti et al. (2016) showed that despite the different variety of foods that are eaten by Pine Martens in different regions, the macronutrient energy ratio in the wide array of diets all provided a similar range of 50–55 per cent of protein, 38–42 per cent of lipids and 5–10 per cent of carbohydrates. Pine Martens are able to achieve this broad annual stabilisation of macronutrient ratios by using a variety of alternative animal foods to compensate for the high fluctuation of some prey species and by sourcing non-protein energy (carbohydrates and fats) from plant-derived foods, particularly fruits.

Remonti et al.'s (2016) analysis of a range of different studies provides evidence that Pine Martens use their dietary versatility as a way of achieving long-term stable macronutrients across populations throughout their wide latitudinal range. Specifically, by using a range of alternative sources of protein (i.e. mammals, birds and insects), Pine Martens can compensate for high fluctuations in the availability of any one prey type, leading to a broad annual stabilisation of the proportional contribution of protein energy to the diet. By having such broad diets based on both animal and plant foods, martens are able to maintain

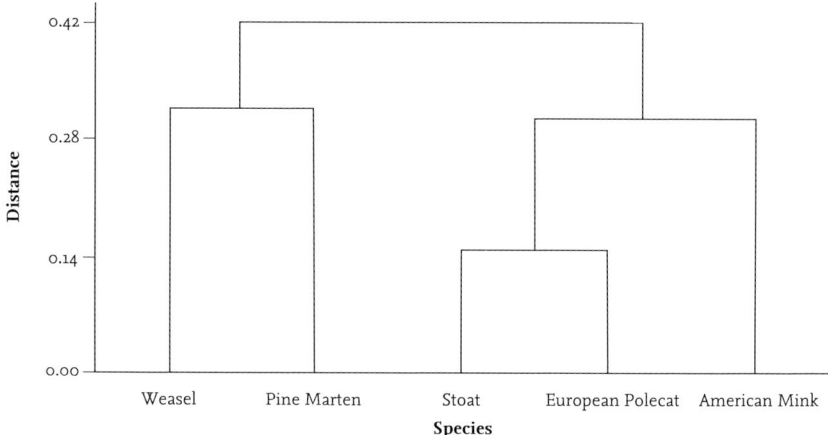

FIG 139. Community dendrogram showing the distances between the dietary niches of the small mustelids living in Britain. We can see that those pairs of species that are closest in dietary niche are not necessarily the ones that are most likely to occur in the same habitat patches or to be foraging at the same time. For example, Weasels *Mustela nivalis* can occupy open grassland, whereas Pine Martens *Martes martes* are more likely to be found hunting in grassland patches within woodland. Nocturnal European Polecats *Mustela putorius* and predominantly diurnal Stoats *Mustela erminea* show temporal segregation, while using a similar prey base. Redrawn from McDonald (2002).

a stable balance of macronutrients in spite of geographic and temporal variation in food availability. Two studies included in their analyses were outliers. This is consistent with Pine Martens showing some risk tolerance in that they may target a particular balance of nutrients but have the ability to survive periods of food shortages by eating otherwise imbalanced foods.

Several carnivores show wide seasonal variation in their diet composition, often related to fruit availability. Stone Martens and Red Foxes are both able to switch from a diet based mainly on animal food (more than 90 per cent) in the cold season to one containing a high proportion (35–50 per cent in volume) of fruit in the warm season (Remonti *et al.* 2012). Such opportunism allows these predators to exploit temporary, highly profitable resources such as fruits and is likely to be adaptive.

The four species of small mustelid native to Britain and the two in Ireland have coevolved over a long period of time and, while their dietary habits and niches appear superficially quite similar, they are sufficiently different (albeit with some overlaps) to enable them all to make a living. Pine Martens eat much more fruit, as well as invertebrates and carrion, than the smaller mustelids, while European Polecats can diversify their diet with amphibians and fish if necessary. Robbie McDonald analysed the dietary niches of all the British and Irish mustelids and found that they were partitioned along several axes (McDonald 2002). The dendrogram in Figure 139 shows the distances between them for the small mustelids in Britain.

WHAT'S YOUR POISON? MUSTELIDS AND RODENTICIDES

As a result of the varying amounts of small mammal prey, particularly rodents, in the diets of Weasels, Stoats, European Polecats and Pine Martens, they are susceptible to secondary exposure to a number of widely used anticoagulant rodenticides. This can occur when they consume exposed prey which are the target of control measures, such as the Brown Rat and House Mouse, or other non-target species that are attracted by the same bait and inadvertently contaminated.

It is estimated that rodents (primarily Brown Rats) cost the UK economy between £60 million and £100 million a year, with the damage caused mainly by spoiling of food and from disease transmission. Anticoagulant rodenticides (ARs) dispensed in bait are commonly used as a way to reduce this damage. They work by

FIG 140. A Ferret *Mustela putorius furo* with a Brown Rat *Rattus norvegicus*.

interrupting the blood-clotting mechanism, and a lethal dose results in death from internal haemorrhaging. Warfarin was one of the earlier ARs and was widely used until the emergence of resistance in rats. This led to the development of a range of second-generation anticoagulant rodenticides (SGARs) with higher toxicity, and these are now used worldwide for rodent and other mammalian pest control.

The levels and risk of secondary exposure in predators can be related to habitat and time of year. Predators that eat rats and other target species have the highest risk of secondary exposure and poisoning by SGARs. As we have seen, Stoats and European Polecats are mainly predators of rabbits, while Field Voles are the preferred prey of both Weasels and Pine Martens. Nonetheless, anticoagulant rodenticides have been found in liver samples from all of these species. In some predators, the cumulative effect of multiple sublethal exposures means that the amount of SGAR residue is greater in older individuals (Ruiz-Suárez et al. 2016).

In a study of rodenticide residues in polecat carcasses collected in Britain between 1992 and 1999, 31 out of a sample of 100 animals had detectable residues of at least one SGAR (Shore et al. 2003). Detection rates were slightly higher in animals that had died during the first half of the year, leading to speculation that this may have been due to the higher frequency of rats in polecat diet during the winter months, when rats can comprise up to 65 per cent of polecat diet (Birks 1998).

A similar study was subsequently carried out using carcasses collected between 2013 and 2016 as part of Vincent Wildlife Trust's national polecat monitoring survey (Sainsbury et al. 2018). In the intervening period between the two studies, polecats had continued to recover and expand their range into parts of Britain traditionally associated with higher usage of SGARs (Packer & Birks 1999, Garthwaite et al. 2012). This gave rise to the concern that overall exposure in the polecat population may have increased. Overall, the prevalence of anticoagulant rodenticide residues in carcasses in Sainsbury et al.'s (2018) study was greater than that reported for polecats that were collected in the 1990s in Britain, even when improvements in analytical sensitivity were accounted for. They identified an increase by a factor of 1.7 in the prevalence of SGAR residues over the 25 years from 1992 to 2016 inclusive. However, they found that rates of detection were similar in polecats both within and beyond their 1990s geographical range, which suggested that the increase in exposure since then had occurred throughout the polecat's distribution in Britain and was not caused by expansion into new areas of higher rodenticide use.

Total concentrations of SGARs were higher in polecats from arable rather than pastoral areas, but also higher in polecats collected in the west compared to the east. This was surprising, as rodenticide usage was thought likely to be higher in the more arable landscape of east England. However, Shore et al. (2003) had also found higher bromadiolone residues in polecats from Wales, the Midlands and the west of England than in animals in the east and the southeast. Residues of difenacoum were also higher in Wales than in east or southeast England. Contrary to Shore *et al.*'s earlier study, Sainsbury *et al.* did not find any variation in exposure at different times of year in the polecats that died in 2013–16. Therefore, there was no evidence that recent exposure in polecats was highest in the autumn and winter, as previously thought; instead, exposure is now similar all year round. Sainsbury *et al.* (2018) concluded that SGAR contamination in polecats in Britain is likely to be greatest in older animals that eat rodents, live in the west of the country and inhabit arable areas. These polecats are probably most at risk of adverse effects. Given that recent data indicate that the high proportion of polecats exposed to SGARs is associated with an occurrence of rodents in the diet that comprises less than 10 per cent of the total, this suggests that even relatively low rates of rodent consumption can result in high rates of secondary exposure.

It has been shown that Stoats and Weasels are also susceptible to secondary consumption of anticoagulant rodenticides. In a study by McDonald *et al.* (1998), carcasses of Stoats and Weasels were collected from gamekeepers on eight shooting estates in central and eastern England where rodenticides were used and, for comparison, one estate in an upland area where rodenticides were not used. The livers were analysed for the presence of six anticoagulant rodenticides (brodifacoum, bromadialone, coumatetralyl, difenacoum, flocoumafen and warfarin). The study found that exposure of Stoats and Weasels to the anticoagulant rodenticides was commonplace on the estates investigated.

The targets of rodenticide application programmes, which include squirrels as well as rats, are only rarely eaten by Stoats and Weasels in Britain. However, Stoats and Weasels are almost exclusively carnivorous and are therefore unlikely to eat grain-based rodenticide baits directly. This suggests that Stoats and Weasels whose livers contain residues of these compounds must have been exposed to rodenticides by eating contaminated small rodents. Given that the use of rodenticides is not restricted to the immediate vicinity of barns and other farm buildings, predators can be exposed to poisoned baits some distance away from farmyards.

All of the mustelids sampled in these studies had survived exposure to rodenticides until they were killed by gamekeepers or road-killed. Therefore, these studies can only reflect the extent of sublethal exposure of small mustelids to rodenticides. However, anticoagulants have a cumulative effect, and several samples in McDonald *et al.*'s (1998) study were sufficiently contaminated to suggest that, were exposure to continue, fatalities were a risk. The same is true of the road-killed polecats that Sainsbury *et al.* (2018) analysed. Many had high SGAR residues present in liver samples, but it is not known if those animals would eventually have succumbed to SGAR poisoning had they not been run over. Equally, it is not known if sublethal levels of SGARS affected these animals in a way that may have contributed to their likelihood of being killed on the road or exacerbated the trauma that they sustained. It is difficult to obtain the carcasses of Stoats, Weasels, European Polecats or Pine Martens that may have died as a direct result of ingesting rodenticides. Most wild mammals keep out of sight when they are sick or dying, so they are rarely seen or found dead except when killed on the road. SGAR residues were more prevalent in female than male Stoats in McDonald *et al.'s* (1998) study. Small rodents are the most likely source of secondary exposure to rodenticides, and female Stoats eat more small rodents than males. This is also true of polecats when rabbits are scarce as they were in the 1950s and 1960s. With rabbit populations currently declining once more, secondary exposure to rodenticides may become a risk that is more severe for female polecats as well as female Stoats. Any effects of sublethal exposure on female reproductive output may, therefore, be of great significance for both of these species with unknown impacts at the population level.

Aside from any negative effects of secondary exposure from eating contaminated prey, the widespread and indiscriminate use of rodenticides around farmland, fields and hedgerows will inevitably reduce the prey base of many predators. Limited food supply is a major factor affecting reproductive success and juvenile survival in small mustelids and can cause local population declines, particularly for Stoats and Weasels.

Pine Martens should be less likely to come into contact with contaminated prey by virtue of their preference for wooded habitats and general avoidance of areas with high human disturbance. However, in this species too, surprising levels of anticoagulant rodenticides have been reported from studies elsewhere in Europe. In France, vole populations are an agricultural pest during periodic outbreaks and are a target of rodent control. A study in the Pyrenees found residues of first- and second-generation anticoagulant rodenticides in liver samples from 10 mammalian predators of voles, including Pine Marten, Stone Marten and Weasel. Red Fox and Stone Marten, both species with a high tolerance for human disturbance, had the most exposure, but even Pine Martens had detectable levels of one or more AR in more than 40 per cent of samples (Lestrade *et al.* 2021). In Finland, a similar study detected at least one AR in 100 per cent of Pine Martens and Weasels and almost 70 per cent of Stoats examined (Koivisto *et al.* 2016). In Denmark, ARs have been found in 99 per cent of Stone Martens, 94 per cent of European Polecats, 97 per cent of Stoats and 95 per cent of Weasels (Elmeros *et al.* 2011), highlighting yet another of the risks faced by these species in our human-dominated landscapes.

CHAPTER 10

Of Mice and Mustelids: Predator and Prey Interactions

There is often a perception that predators, like the small mustelids, have a negative impact on prey species, some of which are of conservation concern or have economic importance for humans. This can lead to conflicts. Dietary analysis in conjunction with population studies of both predators and prey can determine if or to what extent this is an issue and therefore help to better target mitigation measures should they be required. The complicated question of whether predators limit prey populations ('top down' effect) or vice versa ('bottom up') has been debated for many years. In the past it was thought that predators predominantly took the old, sick or otherwise weaker individuals that would have died anyway (the so-called 'doomed surplus') and therefore had little impact on prey populations. This view assumed that predation mortality was *compensatory* in that it was offset by a reduction in mortality from other causes or the remaining individuals in the prey population could compensate for predation by increasing their reproduction.

Most potential prey animals, particularly the very young, die for reasons other than predation. Therefore, studies of the impact of predators have to work out whether predation adds to all these other causes of mortality or is merely a substitute for some of them. That calculation is an extremely difficult one but also depends on where and when you look and on the scale of the interactions studied (Powell 2001). The effects of predation depend on the numbers and behaviour of both predators and prey. It is rare that predation occurs evenly through a prey population: it may be particularly high for some age groups, sexes or in certain areas, and it may also vary through time. Nonetheless, there is now increasing evidence that under certain circumstances, predation can be *additive*, resulting in an increase in overall mortality, and that predators can limit prey numbers (Valkama *et al.* 2005). Predators can be classified as specialists, which feed on only one or a small number of prey species, or as generalists, which will take a variety of prey. Some predators may switch from being specialists to generalists depending on prey availability at different times of year or in different locations, as we have seen from dietary studies of some of the small mustelids.

Mustelids are members of the guild of predators in a variety of ecosystems. Due to their diversity in both size and degree of dietary specialisation, their influence on the dynamics of prey species through predation is thought to range from being negligible in many cases to being major drivers of the population dynamics of some prey species. An oversimplification of this process assumes that it is just about numbers: if predators remove animals from the prey population at a faster rate than they can be replaced, then the prey population will decline. But these relationships are rarely straightforward and never simple, so it is useful to understand some of the ecological principles that describe them. The number of prey that an individual predator successfully kills is a function of prey density and is known as the functional response. As an example, for Weasels this would equate to the increase in the number of voles taken per Weasel as the density of voles increased. The functional response is affected by the availability of preferred (vole) and alternative prey, as well as (intraspecific) competition from other Weasels and (interspecific) competition from other predators. The simplest is the type I functional response in which this relationship is linear. The number of prey eaten per predator increases indefinitely as prey density increases. This assumes that the predator essentially acts like an automaton, capturing and consuming prey at a constant rate. Examples of this are seen in some aquatic filter feeders which filter water at a fairly constant rate. As the density of food or prey in the water increases, so does their consumption of it, with no increase in effort. This is rarely the case for mammalian predators, few of which are efficient prey harvesting machines, even those as supremely well adapted as Weasels. It is very difficult to measure the efficiency of predators in the wild, but in fact one study of Weasels hunting rodents found that fewer than half of all attempted kills were successful (Jedrzejewski *et al.* 1992). The most frequently observed functional response for mammalian carnivores and their prey is what is called the type II response. This describes a relationship where consumption rate increases with prey density until it reaches a plateau beyond which it remains constant, irrespective of further increases in prey density. An explanation for this is that it takes a certain amount of time to find and then to catch, kill, eat and digest each prey item (known as 'handling time'). As prey density increases, it becomes easier for a predator to find; nonetheless, it still takes the same amount of time to catch,

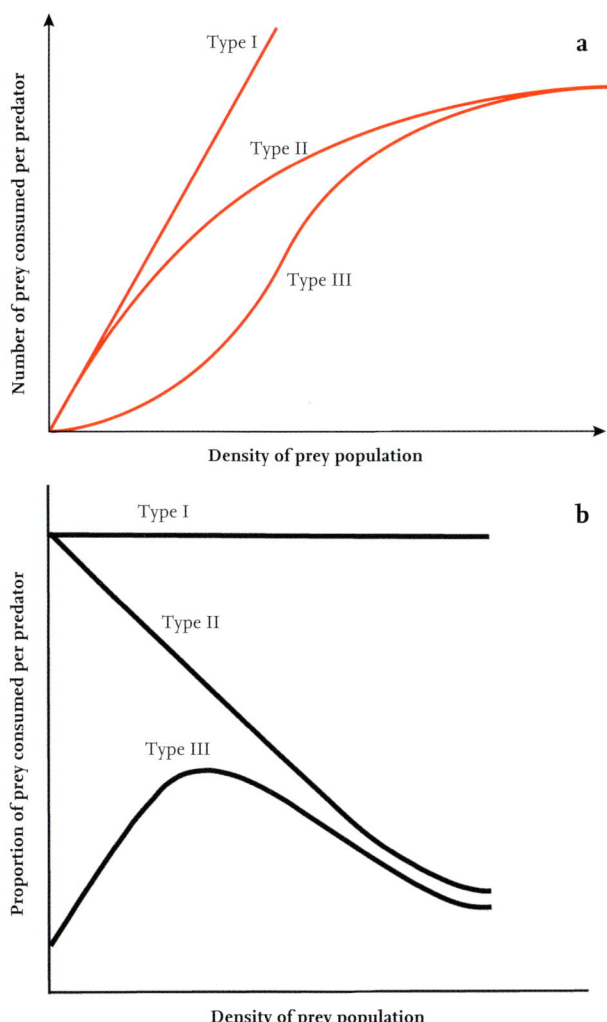

FIG 141. (a) The three types of functional response identified by Holling (1959a). (b) Diagram illustrating the *proportion* of prey consumed per predator as prey density increases for the three types of functional response. In the type I response, each predator consumes a constant proportion of the prey population, but with a type II response, predation rate decreases with increasing prey density. In the type III response, predation rate as a proportion of the prey population peaks at intermediate density and then declines.

kill and consume. Consumption rate therefore reaches a plateau determined by the maximum number of prey items that can be 'handled' per unit of time. Predators can also show a third type of functional response (type III) where capture rates are low when prey density is low, but then increase at intermediate prey density before reaching a plateau. Potential reasons for this type of response can include patchily distributed refuge habitats in which prey are able to reduce their risk of predation in some places. However, as prey population size increases, there are not enough refuges for all prey. Predators form search images when they succeed in catching prey, which increases their hunting efficiency. At low prey abundance, the search images may not be reinforced, making them less efficient at hunting a particular prey. A third explanation is that predators make use of alternative species when their main prey is at low density and only switch back when it becomes sufficiently abundant to be profitable.

Weasels, along with other predators, can kill significant numbers of rodents and they do increase their kill rate as rodent numbers increase. However, the effects of this on rodent populations depend on a combination of the functional response of individual predators and the numerical response of the predator population. It has been shown that when rodent numbers are high, Weasels do not go hunting more frequently but they can make more kills per hunt. Sundell *et al.* (2000) demonstrated this by keeping Weasels and voles together in 0.5-ha enclosures. As the density of voles increased from 4–16 per ha, the number of voles that each Weasel killed also increased but only until it reached a maximum threshold of around three voles per day. While vole density continued to increase up to 100 voles per ha, the Weasels' kill rate remained at the threshold.

In addition to the functional response of individuals, populations of predators can respond to varying prey densities by changes in reproductive rate, mortality, emigration or immigration. This is called their numerical response. Weasels can show an impressive numerical response through increased breeding success when there are increases in vole density and therefore food. In a study in Białowieża, Jedrzejewski *et al.* (1995) observed an increase in Weasels from a pre-breeding density of around 19 per 10 km^2 to over 100 per 10 km^2 in less than three months in response to an outbreak of rodents. They found that over five years of fluctuations, rodent and Weasel numbers were positively correlated during crashes and outbreaks, with no time lag. This is not always the case, however.

The belief that predators must control prey populations is widespread and common. However, most potential prey animals are not killed by small mustelids or other predators but die for a number of other reasons before they ever encounter a Weasel, Stoat, European Polecat or Pine Marten. The ability of a predator to control prey is dependent on the predator's functional and numerical response. Both of these are influenced by a number of factors, including how the predator population behaves in response to its own densities (i.e. emigration/dispersal if density is too high, particularly for territorial animals), prey preferences and the availability of alternative prey, as well as interactions with other predators. It is important to realise that there is a difference between a study of predation and a study of what predators eat. In addition to rates of predation and relative numbers of the predator and its prey, a study of predation must also take account of other predators and competitors and the responses of both predators and prey to varying environmental conditions. Dietary studies can provide information on the food habits of a population of predators, but before drawing any conclusions about predation and its impacts, the densities of both predator and prey must be known, as well as the relative importance of other predators and prey in the same area. Unfortunately, data on the absolute density of small mustelids are very difficult to obtain, due to their elusive nature, low densities and the practical difficulties of detecting them and then calculating their numbers over meaningfully sized areas (as detailed in Chapter 8). Absolute estimates of density are rare, but measures of population abundance repeated over several years are even rarer. Most of what we know about the population dynamics of Stoats, Weasels, European Polecats and Pine Martens has been derived from indices of abundance.

There are a number of ways in which some of the problems of studying the impact of predation by small mustelids such as Weasels and Stoats on their prey have been tackled. The first uses the known daily food requirements of individual animals to estimate how much loss would be accounted for by them eating prey at that rate, or how much loss a given population could impose and how that compares with the amount that would be necessary to reduce prey of a known density by a given amount. Published daily food requirements of some species are shown in Table 13.

An assumption of this method is that small mustelids kill a certain, average number of prey per day, but this is not the case. Small mustelids are adaptable and intelligent predators that kill as many prey as they can. Depending on circumstances, this is sometimes more and sometimes less than they need.

TABLE 13. Daily food requirements (DFRs) of American Mink *Neovison vison*, European Polecat *Mustela putorius*, Stoat *Mustela erminea* and Weasel *Mustela nivalis*.

	DFR (males)	DFR (females)
American Mink	235 g	150 g
European Polecat	100 g[1]	–
Stoat	57 g	33 g[2]
Weasel	36 g	22 g[3]

[1]Data from a captive European Polecat, sex not specified (Blandford 1987); [2]female Stoat DFR increases 2–5.75 times during lactation (Erlinge 1983, Corbet & Harris 1991); [3]female Weasel DFR increases to 180–200 per cent of normal during 4–12 weeks of lactation (Corbet & Harris 1991).
Data from Dyczkowski & Yalden (1998).

Another method is to count the number of prey alive and then count how many of them have been removed. This is highly problematic and susceptible to huge errors of calculation. In one early study, this method resulted in an estimate that Weasels had eaten over three times the number of voles that were present (Golley 1960). It can also result in underestimation from analyses of diet, because small mustelids such as Weasels will sometimes kill and cache a higher number of voles than they eat.

The simplest method is to remove Weasels, Stoats, European Polecats and/or Pine Martens from an area and monitor what happens to the prey populations. This assumes, however, that any subsequent change in prey numbers must have been caused by removal of the predator. Unfortunately, unless the experiment is very carefully designed to control for other variables, it is impossible to eliminate all other possible explanations. As an example, predation might only reduce the number of young produced, whereas adult density is controlled by something else. This was seen in one study in Finland where, following the removal of Stoats and other predators from sites there, both brood size and survival of young grouse improved. However, there was no clear benefit to adult grouse (Kauhala *et al.* 2000). Models can be used to simulate dynamics of predators and prey under a range of scenarios, but these are inevitably simplified versions of the systems that they represent, and their realism is proportional to the amount of data about real animals and populations that they incorporate. All of these methods have been used, often with a combined approach which has advanced our understanding of small mustelids and their prey.

WEASELS

Weasels have evolved to be supreme rodent predators, but paradoxically, the close and finely balanced relationship between Weasels and voles, their preferred prey, has the potential to cause strong population fluctuations in both species. Research into the roles of Weasels in cyclic rodent dynamics has provided more information about how and when predation can have a strong impact on prey populations. It has been suggested that prey population cycles can be a result of predation by specialist predators like the Weasel and this has been studied in several countries, including Britain. Specialist predators are more likely than generalists to cause strong, cyclic fluctuations in prey populations because they are less able to switch to alternative prey when prey populations decline. The many studies in the published literature show that the effects of predation by Weasels are not the same in all situations. In northern boreal forests and

FIG 142. Weasel *Mustela nivalis* with a vole.

BELOW LEFT:
FIG 143. Stoats *Mustela erminea* (and Weasels *Mustela nivalis*) are also prey for larger predators such as this Buzzard *Buteo buteo*.

grasslands where, for most of the year, Weasels have almost exclusive access to small rodents in their burrows and under prolonged snow cover, Weasel predation can have a considerable impact on prey numbers. In contrast, further south in temperate forests and farmland, Weasels are part of a much broader community of predators hunting a variety of prey species. Under these conditions, predation from Weasels typically has much less impact on prey populations.

Many (though not all) northern populations of small rodents, particularly *Microtus* voles and lemmings, fluctuate widely between years. These fluctuations typically follow a three- to five-year pattern, or 'cycle'. Like other species, vole and lemming populations are affected by the availability of food, the weather and disease, as well as predation. It can be difficult therefore to distinguish between factors that might cause cyclic fluctuations. A lot of field experiments, observations and modelling of cyclic rodent populations have been used to test hypotheses about the role that predation might play in these cyclic population dynamics. In the tundra and conifer forests of the far north, Weasels specialise in hunting the different species of voles and lemmings that also occupy these habitats. Voles and lemmings are also hunted by other cold-climate predators, such as martens, foxes, hawks and owls. Many northern populations of these rodent species display cyclic fluctuations in abundance, peaking every three to four years. One way in which this is explained is that in the summers when rodent numbers are increasing and when they are at their peak density, Weasels and other predators are able to take advantage of the glut in food and maximise their breeding success. The combined functional and numerical responses of all these predators takes an increased toll on small mammals, but the predators are not able to outbreed voles and lemmings at that stage, so the rodents can more than replace their losses. During the summer, whilst these other predators are only able to hunt voles and lemmings above ground, the small northern Weasels can follow them through their networks of runways and into their nests. In winter, when the snow arrives, the rodents are no longer breeding, and their food is getting scarcer. The migratory raptors move south, and small rodents are protected from martens and foxes by the snow cover, but this is not a barrier for Weasels. They are able to carry on hunting the diminishing numbers of rodents with deadly efficiency, preventing the remaining individuals from rebuilding their populations. Extended predation keeps rodent populations at low density for several seasons, which allows the vegetation to recover, but as rodents become scarce, Weasel numbers decline through starvation or, if they are able to, by moving elsewhere. This gives the surviving rodents an opportunity to build up their numbers once more. The recovery of their food supplies and a decline in Weasels provide conditions for the whole cycle to begin again. In these cyclic vole populations, it has been observed that the body size of voles varies at different

FIG 144. Weasels *Mustela nivalis* are able to continue hunting small mammal prey under the cover of snow.

FIG 145. High density of burrows and runs observed in Wales during a peak in vole numbers.

densities, being smaller in the decline and low-density phases and largest during peaks in their population cycle. A possible explanation of this (known as the 'Chitty effect', after Dennis Chitty who first described it) is that only the smallest voles survive population crashes because they can escape predation in burrows which even the smallest Weasels are unable to access (Boonstra & Krebs 1979, Sundell & Norrdahl 2002).

Further south, more temperate habitats are home to a much wider range of both predators and prey. Winters can be milder with few (or no) days of lying snow providing cover for voles. This means that not only do Weasels have competition from other larger predators but they must also take care that they do not become the prey. In many temperate habitats, vole populations fluctuate throughout the year, increasing during breeding seasons and then decreasing in between. These seasonal fluctuations are rarely as dramatic as those seen in the far north; nonetheless, they can still make a difference to the hunting prospects of an individual Weasel.

In Kielder Forest in Britain, there has been a lot of research on the cyclical dynamics of voles and their predators, including Weasels (Lambin 2017). Kielder is a 600 km² plantation forest made up mainly of Sitka Spruce *Picea sitchensis*, established in the 1930s. Long-term studies of Field Vole populations here have shown that cyclical fluctuations in their populations drive many ecosystem processes. The Field Vole habitat in Kielder consists of small, fragmented patches of between 5 and 200 ha each. These are areas that have been clear-felled but then within 2–5 years are colonised by grasses. They are connected to each other via a network of grassy river margins and road verges. Shrew species also use these grassland habitats, as well as Bank Voles that use the edges. Bank Vole abundance also fluctuates in response to tree seed production and masting, but their population peaks are not in synchrony with those of Field Voles. In dense Sitka forest, there is virtually no understorey vegetation, and this habitat is only used by Wood Mice at low density. Field Vole surveys have been carried out 2–3 times each year since 1985 based on the presence/absence of vole feeding remains in 25 quadrats per grassland patch. The results were calibrated against density estimates derived from capture-mark-release-recapture. These show that vole dynamics in Kielder fluctuate with a peak every 3–4 years. There appears to have been a clear dampening in the amplitude of fluctuations since the late 1990s, although spring densities have continued to follow a cyclical pattern. Population crashes took place in the summers of 1985, 1988, 1991, 1995 and 2015. Nonetheless, in all years there are still some voles present in Kielder Forest, even though not all grassland patches may be occupied (Lambin 2017). There is also some degree of spatial asynchrony in the dynamics of Kielder Field Voles that means, when numbers are averaged across multiple patches, densities overall during the lowest phase of the vole cycle are always significantly higher than they are in Fennoscandia. A time lag in the numerical response (increase or decrease) of a predator is a necessary condition for it to drive population cycles in its prey. In Kielder Forest, Weasels show a rapid, non-delayed numerical response to changes in vole densities. Graham and Lambin (2002) carried out a five-year removal experiment in Kielder during which they measured the response of Field Vole populations at three sites where Weasels were removed continuously and three control sites where Weasels were present. The removal sites were unfenced, and so continual immigration meant that Weasels were never completely removed from these areas, but continuous trapping meant that they were kept at very low levels. The results showed no differences in patterns of Field Vole density between sites with and without Weasels. From this, Graham and Lambin rejected the hypothesis that a specialist predator is necessary to explain vole cycles. This sparked a lively debate in the scientific literature, with some agreeing that the evidence from Kielder disproved the specialist predator hypothesis (Oli 2003), while others disputed it (Korpimäki et al. 2005), arguing that the vole population fluctuations studied in Kielder were fundamentally different from those seen in Fennoscandia and elsewhere in Europe and that as only Weasels had been removed in Graham and Lambin's experiment, all the remaining predators could have contributed to vole mortality rates. Lambin and Graham (2003) also pointed out that they had been testing only the specialist predator (Weasel) hypothesis and not a more complex model with multifactor interactions.

This was also the case in Korpimäki and Norrdahl's (1998) earlier experiments in western Finland when, during a vole decline in 1992, they removed around

FIG 146. Field Vole *Microtus agrestis* numbers can be very high in peak years.

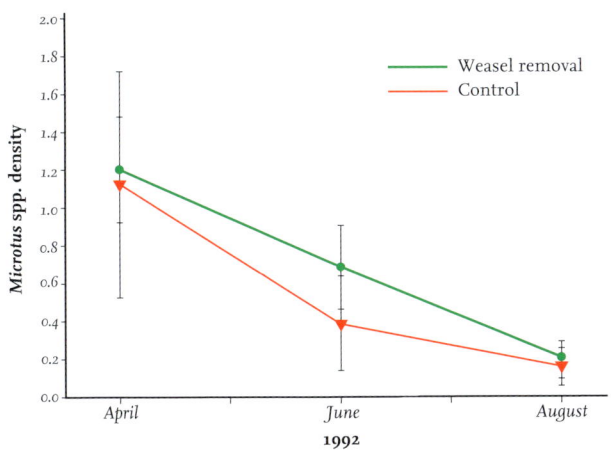

FIG 147. The 1992 decline in the *Microtus* spp. vole population on control plots and on those from which about 70 per cent of Weasels *Mustela nivalis* were removed (Korpimäki & Norrdahl 1998). Redrawn from Krebs (2013).

70 per cent of the Weasels from six paired (control and removal) sites with no effect on the rate of decline of *Microtus* voles.

In Kielder Forest, mobile predators like Weasels can escape starvation after depleting their prey in one area by moving, sometimes several kilometres, to other sites where prey is more abundant. Consequently, Weasel predation does not result in total prey depletion or spatial synchronisation. The emerging picture from the Kielder studies suggests that, here at least, Weasels coexist with cyclically fluctuating Field Voles but have no noticeable impact on their spatial or temporal dynamics. Korpimäki *et al.*'s (2002) experiments showed that in Finland, removing all of the main mammalian and avian vole predators through a population cycle can alter the pattern of population change and increase autumn density. However, other populations of rodents in the far north behave differently, with no cyclicity evident in Norway, close to the coast in regions with mild winters and no permanent snow cover (Strann *et al.* 2002).

One suggestion is that food supply, specifically ground vegetation, is what determines rodent numbers at least as much as predation. This is affected by the length of winter snow cover but is not closely correlated with latitude. Under the dense canopy of a temperate forest, ground vegetation can be as sparse as it is in the tundra, whereas in open grasslands of the temperate zone, it is high. From their studies in Poland, Jędrzejewski & Jędrzejewska (1996) concluded that even when predation accounts for much of the observed rodent mortality, it is not what drives annual or seasonal population cycles in temperate forest. This is determined by winter food supply: in 'mast' years, a sudden, massive drop of seeds in autumn, lying on the ground over winter, results in population increases of both rodents and their predators. In Białowieża Forest, almost all winter mortality of the resident Bank Vole and Yellow-necked Mouse populations is due to predators, including Weasels, but it is variation in tree seed-fall between years that is the cause of dramatic rodent population fluctuations (Jędrzejewski & Jędrzejewska 1996). From their research, Jędrzejewski and Jędrzejewska concluded that rather than predation by vole specialists like Weasels, it is the standing crop of ground vegetation that shapes northern vole cycles. It seems that the potential role of Weasels in driving fluctuations of prey is moderated by other factors and, while predators can modify the shape of a prey population density curve, on their own they cannot generate population fluctuations.

In Britain, the total annual production of Field Voles in England is estimated to be between 677,000 and 982,000 (Dyczkowski & Yalden 1998). This is predominantly from meadow and arable habitats. If none died or were predated, we would very soon be overrun with voles. By extrapolating from a range of data on dietary composition, Dyczkowski and Yalden (1998) calculated the total annual consumption by all predators combined at approximately 980,000 voles. They estimated that two specialist predators, Weasels and Common Kestrels *Falco tinnunculus*, plus two generalist predators, Red Foxes and feral cats, account for 85 per cent of vole predation, with a further 10 predator species dividing the other 15 per cent between them. Around 22 per cent of the voles lost to predators are probably taken by Weasels. During most years, it is thought that the total production of young voles will be roughly matched by total predation, explaining the relatively small changes in the Field Vole populations recorded from year to year in these habitats.

In the real world, the population dynamics of Weasels are almost entirely controlled by the density of prey, whereas the density of prey is affected by many things other than predation, such as food supplies and social behaviour. The rate of reproduction of Weasels is slower than that of any of their rodent prey, and there is a practical limit to the number of rodents an individual can kill in a day. The same is true for Stoats. Consequently, in natural ecological communities, it is well-nigh impossible for predation by Weasels and Stoats to reduce populations of their natural prey, which is why attempts to use them as a biological control agent for European Rabbits failed. One circumstance in which predation by Weasels and Stoats can prevent an increase or start a decrease in their prey is when it occurs at a time when rodents are declining for other reasons and have already stopped adding new recruits to their population. Then the overshoot in numbers of predators caused by the lag in their numerical response means that predation is disproportionately heavy at the time when the prey population is least able to compensate.

STOATS

Although less of a small-rodent specialist than the Weasel, Stoats have also been implicated in driving the population cycles of some species, including lemmings in northwest Canada (Maher 1967) and Greenland (Sittler 1995), as well as fossorial (burrowing) Water Voles in the Swiss Jura (Weber *et al.* 2002). At Sittler's study site in northeast Greenland, the number of lemming nests (used as an index of lemming abundance) went from nearly 3,700 in 1989–90 to just 105 the following year. During the crash, he found Stoats occupying most nests and concluded that the large numbers of newly independent young Stoats were responsible for

FIG 148. A Stoat *Mustela erminea* with a Lemming *Myopus chisticolor*, which can be an important prey for Stoats where the two species co-occur.

FIG 149. Populations of the fossorial Water Vole *Arvicola amphibius* in continental Europe undergo dramatic fluctuations.

pushing down the crashing population of lemmings. The next year, the Stoats had gone, presumably to another area with a higher density of prey. The fossorial form of the Water Vole, found in continental Europe, undergoes population cycles with densities peaking at several hundreds of individuals per ha. Debrot's (1983) studies in the Swiss Jura mountains found that both Stoat and European Polecat numbers responded to Water Vole populations with a one-year time lag. Stoats, European Polecats and Stone Martens also show a functional response to Water Voles (Lachat Feller 1983, Mermod *et al.* 1983). Stoats have lower potential rates of population increase than their prey and therefore have a strong but delayed numerical response to vole densities. This results in delayed density-dependent predation pressure on vole populations.

The effect of predation by Stoats on populations of their prey varies widely. In Britain, it has been estimated that Stoats eat more than 43 million Field Voles a year, contributing just over 4 per cent of the total annual numerical impact of predators on vole populations in Britain (Dyczkowski & Yalden 1998). This level of consumption does not appear to have an impact at either a local or national scale. These days, Stoats are not usually regarded as a serious problem for reared game such as pheasants, but they are still considered a significant predator on populations of wild game birds. On one grouse moor in Teesdale, Curlew *Numenius arquata* populations were found to suffer from high levels of Stoat predation on eggs and chicks. When a simulation model was run using the observed levels of predation, it did result in negative population growth rate (Warren & Baines 2002). However, predation by Stoats has apparently not limited the post-myxomatosis recovery of European Rabbit populations in Britain, whereas Stoat numbers were severely impacted by the reduction in rabbits as a result of myxomatosis. On just one estate in Suffolk, there was a tenfold reduction in the numbers of Stoats killed in the years following the initial outbreak of the disease (Tapper 1992). Stoats were and remain extremely reliant on rabbits (McDonald et al. 2000) and the loss of this important food source was believed to have impaired productivity and survival.

The predator–prey relationships that regulate predation rates by introduced Stoats in New Zealand are very different from those in the northern hemisphere. As discussed previously, rabbits are a significant agricultural pest in parts of New Zealand. Earlier attempts to use introduced predators as a means of biological control were unsuccessful, and now poisoning, shooting and trapping are some of the main methods used. Rabbit control may, however, indirectly have a negative effect by increasing predation rates on some shorebird species. Rabbits are a major prey item for several introduced predators, particularly Stoats, Ferrets and cats. Where rabbits have been at high density and are then reduced, prey switching by these predators may occur, resulting in increased predation on native fauna, including birds. This was seen in the 1940s after rabbit control was interrupted by the Second World War and, as a result, rabbit numbers became very high. Control resumed and intensified in the 1950s, succeeding in rapidly reducing rabbit densities. Unfortunately, the sudden decline of Black Stilts *Himantopus novaezelandiae* recorded at that time was likely a result of prey switching in response to the low rabbit numbers (Pierce 1996). Similarly, on rivers on New Zealand's South Island, nest predation rates on Banded Dotterel *Charadrius bicinctus* were higher in seasons following rabbit poisoning operations (Rebergen et al. 1998). In 1997, when rabbit haemorrhagic disease (RHD) was introduced to New Zealand, prey switching by introduced predators increasingly threatened a number of native bird species. Where RHD reduced high rabbit numbers in the Mackenzie Basin, predation rates on Banded Dotterel nests were similar to those reported immediately after rabbit poisoning and higher than in breeding seasons in which rabbit control had not been carried out. Intensive predator control at some sites resulted in increased breeding success of Banded Dotterels and Wrybills *Anarhynchus frontalis*, but it is not known what the effects of RHD will be in the long term.

For natural communities that evolved on continents or on very large islands, the predator–prey interactions that we see today are a culmination of long-term coadaptation (Sundell & Ylönen 2008). Where predators and some, but not all, of their natural prey have been translocated together to a new environment, it is likely that something akin to that natural evolved relationship will be observed. This explains why the effect of mustelids on rabbits in New Zealand has been no greater than that of their ancestors on rabbits in nineteenth-century Britain. However, when it interacts with some other factor reducing rabbit survival, the results of predation *can* become additive. In the late 1990s, cat and Ferret predation, in combination with RHD, did reduce rabbits on New Zealand's South Island more than when RHD was acting alone (Reddiex *et al.* 2002). The extent to which introduced predators regulate the abundance of introduced rabbits has been the subject of much debate, particularly among the New Zealand farming community, but the evidence points to the fact that it is usually rabbits that drive predator abundance and not the other way around (Norbury & Jones 2015).

The pattern of mustelid predation on rodents in New Zealand differs from that seen in the northern hemisphere. In New Zealand's southern beech forests, Stoats still eat most mice during the short-lived (mast) peak in mouse numbers (King 1983). However, recruitment of juvenile mice stops 8–10 months after the beech mast, at or before their highest numbers are reached in midsummer. By the time the Stoats are producing large numbers of independent young, numbers of mice are diminishing. The mice disappear quickly and would do so regardless of Stoat predation. Jones *et al.* (2011) estimated the functional responses of Stoats to variations in the numbers of mice and rats available in New Zealand ecosystems and concluded that the controlling effect of Stoats on the populations of New Zealand rodents has been overestimated. Rodents can replace themselves faster than the rate at which Stoats can remove them (up to 1.13 mice per Stoat per day and 1 rat per Stoat per 6–7 days averaged over all years). Therefore, as in the case of Ferrets and rabbits, the role of the Stoat in these rodent outbreaks is that of passenger, not driver.

The effect of a predator on its prey is apparently simple because it is a matter of numbers (added to and removed from the prey population) and relative rates of fertility and mortality on both sides. However, all of these factors are variable and affected by a whole host of things, including habitat, season and year. As a result, it is never entirely predictable. It is naive to assume that the subtraction rate imposed by predators will always outpace the multiplication rate of prey. In response to claims that Stoats and Weasels could have an impact on rabbit numbers in New Zealand in the nineteenth century, Richard Henry (a conservationist and one-time caretaker of Resolution Island) pointed out that Stoats were coming to New Zealand from the same place as the rabbits, where the two species had lived and thrived together for thousands of years. Henry concluded that the real natural enemy of any animal is one which removes its food. This is why the rabbit can be a natural enemy of the sheep, but the Stoat is not a natural enemy of the rabbit, unless rabbits are already under stress for another reason, such as drought or a poisoning campaign. At the time when

Stoats and Weasels were being released in New Zealand, rabbits there were already highly abundant, and the small mustelids were released in groups too small to achieve a rate of increase to match that of rabbits. Henry was way ahead of his time in his realisation that predators generally follow rather than control the numbers of fast-breeding prey such as rabbits or rodents.

The bitter experience in New Zealand soon showed that the fertility rates of adult rabbits and juvenile survival rates can be very high in good habitat and other conditions. Rabbit populations have a high natural rate of mortality, which additional losses can rarely exceed. This was eloquently described in a 1920's poem by Arnold Hall, an amateur naturalist and professor of English at Canterbury College, New Zealand:

Where the sheep feed, there feed I,
Depleted lands behind me lie,
Of dogs and guns I take no heed,
I only breed and breed and breed.

At traps and guns and dogs I smile,
I laugh at cats' and weasels' guile,
I frolic, nibble, frisk and feed,
But all the time I breed and breed.

They send my skin to clever folks,
Who turn me into sealskin cloaks,
A sorely hunted life I lead,
But still I breed and breed and breed.

When foes attack I cannot bite,
I have no spirit for a fight,
By other methods I succeed,
I merely breed and breed and breed.

The introduction of mustelids to New Zealand was actively encouraged when Stoats and Weasels were seen as self-perpetuating allies in the war against rabbits. Similarly, the American Mink was welcomed to Britain and Ireland to establish fur farms when these animals were seen as having an economic value. In these and other countries around the world, the profit motive overruled repeated warnings about the risks that the introduced animals might become an irreversible threat to their new environments.

PINE MARTENS

As a result of the acknowledged devastating impacts that introduced mustelids have had on their recipient ecosystems, concerns are often expressed that the recovery or reintroduction of native predators like the Pine Marten to areas from which they have been absent for some time could also lead to negative impacts on populations of their prey. Whilst it is true that introduced, non-native predators can have a devastating effect on naive prey populations, this is much less likely in the case of native predators (Salo *et al.* 2007). When predators and prey coevolve over a long period of time, prey often develop morphological or behavioural adaptations to reduce the rate of encounters with predators or increase their prospects of escape if detected (Lima & Dill 1990). However, where a prey species has already suffered significant declines as a result of habitat loss, fragmentation and other factors, its vulnerability to even slight changes in predation rate may be increased. This means that in some circumstances, the recovery of a formerly very rare native predator could have a negative impact.

Populations of many wild birds in the UK have undergone steep declines over the past 40 years (Baillie *et al.* 2009), including woodland specialists and long-distance migrants (Gregory *et al.* 2007, Hewson & Noble 2009). Probably the most vulnerable life history stage for most bird populations is the egg and nestling stage (Lima 2009). Adult birds can escape predation by flying, but they cannot move their eggs, and so their ability to compensate for predation risk by avoidance is limited once they have committed to a nest site (Cresswell 2011). It has been suggested that increasing rates of nest predation could be a possible cause of the observed declines in UK bird populations (Fuller *et al.* 2005), although studies of individual species (Siriwardena 2004, 2006) and broad-scale surveys of changes in woodland birds (Amar *et al.* 2008) do not support this.

The Pine Marten is an opportunistic, generalist predator with a diet that broadly reflects what is locally and seasonally abundant. It shows some degree of specialisation on voles and has been shown to exhibit a type II functional response (Holling 1959b) to the Bank Vole in Poland (Zalewski *et al.* 1995). There is evidence to suggest that where vole population cycles have pronounced amplitudes, their proportion in marten diet varies more widely than where they are only weakly cyclic (Goszczynski 1986). Where Field Voles are at low density in Scotland, the alternative foods recorded as being taken have been predominantly invertebrates, but also passerine birds (Balharry 1993, Bright & Smithson 1997). However, even during rodent population crashes in Białowieża, Poland, Zalewski *et al.* (1995) found that alternative prey formed a much smaller proportion of the diet than rodents. They therefore considered it unlikely that

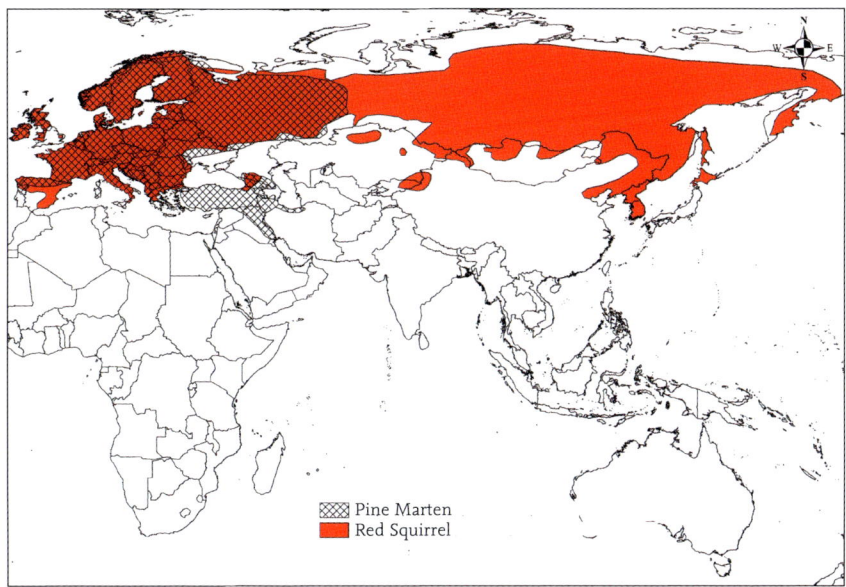

FIG 150. Native distributional ranges of Pine Marten *Martes martes* and Red Squirrel *Sciurus vulgaris*. International Union for Conservation of Nature (2008). Red List of Threatened Species (version 2015-4).

significant declines of alternative prey would be observed, even during times of rodent scarcity.

In the remaining parts of Britain where Red Squirrels still occur, there is often concern that Pine Martens will predate them. As arboreal mustelids, martens are certainly capable of pursuing and killing squirrels in the trees when they are active or by predating adults or young when they are in dreys. Across Europe, the range of the Red Squirrel is overlapped extensively by that of the Pine Marten, as shown in Figure 150.

Red Squirrels have coevolved and coexist with Pine Martens across the continent, and there is no evidence of marten species being a limiting force on sympatric, native squirrel populations (Gurnell 1987). There have been many studies of Pine Marten diet in areas where the Red Squirrel is present as a potential prey item (Yazan 1970, Marchesi *et al.* 1989, Storch *et al.* 1990, Clevenger 1993, Caryl 2008, Paterson & Skipper 2008, Caryl *et al.* 2012b), and studies of Pine Marten diet in those parts of Britain where Red Squirrels occur have found either no Red Squirrel (Caryl 2008, Paterson & Skipper 2008) or only small proportions (less than 1 per cent) of the species in marten diet (Velander 1986, Halliwell 1997).

In Spain and Switzerland, the Red Squirrel was found to occur in 0.5–2.8 per cent of Pine Marten scats (Marchesi et al. 1989, Clevenger 1993). Nevertheless, there are occasional reports of elevated occurrence of Red Squirrels in the scats of Pine Martens, associated with cyclical crashes in vole populations and noted increases in Red Squirrel populations (Storch et al. 1990, Pulliainen & Ollinmaki 1996).

If Pine Martens take rare prey species opportunistically as they encounter them, then prey vulnerability will be related to the amount of time Pine Martens spend in the same habitat as the prey. A species' vulnerability to predation by Pine Martens will also depend on a number of other factors, including its breeding biology, population density, anti-predator strategies and the availability of alternative prey. In addition to this, there will be interactions with other predators, with which Pine Martens are competing for resources as well as avoiding predation themselves. The general perception is often that there will be additional mortality for prey species if Pine Marten numbers increase. However, Pine Martens might have a negative impact on other nest predators and may consume prey that would otherwise have been eaten by other predators. Food webs are highly complex, and predator impacts are rarely as simple as generally perceived. It must be remembered that in order to maintain a stable population, on average each adult bird need only rear one chick to breeding age in its lifetime. Holt et al. (2008) found that predators had a larger impact on productivity than on breeding population and, from a conservation perspective, it is the breeding population that is important. Breeding bird populations can often compensate for increasing predation pressure by mechanisms such as reduced mortality rates as a result of reduced competition for resources, or by increased recruitment of juveniles. While there is some evidence for predators having an impact on populations of passerines, ground-nesting waders and game birds, previous analyses of UK national bird monitoring data, focussing in each case on a single predator species, could not detect any marked effects (Gooch et al. 1991, Thomson et al. 1998, Summers et al. 2004, Chamberlain et al. 2009).

Bright and Halliwell (1999) modelled the potential impacts of Pine Marten predation based on the proportion of birds in the Pine Marten's diet and the proportion of rare birds in communities. The model assumed a type I functional response by Pine Martens, where predation on rare birds would be directly proportional to bird abundance, so the proportional impact of predation was constant for low and high densities of rare birds. The model was based on a high Pine Marten population density and therefore represented a worst-case scenario. The results of the model suggested that Pine Martens would be very unlikely to kill more than 0.8 individuals of a rare bird species per km^2 per year. This would probably not have a significant impact on rare bird populations, even those that were already declining for other reasons. If Pine Marten predation is in direct proportion to bird abundance, then significant impacts are most likely on commoner species such as Blackbird *Turdus merula*. However, the results of Bright and Halliwell's model showed that for there to be any impact on Blackbirds, birds as a whole would need to constitute more than 30 per cent of Pine Marten diet, which is rarely the case. Even under these circumstances, Pine Martens would only predate two Blackbirds per km^2 per year, not enough to have a significant impact on woodland Blackbird populations, which can number 60–100 per km^2 (Bellamy *et al.* 2000).

Human-caused habitat change is a factor that may interact with predation and its effects on prey. Reduced habitat heterogeneity may limit a prey's ability to avoid predation. There can be a high level of spatial variability in predation levels at the landscape scale as a result of heterogeneity in the physical characteristics of the landscape where predator and prey interact. It has been suggested that the declines in farmland birds seen in the UK in recent decades may be partly due to changes in habitat which have left bird species less able to effectively manage their risk of predation (Evans 2004). A lack of mature woodland in many places has resulted in reduced availability of cavities for hole-nesting birds (and other animals), which, in some cases, has promoted the use of nest boxes. Elsewhere in Europe, Pine Martens will predate natural nests of medium-sized hole-nesting birds such as Tengmalm's Owl *Aegolius funereus* and Black Woodpecker *Dryocopus martius*, but they are unable to access natural nest sites of small hole-nesting passerines. This is not the case with nest boxes, however, which are also usually placed at relatively high densities and are distinctive in appearance, increasing their detectability. Predators remember the location of nest boxes where they have found food, and it has been shown that predation by Pine Martens on Great Tit and Blue Tit nest boxes increases with the length of time that the boxes have been in place (Sorace *et al.* 2004). There are, however, simple practical measures that can be put in place to militate against this, which include placing nest boxes on less accessible branches or poles, fitting guards

FIG 151. Blue Tit *Cyanistes caeruleus* on a nest box of the type commonly used for small passerine birds.

or making boxes from predator-proof materials. Pied Flycatcher *Ficedula hypoleuca* is a cavity-nesting bird whose natural nest cavities are too small for a Pine Marten to access. However, where there are nest-box schemes for Pied Flycatcher, these might need to be modified to prevent martens from accessing them. The same is true for the Hazel Dormouse. This species is unlikely to be predated by Pine Martens in natural nest sites or while active, but dormice could be vulnerable to marten predation in a nest box, especially on cooler days when they go into torpor. Depending on the design used, some of these nest boxes can also be slightly modified to prevent a Pine Marten from opening them.

Measures can also be taken to protect game species from Pine Marten predation. An extensive study of Pine Marten diet in Scotland found that the number of Pheasants *Phasianus colchicus* taken by Pine Marten (2.9 per km^2) represented less than 1 per cent of the birds released (Halliwell 1997). This is a small proportion in comparison to other predators, but only relates to free-flying birds. Mammalian predators can cause considerable damage if they get into a Pheasant release pen. However, pens can be protected against Pine Martens and other predators with slight adjustments.

Certain bat species could be at risk from over-predation by Pine Martens in some circumstances. Many woodland bats select roost cavities with entrances too small for a Pine Marten to enter, and previous dietary analysis suggests that Pine Martens very rarely eat bats (Zalewski *et al.* 1995). That being said, bats have been found in the diet of Stone Martens in both Romania and Hungary (Romanowski & Lesinski 1991, Lanszki 2003, Lanszki *et al.* 2009), and in the underground tunnels of Nietoperek in Poland there are examples of Stone Martens preying

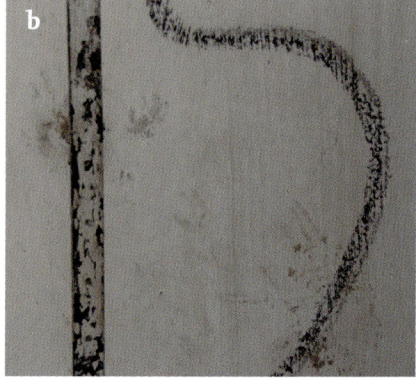

FIG 152. (a) The remains of a bat predated by Pine Marten *Martes martes* in Nietoperek, Poland, and (b) Pine Marten prints on the wall below the evidence of bat predation.

almost exclusively (80 per cent of food biomass) on bats. Recently it has also been found that some individual Pine Martens repeatedly visit these tunnels (Power 2015), and there are examples in the literature of martens frequenting places where they will find bats, like the marl pits in the Netherlands (Bekker 1988).

It seems that while the concentration of bats in colonies reduces the chance of martens finding them, once a colony has been discovered, it provides a readily available food supply. Impact on prey populations from marten predation is only likely to be a risk for species that are nationally rare but locally abundant, or colonial. In the majority of cases, the rarity of a species would mean that martens would be relatively unlikely to encounter it. However, there might be a particular risk relating to some rare species that occupy refuges in areas preferred by martens for foraging.

Black Grouse *Tetrao tetrix* has been highlighted by some as a species of potential concern. Black Grouse is one of the fastest-declining species in the UK and is now only found in the uplands of northern England, Scotland and north Wales. Black Grouse need a fine-scale mosaic of habitats containing relatively small areas of woodland, moorland and grassland. The expansion and growth of conifer plantations, changes in agriculture on marginal land, and increases in deer and sheep numbers are all factors that may have affected Black Grouse habitat (Baines 1996). In Britain, the species is now mostly associated with plantation and forest edges. Grouse populations were not previously considered to be regulated by predation (Lindström 1994), although some effects at a local scale have been reported (Marcstrom *et al.* 1988, Valkama *et al.* 2005). In Scandinavia, Black Grouse breeding success appears to be related to levels of predation on nests and chicks, which is linked to the population cycles of voles as alternative prey. In Britain, there is no evidence for cyclic breeding success in response to predators switching between alternative prey. Despite this, a study by Summers *et al.* (2004) did find between-year differences in Black Grouse breeding success which they thought may have been linked to predation. However, as it was unlikely that predator numbers had changed markedly between the two years of their study, it could have been because chicks are more susceptible to predation during and just after adverse weather conditions (Kastdalen & Wegge 1985). Summers *et al.* (2004) concluded that in years with good weather and reasonable numbers of insects, the impact of predators was probably low and breeding success was similar, regardless of predator management.

Reviews of predator control and its effects on birds have found that predator removal generally has a positive effect on productivity; however, effects on the breeding population are less clear (Côté & Sutherland 1997, Newton 1998, Holt *et al.* 2008). This means that while predator removal can fulfil the primary objective of game managers, which is to generate an artificial surplus of birds and, thereby, larger shooting bags, it may not be necessary to fulfil the rather different objective of conservation managers, which is to maximise stable breeding numbers.

The number of Black Grouse in strongholds in Dumfries and Galloway, Perthshire, Deeside and Speyside in Scotland increased in recent years in areas where Pine Martens are present. It is thought that good weather in 2010, along with major woodland initiatives and conservation efforts, resulted in a very productive breeding season. Capercaillie *Tetrao urogallus*, another forest grouse, also showed increases in some sites in Speyside in the early 2000s, despite reported rises in Pine Marten sightings since 1990 (Birks 2017). This suggests that predation by Pine Marten is not a limiting factor for these species and concurs with findings from a Scotland-wide study which showed that spatial variations in productivity of Capercaillie were related to a combination of impacts from crows and Red Foxes, but not Pine Martens (Baines *et al.* 2004). However, more recent Capercaillie declines have led to calls for urgent action, including removal of Pine Martens from Capercaillie breeding sites. At nature reserves in the Cairngorms, Scotland, managed by the RSPB, predator control was stopped in 2019 and Capercaillie counts have increased. Significant sums of money have been spent on the conservation of this species in Scotland over the past 30 years yet, in common with forest grouse populations in temperate forests across their current range, the overall trend continues to decline. Some researchers suggest that longer-term climate trends may partly explain recent declines of the species, particularly the negative effect on breeding success, but there are a number of other factors, including habitat fragmentation and increased disturbance from human recreation.

Across its range, including in Scotland and elsewhere in Europe, the Pine Marten coexists with many potentially vulnerable rare bird species. Pine Martens are territorial, have large home ranges and live at low population densities, so their impacts on rare birds are likely to be lower than commoner predators such as corvids and foxes. There are many reasons to suggest that recovering Pine Marten populations will not negatively impact other native species. It is, however, important to evaluate specific potential risks and monitor them as and when Pine Martens return to new areas.

The fact that very similar or, in some cases even, the same mustelid species can have vastly different population dynamics in what appear to us to be similar ecological contexts shows that, even now, we understand or know very little of the multiple factors which shape predator population dynamics. As a consequence, it is still very difficult for ecologists to predict when mustelids will drive fluctuations of their prey populations and when they will merely follow along behind.

CHAPTER 11

Pine Martens and Squirrels: Inference and Implications

T he interaction between Pine Martens and the American Grey Squirrel has recently become a subject of much public and media interest, particularly in Britain and Ireland. Anecdotal evidence suggested a potential relationship between Pine Marten recovery, Grey Squirrel declines and resulting Red Squirrel recovery (Carey et al. 2007). This was first noted in 1995 near Stradbally in County Laois and subsequently in 2002 in Tullamore, County Offaly, Ireland, after many decades of Grey Squirrel presence. Red Squirrels were soon being seen in both woodlands after being absent for decades. This recovery of Red Squirrels has been attributed to the corresponding recovery of Pine Martens.

FIG 153. The Grey Squirrel *Sciurus carolinensis* was introduced from North America to Britain and Ireland where it is now an established invasive non-native species.

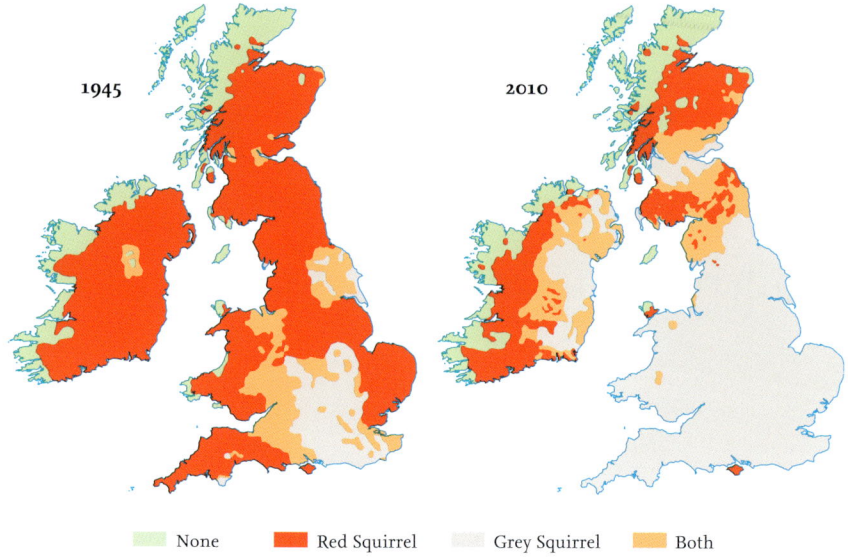

FIG 154. Distribution of Red Squirrel *Sciurus vulgaris* and Grey Squirrel *S. carolinensis* in Britain and Ireland in 1945 and 2010. Redrawn from The Red Squirrel Survival Trust.

The Grey Squirrel is not native to Europe but was introduced to Britain in 1876 and Ireland in 1911, at a time when Pine Marten populations in both countries had undergone significant declines. Grey Squirrel populations established and rapidly spread from their original points of introduction and are now found throughout much of Britain and Ireland, with the exception of the west of Ireland and the north of Scotland. It is worth noting that these are both areas from which Pine Martens were never eradicated. Nevertheless, in the majority of locations where the Grey Squirrel established, the native Red Squirrel disappeared.

INITIAL RESEARCH

An all-Ireland questionnaire survey in 2007 aimed to assess the distribution of all Pine Martens and both Red and Grey Squirrel species in the Republic and Northern Ireland (Carey *et al.* 2007). More than 1,500 postal and website responses were received from across the island. Previous distribution surveys had shown an expanding Grey Squirrel distribution, moving east, north and southwards from the initial 1911 introduction site at Castleforbes in County

FIG 155. The Red Squirrel *Sciurus vulgaris*, native to Britain and Ireland, has been severely impacted by competition and disease from Grey Squirrels *Sciurus carolinensis*.

Longford (Carey *et al.* 2007). The 2007 survey showed that the range of the Grey Squirrel in eastern Ireland had increased. This appeared to have had a negative impact on the distribution of the native Red Squirrel. The survey also revealed that Grey Squirrels were absent from some areas in the Irish midland counties of Laois, Offaly and Cavan where they had previously been common. Some survey respondents anecdotally linked the absence of Grey Squirrels to the natural recovery of the Pine Marten. The results of the survey showed that although patchily distributed, Red Squirrels were still present across widespread areas of Ireland and large vacant areas in Counties Westmeath, Meath, Louth, Kilkenny and Carlow. The area to the west of the River Shannon had mostly remained free of Grey Squirrels, despite the proximity to the introduction site in County Longford. Conversely, the Red Squirrel had persisted in the region, particularly in County Sligo and eastern County Galway and County Clare.

In light of the survey findings, researchers began to investigate the distributional relationship between Pine Martens and the two squirrel species in Ireland. Between 2009 and 2013, a study was carried out to investigate the distribution and population demographics of the three species where there had been an apparent decline in Grey Squirrels in the Irish midland counties of Laois and Offaly and a 30 km radius buffer surrounding these counties (Sheehy 2013, Sheehy & Lawton 2014). The buffer area included much of Counties Westmeath, Meath, Kildare, Kilkenny, Carlow and north Tipperary. Control sites, where the three species were present but a decline in Grey Squirrels had not yet been recorded, were located in woodlands within County Wicklow. The study found a highly significant positive correlation between the distribution (number of sightings per hectad) of Red Squirrel and Pine Marten. There was also a negative correlation between the number of sightings per hectad of Grey Squirrel and Pine Marten, although this was not statistically significant (Sheehy 2013, Sheehy & Lawton 2014). Hair tube surveys also showed a significant negative correlation between detections of Grey Squirrel and Pine Marten within sites. Pine Marten population density (estimated at 3.13 adults per km^2) was much higher in the midland study sites than the 0.1–0.87 per km^2 reported elsewhere in their range in Europe (Marchesi 1989, Balharry 1993, Balharry et al. 1996, Zalewski & Jędrzejewski 2006, Manzo et al. 2012, O'Mahony 2014), and Grey Squirrels were present there at an unusually low density despite favourable habitat (Sheehy & Lawton 2014).

A follow-up Irish survey in 2012 showed that since 2007, the Grey Squirrel had retracted in range (Lawton et al. 2015). This decline was thought to be the result of a negative association between the Grey Squirrel and the recovering Pine Marten population as shown by Sheehy and Lawton (2014). The survey also showed that Red Squirrels had recolonised some parts of Ireland from which the species had previously been recorded as absent. However, this trend was not countrywide. In other areas where Pine Martens were present, the Grey Squirrel had continued to expand its range, extending further in a northwestern, southwestern and southeasterly direction. In the 2012 survey, Grey Squirrels were recorded in 27 out of 32 counties, of which 26 also had records of Pine Martens. There were, however, seven counties with Pine Martens present where no Grey Squirrels were recorded (Lawton et al. 2015). Sheehy and Lawton concluded that Pine Marten density, rather than simply presence, may be a major influence on Grey Squirrel persistence.

FIG 156. The Pine Marten *Martes martes* and Red Squirrel *Sciurus vulgaris* have co-evolved across their native range.

FIG 157. Grey Squirrel *Sciurus carolinensis* remains found at a Pine Marten *Martes martes* den site.

The results of the most recent All-Ireland Squirrel and Pine Marten Survey showed further evidence of range contraction by Grey Squirrels, which were recorded in 38 per cent fewer hectads in 2019 than in 2012, and only recorded present in 22 counties. Conversely, the number of hectads occupied by Red Squirrels had increased slightly, and the number occupied by Pine Martens had increased by 34 per cent (Lawton et al. 2020). The increase in records of Pine Martens and Red Squirrels could in part be due to the greater reach of the 2019 survey. However, the concomitant decrease in range observed for the Grey Squirrel is compelling correlative evidence that the recovering Pine Marten population in Ireland is displacing the introduced Grey Squirrel, to the benefit of the native Red Squirrel.

The mechanism underlying the observed negative correlation between Pine Marten and Grey Squirrel distribution in the Irish midlands has still not been fully ascertained. It is not yet understood whether it is a result of direct predation, avoidance by Grey Squirrels or a more complex interaction of different factors. Being an arboreal species, Pine Martens are able to hunt prey in the trees and can therefore predate squirrels both when they are active as well as resting in their dreys. In the Irish midlands, Sheehy and Lawton (2014) showed for the first time through scat analyses that Pine Martens were predating Grey Squirrels. Evidence of predation has also been captured on camera in Wales following Vincent Wildlife Trust's Pine Marten translocations.

Both Red and Grey Squirrels were present at Sheehy's Irish study sites. Here, the percentage frequency of occurrence (number of scats in which a species was present, expressed as a percentage of the total number of scats examined) of Grey Squirrels in Pine Marten diet was significantly higher than that of Red Squirrels (15.6 per cent compared to 2.4 per cent, respectively). This was also observed in a subsequent study in Northern Ireland (12 per cent greys compared to 4.2 per cent reds) (Twining et al. 2020b) and may be due to differences in the ecology of the two squirrel species. Grey Squirrels occur at much higher densities (2–16 per ha) than Red Squirrels (0.3–1.5 per ha) (Gurnell 1987) and spend more time on the ground. Although substantial niche overlap has been found between Red and Grey Squirrels, with regard to both habitat preference and foraging behaviour, there may be some degree of niche partitioning (Wauters et al. 2002a). Grey Squirrels favour deciduous woodland (Wauters et al. 2002b) and, at 400–720 g, weigh almost twice as much as a 220–355 g adult Red Squirrel (Harris & Yalden 2008). These differences could result in the two species having different levels of vulnerability to predation by Pine Martens.

A review of the literature and a reanalysis of a new dataset confirmed that, in general, Red Squirrels only occur infrequently in Pine Marten diet across

Europe. There were insufficient studies that had investigated squirrel predation by Pine Martens where Grey Squirrels were present to conclude anything from the review, but a re-examination of data from Northern Ireland showed that Grey Squirrels were significantly more likely to occur in Pine Marten scats than Red Squirrels and, interestingly, Grey Squirrels only occurred in scats in spring and summer (Twining *et al.* 2020b). The authors suggested that Pine Martens may be predating Grey Squirrels when young are confined to the drey and hypothesised that by taking adults and young during this period they could be having a marked effect on population recruitment, supporting the suggestion that predation is driving the observed negative association between Pine Marten and Grey Squirrel distributions.

Although direct predation is the most obvious interaction to investigate between the Pine Marten and Grey Squirrel, it has also been suggested that a less direct relationship may exist through a 'landscape of fear' (Laundré *et al.* 2001, van der Merwe & Brown 2008). This proposes that presence of a predator may have detrimental impacts on its prey through indirect mechanisms such as initiation of fear-mediated and stress responses. If Grey Squirrels do perceive Pine Martens as a threat, they may alter their behaviour in order to reduce predation risk. Having to spend more time in vigilance behaviours due to the presence of a predator may have a greater negative impact on Grey Squirrel fitness compared to that of Red Squirrels. This could impact on both breeding success and survival. The difference in body mass between Red and Grey Squirrels means that the daily energy expenditure of Grey Squirrels is significantly greater than that of Red Squirrels. Consequently, to maintain a positive energy balance, Grey Squirrels need to consume more food and spend more time feeding than Red Squirrels (Bryce *et al.* 2001). Sheehy and Lawton (2014) found that Grey Squirrels in their midlands study area had significantly lower body mass and condition in a site where Pine Martens were present at high density.

NAIVETY

Another study (Sheehy 2013) investigated the possibility that Grey Squirrels may be more susceptible to predation due to prey naivety. Pine Martens have never occurred within the native range of the Grey Squirrel in North America, and nor does the native distribution of Grey Squirrels have very much overlap with the two comparable North American *Martes* species, American Marten and Fisher, as shown in Figure 159.

FIG 158. American Marten *Martes americana* (a) and Fisher *Pikania pennanti* (b) are the two relatives of the Pine Marten *Martes martes* whose native ranges overlap with that of the Grey Squirrel *Sciurus carolinensis* in North America.

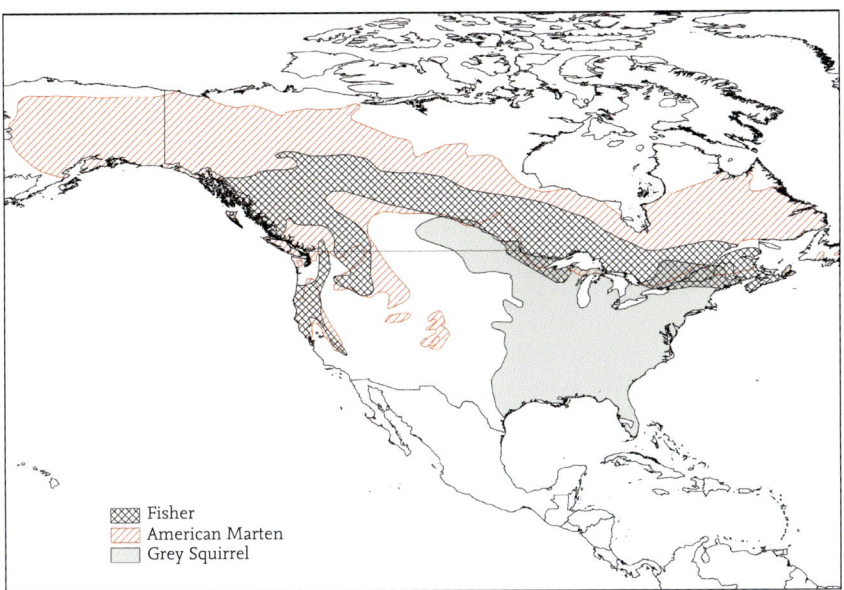

FIG 159. Native distributional ranges of Fisher *Pekania pennanti*, American Marten *Martes americana* and Grey Squirrel *Sciurus carolinensis*. International Union for Conservation of Nature (2008). Red List of Threatened Species (version 205-4).

An experiment by Sheehy (2013) explored the response of Grey Squirrels in Ireland to the scent of both a natural predator, the Red Fox, and a 'new' predator, the Pine Marten. Since Grey Squirrels and Pine Martens evolved on different continents, it was expected that the innate recognition and negative response to marten scent would not be displayed by Grey Squirrels, but that Red Fox scent, being that of a coevolved species, would result in avoidance or more wary behaviour. The results of this study were inconclusive, but a subsequent study carried out at 20 locations in Northern Ireland in 2015 demonstrated a difference in behavioural responses by Red and Grey Squirrels to Pine Marten scent. Standard flip-top squirrel feeders were set up in a number of either predominantly Grey Squirrel or Red Squirrel sites. These ranged from gardens to commercial forestry within Counties Antrim, Down and Armagh, where Pine Martens had not been recorded for 40–70 years. Feeders were baited with peanuts and monitored with trail cameras set to record video. Once visits began, behaviour at feeders was recorded for one week, after which Pine Marten scent (a scat dissolved in a water solution) was sprayed over the wooden feeders and then behaviour recorded for the following week. There was a marked decline in Red Squirrel activity at feeders after the application of marten scent, but no difference in activity by Grey Squirrels (Twining *et al.* 2020a). A suggested explanation for the observed differences in response to Pine Marten scent is that Red Squirrels, which have coevolved with Pine Martens, recognise the scent as a risk and decrease their activity accordingly to reduce the risk of predation, whereas the evolutionarily naive Grey Squirrel does not. Consequently, the lack of an appropriate anti-predator response by Grey Squirrels may mean they are more vulnerable to predation by Pine Martens than are Red Squirrels.

The first year of conservation translocations comprising an initial 20 wild-caught Pine Martens from Scotland to mid-Wales in 2015 (MacPherson 2017) provided an opportunity to investigate the effects of exposure to martens on the survival and space use of resident Grey Squirrels. Ranging data from 29 radio-collared squirrels were used to determine whether different levels of 'marten exposure', a surrogate for marten density, influenced the 'apparent survival' (recapture probability) or the ranging behaviour of the radio-collared squirrels. No effect on apparent survival was detected, although the small sample size and relatively short time frame (a mean of 16 days) may have masked any effect, if there was one. However, home range size and daily distances travelled were both positively related to marten exposure (McNicol *et al.* 2020). The authors offered two possible explanations: the observed results could be due to predation of un-collared individuals in the study area and/or Grey Squirrels in surrounding areas, influencing the ranging behaviour of the collared individuals in the study area. An alternative explanation, contrary to the findings of Twining *et al.* (2020a), is that the martens had created the aforementioned 'landscape of fear', causing the squirrels to extend their home ranges in an effort to avoid predation, and that this effect was apparent almost immediately following exposure to the novel predator. If that is the case, it is not known whether this is a short-term effect or if it is likely to persist.

Despite the widespread presence of Pine Martens in Ireland, Grey Squirrel populations continue to thrive. Grey Squirrels have spread south and west (Goldstein 2014) and achieved typically high population densities in counties such as Wicklow. The potential effects of martens observed in the Irish midlands may be dependent on marten density rather than just presence alone. It has been suggested that Pine Marten densities less than or equal to one per km^2 may have no impact on Grey Squirrel populations, and that densities need to be at least three martens per km^2 to have a negative effect on Grey Squirrels locally (Sheehy & Lawton 2015). So far, this density of Pine Martens has only been recorded in some parts of Ireland.

In 2015, a citizen science survey was carried out in Northern Ireland using camera traps at feeders baited with peanuts to investigate habitat suitability and interactions among Pine Martens, Red Squirrels and Grey Squirrels (Twining *et al.* 2021). Although the results provided more correlative evidence of Pine Martens suppressing Grey Squirrel populations with the probability of Grey Squirrel occupancy being dramatically reduced in the presence of Pine Martens, Grey Squirrels had the highest probability of occurrence in areas of human habitation, where Pine Martens had the lowest probability of occurrence. Consequently, Grey Squirrels may persist in parklands in and around city centres, and therefore these urban populations could act as a source for reinforcing Grey Squirrel populations in the wider landscape. A further two citizen science surveys were carried out in 2018 and 2020, and the results of all three surveys used to investigate whether the impact of the recovering Pine Marten population in Ireland was being mediated by habitat. Pine Martens appeared to suppress Grey Squirrels in both coniferous and broadleaf forests where both species occur, but as found in the earlier study, Pine Martens remained negatively associated with areas of human habitation (Twining *et al.* 2022). The 2015 survey showed that in the county (Fermanagh) with the highest rate of Pine Marten detections (38 per cent of sites surveyed), Grey Squirrels were detected at the fewest (2.8 per cent) sites. Conversely, in Counties Londonderry and Antrim, where Pine Martens were detected at the lowest percentage of surveyed sites (4 per cent and 8 per cent, respectively), Grey Squirrels were found at 53 per cent and 31 per cent, respectively. This is similar to the observation by Sheehy *et al.* (2014). However, this trend is not consistent, and in Counties Armagh and Down, where detection rates for Pine Marten (at 35 per cent and 24 per cent, respectively) were only slightly lower than in Fermanagh, Grey Squirrels were also found to be present at 38 per cent and 31 per cent of survey sites, respectively (MacPherson *et al.* 2016).

In 2016, a further study took place in Scotland to investigate the correlation of predator (Pine Marten) density on Grey and Red Squirrel population densities. This was hoped to clarify the predation-competition hypothesis developed from the previous Ireland study (Sheehy 2014) involving all three species. The study was carried out at sites in the Highlands, centre and Borders of Scotland with long-term, recent and very recently established marten populations, respectively. Not only did it look at the density of each species but also at how intensely Pine Martens used parts of the landscape and, in turn, how these measures related to variation in Red and Grey Squirrel occurrence. The study found that increased exposure to Pine Marten activity was correlated with suppressed density of Grey Squirrels but that there was a positive effect of Pine Marten 'connectivity' on Red Squirrels in all three regions.

COULD PINE MARTENS PROVIDE BIOLOGICAL CONTROL FOR GREY SQUIRRELS?

With these studies in mind, there does appear to be a correlation between marten density and squirrel density; however, the processes underlying this pattern are yet to be fully understood. All of these highly publicised studies have led to significant interest in using translocated or reintroduced Pine Martens as a means of biological control for Grey Squirrels. However, if the lessons of the past have taught us nothing else, we have learned that we should be very careful about managing the expectations as well as considering the potential implications of pursuing this.

Biological control is defined as 'direct human manipulation of one species [an agent or natural enemy] in order to achieve a negative impact on another species [a pest]'. For this to succeed requires certain conditions are met. Crucially, the predator (control) and prey (pest) should have matching population dynamics, capacity for dispersal and 'operational' areas. It is rare that these conditions are met but, if they are, then the specific characteristics of the control agent will determine the detail of the results. A classic example of successful biological

FIG 160. Vedalia Ladybird Beetle *Novius cardinalis* eating Cottony Cushion Scale *Icerya purchasi*.

control is that of the Vedalia Ladybird Beetle *Rodolia cardinalis* (previously *Vedalia cardinalis*) and its use in controlling the Cottony Cushion Scale *Icerya purchasi* in the late nineteenth century. The introduction of Vedalia Ladybird Beetles from Australia to California to control Cottony Cushion Scale on citrus was an immediate success (although later it was thought that the parasitoid fly *Cryptochaetum iceryae* also played a part, being more active in cooler seasons). Within 18 months, the pest was under control. At around the same time, Stoats, Weasels and Ferrets were being released in New Zealand with the hope that they would achieve the same effect on rabbits. Unfortunately, as we know, they did not. With hindsight, the differences between the two examples are obvious. Vedalia Ladybird Beetles are able to equal or even exceed the reproductive and dispersal rates of Cottony Cushion Scale insects. The flying beetles have no territorial restrictions and their aerial searching and dispersal are both quick and efficient. Scale insects remain stationary, clumped together attached to a plant, and are therefore easy prey to find and to kill, posing no risk to the predator. The beetles are able to produce more offspring immediately when numbers of scale insects increase, allowing no time for the prey population to recover. In stark contrast, neither Ferrets nor Stoats can match the breeding rate of rabbits, nor can they match their running speeds, and hunting rabbits underground is relatively slow and inefficient. Rabbits are able to hide in and escape from the multiple exits of their complex burrow systems and can rapidly reinvade an area after it has been cleared. They are alert and well able to fight back so are a relatively risky adversary for a small mustelid in comparison to other equally palatable and, in New Zealand, naive native prey, such as lizards and flightless birds. Both Stoats and Ferrets will stop to eat after killing one adult rabbit or a litter of kits and only resume hunting when they are hungry again. The rabbits carry on multiplying while the mustelids rest up rather than reproducing. Experience showed that even in the huge numbers released, there were still too few mustelid predators to affect the rabbit populations in New Zealand, mainly because they could never remove rabbits at a faster rate than they could be replaced. This is also true for Pine Martens and Grey Squirrels.

Pine Martens are slow breeders, having only one litter of 1–5 kits a year and with a reproductive cycle that includes a period of delayed implantation. For this reason, it is not possible for martens to produce a quick numerical response to rapid increases in prey. The limiting factor is food supply when they are rearing young, so a significant increase of marten numbers the year after a peak in prey/rodent abundance is also prevented, if prey abundance is declining by then. The impact of predation by martens on their prey (expressed as the number of rodents removed per unit area) has been shown not to vary much between years, regardless of rodent abundance (Zalewski *et al.* 1995). Dramatic fluctuations of marten numbers are prevented by the Pine Marten's adaptation to use alternative resources. Grey Squirrels can breed twice a year, producing litters of up to seven kits. With densities of 200–1,600 per km^2, Grey Squirrels are certainly an abundant prey for Pine Martens (and other predators) although, by comparison, in a good year, Field Vole densities equivalent to 76,500 per km^2 have been reported in Britain (Ruffino *et al.* 2018), making them an attractive alternative when they are available. Grey Squirrels are fast and agile prey but can perhaps be taken by surprise in their dreys, as suggested. In Ireland, there is a reduced prey community and Field Voles are absent, which may explain the elevated levels of marten predation on Grey Squirrels seen in the region. Other differences are that Pine Marten densities in some parts of Ireland are much higher than those reported anywhere else in their native range. Throughout Britain and continental Europe, marten densities have been estimated as being between 0.1 and 0.87 per km^2 (Marchesi 1989, Balharry 1993, Balharry *et al.* 1996, Zalewski & Jędrzejewski 2006, Manzo *et al.* 2012). Even elsewhere in Ireland, Pine Marten density is reportedly lower (0.46–2 per km^2) (Mullins *et al.* 2010, O'Mahony 2014) than in midland sites studied by Sheehy *et al.* (2014). At approximately 10 per cent, Ireland has the least overall forest land cover of any country in the Pine Marten's range, as well as extensive habitat fragmentation. Recent studies on the effects of fragmentation on the Pine Marten have suggested that intermediate levels may actually result in smaller home ranges, and therefore higher densities, such as those observed (Mergey *et al.* 2011, Caryl *et al.* 2012a). In addition, species diversity of other predators, particularly small carnivores with which Pine Martens would compete for resources, is also lower in Ireland than in the rest of the Pine Marten's range. This would enable population increases by martens in the absence of direct competition.

PINE MARTENS AS A MANAGEMENT TOOL

All of the Pine Marten translocations that have been carried out in Britain so far have Pine Marten conservation as their primary objective. However, there is hope that the Pine Marten may at least contribute to Grey Squirrel population management through the currently unknown mechanisms that appear to drive Grey Squirrel declines in its presence. Whilst a natural solution to aid in the reduction and eradication of the invasive Grey Squirrel is highly desirable, studies so far in the UK and elsewhere show that the interactions between martens and squirrels are clearly complex and influenced by a number of factors. These include the abundance and types of alternative prey and food sources, as well as habitat and the densities of both martens and squirrels. The negative correlation between Pine Martens and Grey Squirrel density observed in the midlands of Ireland and Scotland is encouraging. However, much more research is needed under a range of conditions before regarding Pine Martens as a panacea for the problems caused by the Grey Squirrel in Britain. A clearer understanding of the mechanism underlying these correlations is still required and it is possible that any negative effect of martens may only be temporary if Grey Squirrels are able to adapt over time to either cope with the presence of or evade predation by Pine Martens. The functional response of Pine Martens means that predation on Grey Squirrels is theoretically unlikely to result in their eradication. When they are at very high densities, predation losses as a proportion of the population will decrease, and when they are at low density, then it is likely that prey switching by martens to more abundant or available prey will occur. Nevertheless, in circumstances where Grey Squirrels are reduced (e.g. by control operations) and are at intermediate density, then Pine Marten predation may contribute to suppressing their numbers below a threshold at which they have a negative impact on commercial forestry and native species. Pine Martens have been unable to achieve their natural carrying capacity in large parts of Britain due to persecution in past centuries. Therefore, it will be some time before we are able to establish what densities Pine Marten populations could achieve in mainland Britain, and under what conditions, if any, they might have a negative effect on Grey Squirrels at the population level. The fact that there are numerous examples where very similar or, in some cases, even the same mustelid species can have vastly different population dynamics and interactions with prey in what appear to us to be similar ecological contexts shows how little we know of or understand the multiple factors involved.

One unwelcome consequence of the publicity around the Pine Marten's potential impact on Grey Squirrels is that there has been an increasing number of lone martens turning up in, usually, unsuitable locations at unfeasibly large distances from the nearest viable population. The assumption has to be that these are a result of well-meaning but ill-informed, not to mention unlicensed, translocations and releases. As discussed in Chapter 6, wildlife translocations are highly complex and high-risk endeavours and should not be undertaken lightly or without adequate consideration both for the welfare of the animals themselves and the potential impacts (such as disease risks) at their release site. We also know that viable Pine Marten populations are vanishingly unlikely to result from the release of these very small numbers of animals and that a single Pine Marten (or, as is often requested, 'a breeding pair', revealing a lack of knowledge about the basic ecology of these animals) will not stay put where it is released and obligingly gobble up all the Grey Squirrels in the vicinity. In fact, the most likely outcome is that the unfortunate individuals will suffer high levels of stress and be motivated to make long-distance dispersal movements on release in search of others of their species. This further increases the risk to those animals of mortality and, indeed, we are seeing an increase in reported Pine Marten road casualties in these outlying areas.

To date, trapping remains the most commonly implemented method of reducing Grey Squirrel numbers to protect remaining Red Squirrel populations and minimise tree damage. Recent translocations of Pine Martens have resulted in establishing populations in Wales and Gloucestershire. Provided translocated individuals continue to thrive, these new populations will increase and expand through reproduction and juvenile dispersal. Other Pine Marten translocation projects are currently being developed and implemented elsewhere in England, adding to this species' presence south of the border. Furthermore, the Scottish marten population continues to spread southwards, with recent records of individuals in some of the northern counties of England. With the gradual expansion and recovery of the Pine Marten population, a species which has until recently been very rare, it is now possible that these animals could be accidentally caught in traps set for Grey Squirrels, so it is important that landowners and land managers are made aware of the implications.

CHAPTER 12

Mythology, Monarchy and Mustelidae

The belief, currently held by many in Britain and Ireland, that the Pine Marten is the nemesis of the Grey Squirrel and consequently the saviour of the native Red Squirrel (not to mention commercial forestry) may or may not turn out to be true. If it does not, it will nevertheless join a host of other myths and legends, not all of them flattering, that have surrounded various members of the mustelid family for centuries, perhaps even longer. The earliest known depiction of a Weasel dates to the Paleolithic and was found, along with other cave paintings, at Niaux in southwestern France. It is thought that the Weasel may have been depicted here as an indication that it was admired for its stealth and perhaps its fur, but this is the only known example from the period.

FIG 161. Paleolithic depiction of a Weasel *Mustela nivalis* from the Niaux Cave, near Arieges, France.

Many ancient beliefs and stories about mustelids, particularly the Weasel, have some foundation in observation, however misinterpreted. Many of the Weasel's characteristic traits can be found in representations of the Weasel in the past, with even the most fantastic and bizarre claims about the animal being based on at least a fragment of truth. Many scholars thought that tame Weasels were kept as household pets in ancient Greece and Rome, although the term 'weasel' was used to refer to several of the small mustelids. I know a couple of people who have taken in orphaned Weasels and kept them as pets up to a point, but in both cases, the Weasels reached an age where they were no longer amenable to being 'domesticated'.

In Aristotle's *History of Animals* (written around 350 BCE), he helpfully gives physical descriptions of each mustelid species and, while not mentioning domesticated Weasels, he does state that the Pine Marten (*iktis*) can readily be tamed. Pliny's *Natural History* (completed *c*.77 CE) mentions two kinds of 'weasel': 'one that lives in the forest at a great distance, which the Greeks call the *iktis*, and the other which wanders about in our houses'. The description of the latter suggests that these Weasels were not truly tame or domestically bred by humans. Nevertheless, these 'house weasels' were almost certainly commensal with people, and so the Greeks and Romans would have been very familiar with the animals and their habits.

From literature, historical accounts and art, we know that the European Polecat was domesticated as the Ferret and earned its keep by protecting

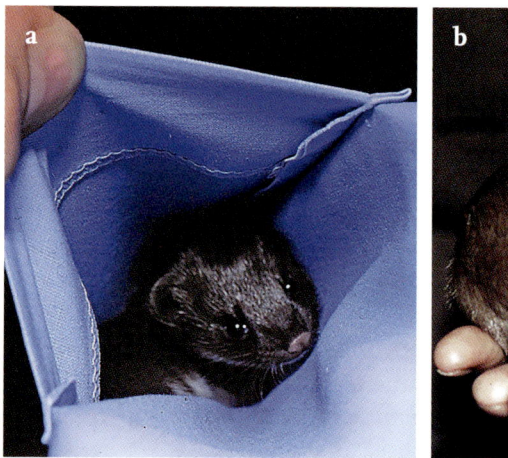
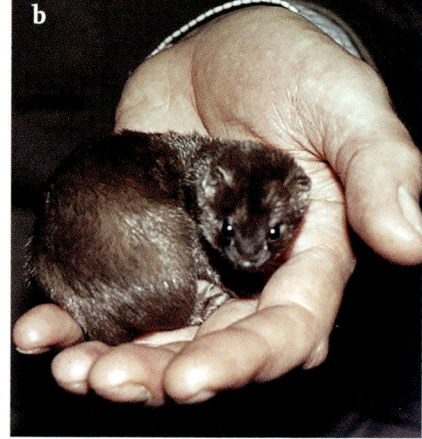

FIG 162. (a) A baby Weasel *Mustela nivalis* kept in a pocket, and (b) a Weasel kept as a pet by Clifford Ownes.

grain and other food supplies from rodents. Pliny describes how people used Ferrets to hunt rabbits by sending them 'into the holes which have many openings ... and they catch the rabbits when they are driven up out of their holes'. In the comedies of the playwright Aristophanes (450–388 BCE), 'weasels' (or more likely Ferrets) are often mentioned, with allusion to the fact that they release a terrible stench when alarmed. Both Weasels and Ferrets feature very early in Greek mythology. In the tale of Alcmene (mortal granddaughter of Perseus and Andromeda), she is seduced by Zeus disguised as her husband. That old trick! Later on, when Alcmene is in labour with Zeus's son, Heracles, the jealous goddess Hera asks the Moirae (Fates) and Eileithyia (the goddess of childbirth) to prevent the birth. Alcmene's wily friend, Galinthias, rushes in with the false report that Alcmene has already given birth, breaking the spell. To punish Galinthias for having tricked them, the Moirae turn her into a Weasel. In the Roman version recounted by Ovid, Lucina, the goddess of childbirth, tries to prevent the birth at Juno's request. Lucina is also deceived by Galinthias, who is transformed into a Weasel but then continues to live with Alcmene, whereas in the Greek version, the more vengeful Moirae condemn her to live in holes. Weasels do nest in holes and burrows, as well as in the nooks and crannies of stone walls, and this ability to slip in and out of extremely narrow places was noted by ancient observers. Pliny said that the Weasel is the only animal capable of killing a basilisk (*Harry Potter* fans take note) because it can get into the holes where the basilisk lives and kill it there.

MUSTELID MIDWIVES: WEASELS, FERTILITY AND CHILDBIRTH

Ad-Damiri (1341–1405), author of the first systematic text on Arabic zoological knowledge, wrote that the Weasel was the enemy of the crocodile because it was able to slip inside the crocodile's open mouth, get into its stomach and eat its way out from the inside. A similar talent is ascribed to the Weasel by the Ojibwe people of Wisconsin, who believed that the Weasel could kill the dreaded wendigo (a cannibalistic monster) by rushing up its anus! (Barnouw 1977) This ability to slip in and out of small spaces may be why it became a talisman for easy childbirth. Popular traditions in a number of regions associate the Weasel not only with childbirth but also with fertility. In Carinzia, Austria, it is said that a woman who is pregnant has been 'bitten by a weasel', whereas if she is unmarried and pregnant then it is, rather more coyly, said that 'a weasel breathed on her'. If we go back to the ancient world and Pliny again, he describes a revolting aid to childbirth if a woman 'drinks the liquids which flowed out of the uterus of a weasel through its genitals'. Presumably the rationale was that this could have a positive influence on the 'flow' of labour and birth. The Hopi of southwest America believed that eating the meat of a Weasel (or an American Badger *Taxidea taxus*) could help a pregnant woman deliver her baby for the reason that these animals are able to burrow into the ground if trapped by hunters and therefore can help the baby come down quickly. This is slightly more palatable and almost certainly easier to obtain than the remedy suggested by Pliny. It is easy to see how the physical characteristics and behaviour of the Weasel make it a suitable symbolic expression for childbirth. It is slightly more difficult to imagine the origin of another Roman perspective on the powers of the Weasel described by Aelian that 'the testicles of a weasel, placed upon a woman by trickery or with her consent, prevent her from becoming a mother and restrain her from intercourse'!

Other legends focus on the Weasel's maternal prowess. Female Weasels take good care of their kits, which are born hairless and helpless. Unweaned nestling Weasel kits are unable to maintain their own body temperatures and so, when their mother is away from the nest, they can go into a temporary cold torpor to conserve energy. When the adult female returns, this is reversed, and the kits rapidly 'come back to life' to huddle up close to her and use her warmth so that they can channel all of their energy into growth. This aspect of the Weasel's

FIG 163. Weasel *Mustela nivalis* mother breathing life back into her dead young. Bodleian Library (MS. Douce 308, folio 096v).

biology may have given rise to the legend about its ability to revive its dead kits, which has been further embroidered in various European folktales where the Weasel is able to bring her dead offspring back to life by making use of a miraculous herb.

By slowing their metabolism when they are left, the kits are able to maximise their growth rate when their mother is with them, and when she is away, they reduce the chances that they will run out of energy altogether before she comes back. The disadvantage is that if the mother has to spend a lot of time away from the nest hunting, the young have few opportunities to grow and may end up being permanently small or dead. Unweaned Weasels spend most of their time asleep and do not leave the nest. However, the mother may move them if there is a shortage of prey in the immediate vicinity or if the family is threatened by any interference. Then she carries each one in turn to the safety of a new den. When moving the kits, the mother Weasel carries them in her mouth in the typical carnivore way. This behaviour was well known to the ancient Greeks and Romans and is referred to by both Aristotle and Pliny. Aristotle thought that it was the explanation for the popular belief that Weasels gave birth through the mouth and wrote, 'This notion is really due to the fact that the weasel produces very tiny young ones ... and that it often carries them about in its mouth.'

Antoninus and Ovid both refer to the belief that the Weasel conceives through the ears and gives birth through the mouth, but according to Aristotle, this

FIG 164. Weasel *Mustela nivalis* with young in its mouth. Chalon-sur-Saône Public Library, Virtual Library of Medieval Manuscripts (MS. 14, folio 84v).

a

b

FIG 165. (a) and (b) Manuscript illuminations depicting Weasels *Mustela nivalis* having mouth-to-mouth intercourse and giving birth through their ears. (a) National Library of France, Manuscripts Department, Francais 14969, folio 46, and (b) Queen Mary Psalter, British Library, MS. Royal 2 B VII, folio. 112v.

legend dates back at least to the pre-Socratic philosophers. Plutarch interpreted the myth that Weasels conceive through their ears and give birth through their mouths as an allegory for language. In the book of Leviticus in the Bible, the Weasel is included in a list of animals that are impure and prohibited from being eaten, although it is not clear why it would have occurred to anybody to eat Weasels in the first place. In the *Physiologus*, a Christian text dating from the second century CE describing animals, birds, and fantastic creatures along with moral content, this myth is inverted. The law says: 'Do not eat the weasel nor what is like her. Concerning the weasel the *Physiologus* has said that she has the following nature: she conceives from the male with her mouth and when she becomes pregnant she gives birth through her ears.' The Latin versions of the *Physiologus* add further details, such as the fact that if the weasel gives birth through her right ear, the offspring will be male, and if she gives birth through the left ear, female. In *Women and Weasels: Mythologies of Birth in Ancient Greece and Rome*, Maurizio Bettini suggests that this may have been because of Christian belief that the Virgin Mary conceived through her ear (by receiving the word of God), and so an impure animal such as the Weasel could not do the same.

WEASELS AS WITCHES

In contrast to tales about the Weasel's benevolence, there are equally as many in which the animal is a malevolent creature. In some ancient versions of the story of Alcmene, the witch Hecate takes pity on Galinthias after she is changed into a Weasel and appoints her as her sacred servant. In these accounts, although the Weasel defeats the witches who are the enemies of Alcmene, she then becomes a witch herself. Aelian describes 'the land-dwelling weasel' as an animal that 'used to be a human being ... She was a witch and an enchantress, and she was terribly licentious and afflicted with abnormal sexual desires; those things have reached my ears, and also that the wrath of the goddess Hecate turned her into this accursed animal – that has not escaped my attention. May the goddess be gracious to me! I leave to others myths and their telling. But the weasel is a very malicious animal: it is known that they come around dead bodies and attack those that are not guarded, and they pull out the eyes and eat them.' This image of the Weasel-witch extends beyond the ancient world and can be found throughout European folklore. In some languages and dialects, the name for the Weasel translates as 'witch' or 'enchantress', further indication of how the Weasel has become linked to the world of magic and witchcraft. Knowing that the Weasel can be identified as a witch makes some of the other folklore more understandable, such as the fact that the Greeks considered the Weasel's cry to be sinister or disturbing, likening it to a piercing laugh or cackle. It also explains why in Greek culture it was considered a bad omen if a Weasel crossed your path. Nonetheless, there are also many regions where this is thought to be a good sign. In ancient Macedonia, if a woman suffered a headache after washing in water that was drawn overnight, she would think that a Weasel had used the water for a mirror, but it was considered unlucky to say the animal's name for fear that it would destroy clothing (Abbott 1903). In southern Greece, popular superstition has it that the Weasel is a failed bride who will jealously destroy the wedding dresses of other brides. In one of the stories in *Aesop's Fables*, a Weasel falls in love with a handsome young man and asks Aphrodite to make her human so that she can marry him. The goddess transforms the lovestruck Weasel into a beautiful girl and the young man immediately falls in love with her and they are married. However, when the couple enter the bridal chamber, Aphrodite lets a mouse loose in the room to see if the Weasel has truly changed. The young woman leaps from the bed to chase the mouse, enraging Aphrodite, who changes her back into a Weasel. This is probably the basis of a Greek tradition, where a bride's family leaves an offering of honey and perfume with the bride's trousseau to stop the Weasel from destroying it.

The Weasel's ability to get into small spaces made it a mortal enemy to poultry and it was (and by some, still is) generally thought to be an especially vicious predator that would suck the blood of any victim that was too large to be carried out of the hole through which the Weasel had entered. Weasels do kill their prey with a characteristic bite to the back of the head or the neck, leaving two puncture wounds from the canines. They may then lick any resulting blood on the prey before eating it, but adult Weasels are physically incapable of sucking blood or anything else. Nevertheless, there was also a widespread belief that they sucked the yolks from birds' eggs, as immortalised by Shakespeare. In *As You Like It*, Jacques says to Amiens, 'I can suck melancholy out of a song, as a weasel sucks eggs', and in one of Westmoreland's speeches in *Henry V*, while England is described as an eagle, he warns that 'To her unguarded nest the weasel Scot comes sneaking and so sucks her princely eggs'! Shakespeare was equally derogatory about the polecat, using it as an insult in *The Merry Wives of Windsor* (Ford: 'Out of my door, you witch, you hag you baggage, you polecat, you runyon!'). The Weasel has long been maligned in the English language, where 'weasel words' are defined as statements that are intentionally ambiguous or misleading. 'Weasel' is also used to describe someone who is conniving or duplicitous, as in the idiom 'to weasel out of something' (an obligation or an agreement). This is clearly a metaphor for the way in which Weasels are physically able to get out of almost any tight spot. Similarly, using the Weasel as a symbol for a nosy person or trying to weasel something out has probably arisen from observation of the habit of both Weasels and Stoats, especially when hunting, of standing upright on their hind legs in order to get a better view.

Ancient Greek beliefs about the Weasel's sexual proclivities seem bizarre but they persisted until as recently as postwar modern Greece, when the Weasel was still widely believed to have exorbitant sexual desires. Again, some of this reputation is probably grounded in observation of the animals' mating behaviour. Weasels are sexually mature in their first year and can mate from the end of

February through until September. In the prelude to mating, the two Weasels make excited trilling noises (which the Romans described as *drindrare*), and a female, if she is receptive and decides to accept the male's advances, leaps and dances around him. The male then seizes the female by the neck, while she assumes a passive pose. The act itself is quite vigorous as it is this that stimulates ovulation in the female, and copulation can last several hours, interspersed with occasional breaks for rest. When the couple finally decides that they have had enough, they may stop and rest, but over the next two or three days, they will repeat the entire performance several times. During this period, however, the female will also accept other male partners, resulting in litters with multiple paternity. Ovulation usually occurs on the second or third day after mating, so the later males are at an advantage. If there is an abundance of food that season, the female Weasel will go into heat a second time as early as May once her first litter is weaned. Although a second litter can be born in July or August, these late litters are often lost. In a sample of adult Weasels collected by McDonald and Harris in Britain, none of the females trapped after 27 August were lactating, suggesting that very few late-born young survive (McDonald & Harris 2002).

FIG 166. Stoat *Mustela erminea* on hind legs.

The Stoat is the most sexually precocious of the small mustelids. Not only will the female Stoat go back into heat immediately after her kits are born, but when a male gets into the nest where the female is with her litter, he will impregnate not just the mother but also the female kits, who respond with soft trills and chuckles, similar to those of adult females. As Carolyn King concludes, 'If there were a prize for the sexiest animal, the Stoat would surely win it.' It is ironic, therefore, that the Stoat (in its white winter coat of ermine) has long been seen as a symbol of moral purity and of errorless judgement. However, the poor old Weasel was the one that became a symbol of deviant female sexuality.

WEASELS IN THE RENAISSANCE

In the sixteenth century, mustelid pelts (including ermine, sable, marten and mink) often appeared in Renaissance portraits of high-ranking noblewomen. All of these furs were historically very valuable, so including them in a woman's portrait was a way of communicating her wealth as well as her high social status.

These pelts were also believed to draw fleas away from biting the skin and were sometimes known as flea furs, or *zibellini*. Many were ornamented with special metal or glass heads, and sometimes feet, adorned with precious gems, as seen in the portrait of an unknown lady attributed to William Segar (Fig. 167a) and a surviving example from the Walters Art Museum, Baltimore (Fig. 167b).

However, underlying the portrayal of these mustelids was a fascinating language of sexual symbolism (Musacchio 2001). In the Renaissance, a painting of a young bride with an ermine, sable or mink fur was thought to bring good luck with her fertility, and the brides can be seen touching their wombs while holding a mustelid fur. Examples include the portrait of Lucina Brembati, by Lorenzo Lotto (Fig. 168a), and the portrait of Antea, by Parmigianino (Fig. 168b). It was also believed that wearing the furs of these animals directly on the skin could help with childbirth (harking back to the earlier associations of Weasels). In a portrait by Paolo Veronese, the pregnant Countess da Porto is depicted with the pelt of a marten draped over her arm (Fig. 168c). Martens were thought to be particularly powerful in this respect, so when her daughter, Catherine, was thought to be pregnant, Christina de Medici sent her an ermine or marten belt which she claimed had helped her during childbirth, as described in letters between the two women.

FIG 167. (a) *Unknown Lady*, c.1598, attributed to William Segar (City of Kingston upon Hull Museum, UK), and (b) the head of a Pine Marten *Martes martes* adorned with gold, rubies, garnets and pearls, c.1550 (the Walters Art Museum, Baltimore).

FIG 168. (a) *Lucina Brembati* (c.1518), by Lorenzo Lotto, (b) *Antea* (c.1520), by Parmigianino, and (c) *Portrait of Countess Livia da Porto Thiene and her Daughter Deidamia* (1552), by Paolo Veronese.

OPPOSITE: FIG 169. *Lady with an Ermine* (c.1489–90), by Leonardo da Vinci.

MYTHOLOGY, MONARCHY AND MUSTELIDAE · 287

Leonardo da Vinci's *Lady with an Ermine* (Fig. 169) is probably the most famous Renaissance mustelid painting. The woman has been identified as Cecilia Gallerani, mistress of Lodovico Sforza of Milan. In 1490, when the portrait is believed have been painted, Cecilia was only 17 and probably pregnant, giving birth in May 1491. The depiction of the apparently live mustelid and its prominence in the painting has long been debated. It would have been

an unsuitable pet and therefore it is more likely that it should be considered symbolically. The white ermine was a symbol of purity, chastity and moderation, and in a bestiary later compiled by Leonardo, he recorded that 'the ermine, out of moderation never eats but once a day and it would rather let itself be captured by hunters than take refuge in a dirty lair, in order not to stain its purity' (Beck 1993). A similar illustration in Henry Peacham's book of emblems (1612) shows an ermine with the motto '*Cui candor morte redemptus*' (Purity bought with his own death), associating the animal's self-sacrifice with impeccable morality.

STOATS AND STATUS

Wearing ermine was restricted to royalty and high nobility, a symbol of status which persists to this day in the tradition of the ermine-trimmed coronation robes still worn by the monarch and peers. These were detailed in 1614 by a barrister, John Selden, in *Titles of Honour*. Selden's descriptions still match, almost to the last detail, the robes worn by today's peers. Each rank of the modern peerage (baron, viscount, earl, marquess and duke) has its own ceremonial dress, worn at the coronation of a sovereign and at the state opening of Parliament. Robes, which are handed down the generations, are made of crimson silk velvet with rows of ermine extending around the cape. The rows of ermine spots denote a peer's rank. A duke has four rows, a marquess three and a half, an earl three, a viscount two and a half and a baron only two.

The same code signifies the rank of a peeress but, instead of a loose cape as worn by peers, a peeress's robe is close fitting, and a small cape is worn across the shoulders. In addition to rows of spots on the cape, the peeress's symbols of rank are the width of the white fur edging to her dress and the length of the train.

In early modern England, the monarch traditionally wore the Robe of State, an ermine cloak, or mantle, during his or her coronation and when sitting in Parliament. This began in the late medieval period and confirmed the role of ermine as a material of ritual and ceremonial significance. The traditions surrounding a royal coronation have been relatively unchanged for a thousand years. However, unlike peers' robes which are handed down, a new robe is usually made for the monarch at his or her coronation. Ermine has been a feature of coronation robes for much of this time. George III's coronation robes (1761) are the oldest surviving and are now held in the court dress collection at Kensington Palace. In 1937, 50,000 ermine skins were sent from Canada to make robes for the coronation of King George VI. Queen Elizabeth II's (1953) robe followed the design guidelines of previous coronations and featured 'a six-yard train in best quality handmade purple silk velvet trimmed with best quality Canadian ermine 5" on top and underside ... complete with ermine cape and all being tailed ermine in the traditional manner'.

At his coronation in 2023, King Charles III chose to wear the ermine-trimmed robe of state originally worn by King George VI in 1937, doubtless to the relief of stoats everywhere.

Despite the long association of ermine with the English monarchy, Henry VIII appears to have favoured sable over ermine which, it is suggested, was to distinguish himself and his reign from that of his father, Henry VII, who was often portrayed in ermine robes. In 19 out of 24 portraits of him, Henry VIII is depicted wearing sable rather than ermine, although surviving warrants from Henry VII's reign show that he ordered a set of formal robes for Henry VIII when he was prince. The details of this illustrate the huge number (and expense) of ermine skins that were needed to fully line a garment. In 1498, a long ermine gown with 2,800 'powderings' (black ermine tails) was made for Prince Henry at the king's request. In doing so, Henry VII used the symbolism inherent in the ermine as a way of signalling that Prince Henry was the heir apparent and future king. The finer details of how ermine trimming was meant to be worn by different degrees at court were recorded in an unpublished handwritten manuscript titled 'Memorandum that all manner of Estates shall ware there Apparell Powdred As ys Abouesade'. This was written in the sixteenth-century hand and most likely dates from Henry VIII's reign. The manuscript includes two pages of diagrams recording the placement of ermine tails, the number of tails

FIG 170. Tailors pinning and measuring the ermine coronation robes for King George V and Queen Mary in 1911.

used in a row, and their distance permitted to each rank. For example, the queen was allowed to wear her ermine tails at quarter-inch intervals on her sleeves and gown, a baronette's wife could wear two rows an inch apart on her sleeve, whereas a knight's wife could only wear one row of ermine tails. It is not clear whether this was legally implemented or acted merely as an unpublished guide to sartorial etiquette at court.

Scarcity made ermine sufficiently rare that it was beyond the means of the majority of people, which explains its absence in the Tudor Acts of Apparel. There is no mention of ermine in the sumptuary legislation passed during the reigns of Henry VIII or his daughters, Mary I and Elizabeth I, but this is most likely because its very limited availability and affordability meant that it was not subject to misuse. The best ermine fur is only produced by Stoats in regions where there is snow on the ground for a minimum of 40 days per year, and the animals have white fur only for the duration of the snowy season. Whereas mink and Sables remain the same colour and can be harvested all year round, white ermine fur is the product of a very limited season and was therefore far less common and relatively much more expensive than most other furs. Despite a timber (40 skins) of ermine costing just £1 compared with £13 6s 8d for a timber of the worst sables, because the ermine skins were significantly smaller than sables, many more ermine skins were needed to make a mantle.

In a technique called 'powdering', actual ermine tails mixed with black lamb's wool were inserted into the slit of 'pinked' white pelts. A pelt of ermine skin was no larger than the palm of a hand, and so thousands of pelts and tails were needed to make a single mantle, resulting in a garment of enormous expense and rarity. In the *Coronation Portrait of Elizabeth I* (copied from a lost miniature from 1559), the painter emphasises the vast number of ermine pelts needed to construct Elizabeth's mantle by condensing the spacing between the black tails. The quantity of ermine tails used in the queen's mantle also signalled her royal status, as only the monarch and royal family could wear ermine trimmings with tails at such small spacing.

FIG 171. Detail from *The Ermine Portrait* of Elizabeth I (c.1585), attributed to William Segar.

The Ermine Portrait of Elizabeth I (Fig. 171), thought to have been painted by William Segar, is so called because of the ermine perched on the queen's left arm, symbolising purity and royalty. The animal even wears a crown around its neck to reinforce the message. The ermine's coat is speckled with black dots which, rather than suggesting that the painter had never seen a live Stoat (although that may be true), is probably artistic licence in which the live ermine is conflated with its already powdered fur. The ermine and its coat combine in a heraldic symbol of the Virgin Queen's royal lineage and purity.

MUSTELID FOLKLORE IN IRELAND

In Ireland's mild climate, Stoats rarely change colour in winter and so the 'pure' white ermine does not feature in Irish folklore. Irish Stoats were represented as very intelligent but also vengeful. Among their powers, Stoats were believed to understand human speech, so if you met one, it was essential to greet them politely. In County Clare, it was customary to also tip your hat or even to bow respectfully, although some people would spit and cross themselves for protection instead. Anyone who insulted a Stoat or threw stones at it could expect to lose all their chickens shortly afterwards in retaliation. Deliberately killing a Stoat was very unwise and would cause all of its relatives to descend on the murderer's house and attack them. The only way to make amends was to kill one of your own hens as compensation. The Stoat's saliva was (wrongly) believed to be poisonous, because they had an affinity with serpents (echoing beliefs elsewhere about the Weasel and its basilisk-slaying abilities). It was believed, therefore, that anyone who came in contact with Stoat spit would get blood poisoning. However, according to one popular story, the Stoat was fair-minded. Some men were mowing a field close to where a Stoat had her nest. Thinking that her young were going to be killed, the Stoat spat poison into the workmen's tea can. However, on discovering the nest, one of the workmen carefully put it safely to one side out of harm's way. The men then stopped for their tea break, but as they approached the can, the Stoat ran and knocked it over so that they would not drink the poisoned tea after all. There are various other superstitions surrounding the Irish Stoat.

FIG 172. A procession of Stoats *Mustela erminea*.

Observations of the animal occasionally moving in packs (probably a mother and kits) may be the origin of one belief that Stoats held funerals for their dead in much the same way as humans. It was said that two of the chief mourners would go in front carrying the body of the deceased between them while a procession of mourners followed behind.

Another widespread belief was that it was lucky to own a Stoat-skin purse as it would always contain money. However, it was vital that the purse should be found and not made because deliberately killing a Stoat for this purpose would bring extremely bad luck. Seeing a Stoat when setting out on a journey was also thought to be a bad omen. In County Donegal, to see a Stoat eating grass meant that rain was on the way, while to hear the sound of a Stoat screeching was a sure sign that it was going to be misty. As well as being credited with forecasting the weather, the Stoat features in some Irish folk medicine. Stoat skin was thought to cure rat bites, while the testicles of a male Stoat could be used for contraception (sound familiar?) if an Irish woman was able to castrate an unfortunate Stoat while it was still alive and then stitch the testicles into a piece of skin from a gander. Hung around her neck at all times, especially during intercourse, this was thought to prevent pregnancy, although it is more likely that it would prevent intercourse in the first place. Irish names for the Stoat include the *whutret* or *whitrick* in Ulster, believed to derive from 'white rat', and *easóg* elsewhere in Ireland, from the Stoat's eel-like shape (*eas* being Irish for 'eel'). In Cormac's glossary from the eighth century, the Stoat's Irish name came from *ní fhois*, or 'unquiet', describing its constant restless movement. Curiously, there are very few Irish place names that refer to the Stoat, though one example, Drumaness (*Droim an Easa* – ridge of the Stoat), can be found in County Down. In Irish mythology, the tale of Gaoine, a hero of the Fianna (warrior-hunter bands), tells of how, as an infant, he proved his worth by killing a Stoat. When the Stoat ran across his mouth looking for food, young Gaoine caught it in his hands, gripping it so firmly that it was later difficult to retrieve. A common notion in Irish folklore was that the Stoat was the cat of the Norsemen, brought to Ireland by them as pets (Mac Coitir 2015).

In Ireland, the Pine Marten too was likened to a cat. Indeed, its Irish name, *cat crainn*, literally translates as 'tree cat'. The 'cat' in the names of many places in Ireland, such as Craignagat (*Creag na gCat* – 'rock of the cats'), in County Antrim, and Poulnagat (*Poll na gCat* – 'hollow of the cats'), in County Westmeath, probably refers to the Pine Marten rather than to the Wildcat. There is little evidence of Wildcats in Ireland except between c.9,000 and c.3,000 years ago, most likely from human introductions (Montgomery *et al.* 2014). But there are some curious Irish beliefs about cats, some of which may have been transferred to the *cat crainn*. Cats were believed to have once been serpents and to still have a serpent's tooth which could poison anyone who was scratched or bitten by a cat. A modified version of this might be the origin of the belief in Ireland that the Pine Marten's tail concealed a claw or poisonous spike that could be used as a weapon.

In the Brehon Laws of early medieval Ireland, Pine Martens are referred to only as pets. Queen Meadhbh had a pet Pine Marten around her shoulders, which was killed by the warrior Cúchalainn with his slingshot. In a similar story to that of Gaoine and the Stoat, the Gaelic hero Fionn Mac Cumhaill's first 'heroic' act while he was still an infant was to strangle a Pine Marten when it approached the hollow tree where he lay, attracted by the smell of the meat that Fionn was eating. The Pine Marten was skinned and its pelt wrapped around Fionn as a blanket. The heroic wearing of marten skin is also mentioned in one of the oldest-surviving Welsh lullabies, 'Pais Dinogad' ('Dinogad's Smock'), in which a mother tells her son all about his father's hunting prowess, listing the animals he kills whilst wearing his marten-hide smock.

WEASEL BELIEFS AROUND THE WORLD

Elsewhere in the world, some of the folklore in which Pine Martens feature includes the belief in Romania that Pine Martens, as well as Weasels, are responsible when horses' manes are tangled because they come and plait the manes during the night (when you would think they had better things to do). In Croatia, where (until January 2023) the *kuna* (pine marten) was the unit of currency and bore the image of a Pine Marten, the animals are valued for their help with rodent control.

In the Great Lakes region of North America, the Ojibwe First Nations people have several clans, one of which is the Waabizheshi Odoodeman, or Marten Clan. Members of the Marten Clan are presumed to carry the same characteristics as those with which the animal itself is credited: lithe, ferocious and an excellent hunter. They are, therefore, the strategists, warriors and builders within their community. As well as being clan animals of the Ojibwe, American Martens also feature in the mythology of other Native American tribes. Whereas the Weasel and Wolverine are usually the villains, martens are generally portrayed as lucky spirits, brave heroes and skilled, determined hunters. The marten has a special meaning to the Mi'kmaq creation story as the first animal to sacrifice itself as food for the human race. In a book of stories intended to educate young tribal members compiled by elders of the Kootenai from western Montana, Marten learns a valuable lesson in obedience during a run-in with a bear that leaves his fur spotted.

Interestingly, there is little folklore associated with European Polecats other than the use of their name as an insult or simile for unpleasant odour. In his book about polecats, Johnny Birks (2015) recounts some of the tales he was told by elderly farmers when he was tracking polecats in Herefordshire in the 1990s. These included that they chew the ears off sheep as they lie sleeping in the fields at night and that they spring up to bite grown men on the back of the neck, causing paralysis and death. There are many more examples of the weird and wonderful tales told about the mustelid family, but, as Aelian (170–c.235) explains in the epilogue to his massive work *On the Nature of Animals*, I hope that after reading this chapter dedicated to ancient beliefs and folklore about the weasel, the reader will at least be surprised by the extraordinary quantity of unusual habits associated with the weasel [family], its symbolic meanings and connotations, and all the stories and legends in which this little animal was involved in years gone by (Aelian, translation by Scholfield 1958).

CHAPTER 13

Mustelid Mysteries: What We Know and Don't Know

Despite the small mustelids being generally described as 'common and widespread' and of Least Concern according to IUCN Red List criteria, there is still much that we do not know or understand about these enigmatic little predators. They are not alone in this. There is a relative dearth of information on many small carnivores, compared with large predators and species that are of commercial interest or feature frequently in human–wildlife conflict. As elusive as they are, small mustelids are not often studied to address general ecological questions. Nonetheless, we know that mustelids can be highly vulnerable to impacts from anthropogenic activities and many have already undergone numerical or distributional declines, as we have seen in earlier chapters. As a result of going through population 'bottlenecks' in Britain and Ireland, some species, such as the Pine Marten, have low genetic diversity in comparison to populations elsewhere (O'Reilly *et al.* 2021), which may reduce their potential for adaptation in response to future environmental change or new emerging disease risks. Stoats in Britain also show lower genetic diversity than their invasive cousins in New Zealand.

Populations of invasive non-native species generally show major reductions in their genetic diversity as a result of the demographic bottleneck they often go through during colonisation (in cases where founding populations originate from relatively small numbers of animals). However, a recent study by Veale *et al.* (2015), comparing mitochondrial genetic diversity of the Stoat in New Zealand with that in the species' native range in Britain, actually showed the opposite effect. Results showed that the introduced and invasive Stoat population in New Zealand is significantly more diverse than that of native Stoats in Britain. Five

FIG 173. The author extracting DNA from marten scat in the lab.

mitochondrial DNA haplotypes were found in the New Zealand Stoat population (from a sample of 80 New Zealand Stoats), whereas only one of those haplotypes was detected in British Stoats. (A haplotype is a set of DNA variants along a single chromosome that tend to be inherited together.) All of the New Zealand haplotypes differed by only one base pair of DNA and were therefore more similar to the remaining British haplotype than they were to those found in Stoats from mainland Europe or Ireland. This supports the theory that the New Zealand Stoat population originated solely from one or more British sources, as historical records indicate. After running simulations to investigate what severity of population decline in British Stoats would result in the observed difference in haplotype frequencies between the two populations, the authors concluded that it would require an extremely severe bottleneck of less than 1 per cent of the original population lasting for several years. These data suggest that the Stoat population in Britain underwent a very significant genetic bottleneck sometime after Stoats were introduced to New Zealand. Trapping records from the 1950s show that British Stoat populations declined significantly during this period as a result of myxomatosis in rabbits (Jefferies & Pendlebury 1968, King & Powell 2007). For the 10 years following the outbreak of this disease, the average number of Stoats caught annually on game estates decreased by 84 per cent, with some estates trapping no Stoats at all in some years, with no reduction in trapping effort (King & Moors 1979a). Throughout Britain, Stoats remained scarce during a demographic bottleneck that lasted between 10 and 15 years (Veale *et al.* 2015).

Whilst the New Zealand and British Stoat populations underwent similar population bottlenecks at different times, there are differences in the demographic histories of each, which probably influenced the severity of the genetic consequences for each. Although only a relatively small number of Stoats were introduced to New Zealand (albeit still in the thousands), there was a rapid population expansion following the initial releases. This would have minimised any potential loss of genetic diversity. In contrast, the severe decline in numbers of Stoats in Britain resulting from the loss of their main rabbit prey caused a bottleneck which lasted for several generations and was slow to recover. Under these circumstances, a loss of genetic diversity is highly likely (Allendorf *et al.* 2012). Veale *et al.*'s (2015) simulations suggest that for this demographic bottleneck to result in the observed differences between the British and New Zealand Stoat populations, the decline in Stoat numbers in Britain must have collapsed to an effective population size of only a few hundred females and lasted for several years. These population bottlenecks that some species of small mustelids have gone through in Britain and Ireland and their impacts on genetic diversity may limit their potential for adaptation in response to future environmental change or new emerging disease risks. Although we know a great deal about the continuing threats to some members of the mustelid family (Bright 2000), for others, particularly the smaller mustelids, their naturally low densities, elusive behaviour and lack of study mean that there are still significant gaps in our knowledge about them (Westra 2019). Therefore, there is a very real danger that some species could be on the verge of extinction before we even realise it.

The difficulties of surveying small mustelids (as discussed previously), as well as their highly variable home range sizes, densities and a lack of information in many cases on key predictors of these, make it problematic to extrapolate numbers across large areas. Our current estimates of population size are based on assumptions which may not be met and have low precision. In a report published in 2018, the Pine Marten population size in Britain was estimated at 3,700, comprising 39 individuals in Wales as a result of translocations (which were still in progress when the analysis was carried out), with the rest of the population in Scotland. Estimates for Scotland were extrapolated from published densities and based on the assumption that Pine Martens are present in all woodlands within the species' geographical range. However, the authors acknowledged that there is a lack of information on rates of occupancy for Pine Martens, so if this assumption is incorrect, the population size may be an overestimate (Mathews *et al.* 2018). There are similar uncertainties with the Pine Marten population estimates for Ireland. These were based on estimated area of habitat occupied within the known distribution for each jurisdiction and extrapolated from mean territory size. The estimate for the Republic of Ireland was 2,740 and for Northern Ireland 320 (O'Mahony *et al.* 2012).

Pine Martens, European Polecats, Stoats and Weasels, in common with many other solitary carnivores, are all territorial. They show intra-sexual competition, where animals actively avoid others of the same sex but will tolerate an overlap in their home range with members of the opposite sex. Typically, males have larger territories than females, and a single male's territory may overlap those of multiple females.

Pine Martens in Britain show huge variation in their mean home range sizes, which is linked to the availability of resources such as food (prey), shelter (den sites) and potential mates. For territorial animals, it is broadly thought that home range size decreases with increasing habitat quality. Smaller home ranges should be preferred if the habitat is good enough as they are less costly to defend (Moll *et al.* 2016). This has been shown in Pine Martens, where female home range sizes reduced in response to a greater abundance of rodents (Zalewski & Jędrzejewski 2006). In Scotland, the smallest home range sizes have been recorded where woodland cover is between 25 and 30 per cent of the landscape, although Pine Martens can still establish in areas with only 4.1 per cent woodland cover (Caryl *et al.* 2012a). A study from Poland concluded that approximately 2 km² of woodland is needed for a Pine Marten to establish a territory (Zalewski & Jędrzejewski 2006). Based on a maximum female home range of just under 10 km², this equates to a minimum woodland percentage of about 20 per cent. The implication is that to support a viable population of Pine Martens, at least 20 per cent of the landscape needs to be woodland comprising patches of forest that are at least 0.25 km². Better habitat provides sufficient quantity and diversity of resources within a smaller area compared with poorer-quality habitat, where animals may need to range more widely to acquire what they need. This can be seen when comparing the mean home range sizes of radio-tracked Pine Martens from a study in an upland conifer plantation in Scotland, which averaged nearly 33 km² for males and 11 km² for females, with those of animals in lowland mixed forests, which ranged from 3 km² for males and less than 1 km² for females. These are summarised in Table 14.

Male home ranges are usually between one and a half to three times larger than those of females. The bigger, heavier males require more resources than females, but also try to maximise their reproductive success by having access to as many females as possible within their home range. In continental Europe, mean home range size for Pine Martens is between 1.8 km² and 28.6 km² for males and 1.4 km² and 9.8 km² for females, across a variety of habitats from Spain to Poland (Zalewski & Jędrzejewski 2006), whereas in Ireland, one of the least forested countries in Europe, Pine Marten home range sizes are the smallest recorded throughout their geographical distribution. In the 1970s, Paddy Sleeman studied Pine Martens in Dromore, a lowland broadleaf woodland in County Clare, where the largest male home range was just 0.44 km². It has been suggested that the

FIG 174. Radio tracking is one way of collecting data on the size of Pine Marten *Martes martes* home ranges.

reason Pine Martens are able to make a living in such small territories in Ireland, compared with the rest of Europe, may be related to Ireland's mild westerly climate and the absence of potential competitors such as the Weasel, European Polecat and Stone Marten. There are also fewer small mammal species, notably a lack of native voles (Macdonald 2009). As a result, Pine Martens in Ireland appear to occupy a slightly different dietary niche than in Britain, eating fewer small mammals and significantly more earthworms, invertebrates and fruit (McDonald

TABLE 14. Mean home range sizes of Pine Martens *Martes martes* from studies in Scotland, Ireland and Poland.

Location		Habitat type	Males (km^2)	Females (km^2)
Scotland	[1]Kinlochewe	Fragmented upland conifer	23.63	8.83
	[1]Strathglass	Lowland mixed conifer	6.28	3.57
	[2]Galloway (Minnoch)	Upland spruce monoculture	32.86	9.78
	[2]Galloway (Glen Trool)	Lowland mixed conifer	8.36	4.52
	[3]Novar	Lowland mixed conifer	3.04	2.01
	[4]Morangie	Lowland mixed conifer	5.63	0.63
Ireland	[5]County Clare	Lowland broadleaf	0.42	0.2
	[6]County Leitrim	Upland conifer plantation	1.71	0.99
Poland	[7]Białowieża	Lowland broadleaf	2.23	1.49

Data from [1]Balharry (1993), [2]Bright & Smithson (1997), [3]Halliwell (1997), [4]Caryl (2008), [5]O'Sullivan (unpublished data, cited in Birks 2017), [6]O'Mahony (2014) & [7]Zalewski *et al.* (1995).

FIG 175. The author searching for a radio-collared Pine Marten *Martes martes* in its daytime den.

2002, Lynch & McCann 2007). This is possibly how they are able to make do with the very small home range sizes observed there.

Pine Martens of both sexes show seasonal variation in space use within their home ranges, which is larger in summer and much reduced in winter (Zalewski 2000). This ties in with them being more active in summer, when they are increasingly diurnal (active in daylight), compared with winter, when they are almost strictly nocturnal. A study in Poland of 14 Pine Martens radio tracked over the course of one year recorded a reduction in marten activity from an average of 12 hours a day across June and July to less than 3 hours a day in February and March. The Pine Martens also showed a strong behavioural response to avoiding the energetic demands of being active in cold weather, decreasing their activity from 13 hours a day on warm days to 2.5 hours a day when it was cold. When the animals were inactive in cold weather, females were inactive for longer than males, which is not unexpected given their smaller body size and greater susceptibility to cold (Zalewski 2000).

PINE MARTEN POPULATION DENSITIES

Related to variations in home range size, Pine Marten population densities can also vary widely. Across a range of sites in continental Europe, mean population density was 0.2 per km^2, with a maximum population density of 0.865 per km^2 (Zalewski & Jędrzejewski 2006), whereas in Scotland, densities of between 0.12 per km^2 and 0.82 per km^2 were recorded across four different populations (Halliwell

1997). As well as having the smallest home ranges, Pine Martens in Ireland have the highest recorded population densities at 0.5–3.13 per km² (Sidorovich et al. 2006, O'Mahony et al. 2017), with a widespread broad estimate across Ireland of 1.6 per km² (Virgós et al. 2012). This may be due to reduced competition, as previously discussed, in combination with warm winters. It has been shown that Pine Marten population density in continental Europe is related to mean monthly winter temperatures (November–March) and seasonality. Other known factors include the abundance of voles, especially in winter, and the availability of carrion (Brainerd 1990, Zalewski & Jędrzejewski 2006).

Although Pine Martens have always been thought of as old-forest specialists (Balestrieri et al. 2016), further research has shown that they are highly adaptable and can make use of a range of habitats (Pulliainen 1981), although forest cover of some kind seems to be a requirement of good Pine Marten habitat (Pulliainen 1981, Zalewski 2000, Brainerd & Rolstad 2002, Balestrieri et al. 2015). However, scrub, tussocky grassland and hedgerows are often excellent foraging habitats, being rich in voles and other small animals, and are often used by Pine Martens (Caryl et al. 2012a, MacPherson et al. 2014). For this same reason, forest edge and riparian woodland along watercourses are key components of good-quality Pine Marten habitat. At the time when Pine Marten populations were at their lowest in Britain, they were restricted mainly to the north and west of the country, in areas dominated by rocky, upland habitats. Conventional belief is that these Pine Martens were able to persist in these regions because they had the lowest persecution by people. But an alternative suggestion is that these rocky landscapes provide similar structural (particularly vertical) features to those found in woodland and that they might have been actively selected by Pine Martens as an alternative habitat

FIG 176. Quickly climbing a tree is one of the ways in which a Pine Marten *Martes martes* can escape from a predator such as a Red Fox *Vulpes vulpes*, and opportunities to do this decline with increasing distance from woodland cover.

(Birks & Messenger 2010). The Pine Marten's similar retreat to rocky western parts of Ireland, particularly to the craggy limestone pavements and scrub of the Burren in County Clare, where the species still thrives, shows that Pine Martens can certainly make a living in such places. However, they usually only venture a short distance away from cover of some description, be it woodland, scrub or rocky overhang. It has been shown that males will travel further, on average, into open habitats than smaller, more vulnerable females (75.1 m for males and 30.4 m for females) (Caryl *et al.* 2012a). Pine Martens do avoid some habitat types, such as agricultural land, heathland, cleared and open areas. They also tend to keep away from urban and built-up areas, although Pine Marten use of and denning in the roof spaces of occupied buildings is not unheard of, particularly in Ireland, where suitable denning sites may be scarce – all of which illustrates that Pine Martens can live in some surprising places. Our knowledge of Pine Martens in non-wooded habitat is based on very few studies, and our assumptions about them, particularly for the purposes of estimating density and population sizes, may be a considerable oversimplification.

SPACE USE BY WEASELS AND STOATS

When it comes to Weasels and Stoats, a recent Red List assessment of British mammals (Mathews & Harrower 2020) noted that both were extremely data deficient, with estimates of population size and changes being extremely uncertain. It concluded that further information is required to permit a reassessment of this species. For Weasels, a number of independent studies, documenting home range and densities, have shown that, as with Pine Martens, there are important differences among populations. Unlike Pine Martens, Weasels have small home ranges, measured in hectares rather than square kilometres, and they sometimes tolerate partial overlap with the home range of their neighbours. When and where there is an abundance of prey, Weasels can reach relatively good numbers, but in areas where or years when prey are scarce, then very few (or no) Weasels will remain. As is the case for Pine Martens, studies show that male Weasels' home ranges are larger than those of females, often being double or more in size, as can be seen in Table 15 (adapted from King & Powell 2007).

Male Weasels need more space than females partly because of their larger body size. However, male home ranges would seem to be disproportionately large compared with females than the difference in their food requirements would predict. Other mammalian predators follow this same pattern, including Pine Martens (Powell 1994). Recent research by Yamaguchi and Macdonald (2003)

TABLE 15. Summary of studies of home range sizes of Weasels *Mustela nivalis* in different habitats.

Country (years)	Habitat	Males (ha)	Females (ha)
[1]Scotland (1960–63)	Young plantation	1–5	<1
[2]England (1968–70)	Deciduous woodland	7–15	1–4
[3]England (1991–92)	Farmland	21–192	4–29
[4]Scotland (1971–73)	Farmland	9–16 (winter) 10–25 (summer)	c.7
[5]Scotland (1977–79)	Farmland	2	1
[6]Poland (1990–91)	Deciduous forest (rodents high)	24	
	Deciduous forest (rodents low)	167	

Data from [1]Lockie (1966), [2]King (1975), [3]Macdonald et al. (2004), [4]Moors (1975), [5]Pounds (1981) & [6]Jedrzejewski et al. (1995).

provided an explanation for this with American Mink, by showing that decreased availability of prey resulting from home range overlap means that male mink need extra-large home ranges. Because they have less food when overlapping territories are taken into account, the home ranges of males are not really disproportionately large. Nonetheless, it is likely that differences in home range sizes between males and females arise from a complex interaction of a range of factors, including food requirements and different prey preferences, as well as the males' need to access females and to know where they are located ahead of the breeding season.

Estimates of home range sizes may differ where different methods have been used. Earlier studies were valuable but they often used trapping, which inevitably interferes with the movements of animals around their home range. Advances in technology meant that radio telemetry could be used to collect data on the ranging behaviour of free-moving animals, but limitations on size, weight and therefore battery life of transmitters mean that these data can only be collected over the short term. Studies by Moors (1974) and Pounds (1981) of Weasels in the same study area and habitat but using trapping and radio tracking, respectively, resulted in quite different estimates of home range size, as shown in Table 15. Both studies showed that Weasels were mainly active in the thick vegetation along walls and hedgerows, avoiding agricultural fields. However, each author made different assumptions on how far Weasels would travel from the cover of stone walls and hedgerows, which resulted in different boundaries and areas being attributed to each animal.

Both King (1975) and Macdonald et al. (2004) used the same study area for their work on Weasels, but with different methods and focussing on different habitats. King's study used trapping records collected over more than two years in woodland (with between 3 and 109 recaptures per individual animal), whereas Macdonald et al. radio tracked weasels in farmland, obtaining more than 5,000 fixes over 103 days. Both studies resulted in very different estimates of Weasel home range areas, despite both using minimum convex polygons (MCPs), in which the smallest polygon is drawn around points with all interior angles less than 180 degrees. MCPs are common estimators of home range but actually just describe the extent of distribution of locations of an animal and can potentially include a lot of space not used by the animal and therefore overestimate the home range – all of which shows that the limitations of various available methods mean that there is still a lot that we don't know about these ostensibly common and familiar creatures.

There has been a lot of field research on Stoats, probably more than on any of the other small mustelids. Early studies of their home ranges in Finland, Russia and Sweden were done using snow tracking (Nyholm 1959, Vaisfeld 1972, Erlinge 1977b), as well as trapping and radio telemetry. Stoats can range widely over very large areas relative to their size. In studies of radio-tagged Stoats, their movements consist of bouts of intense activity when they are hunting for prey in a small patch and then consuming all or part of the caught prey, interspersed with quick movements to other patches to a den or rest site. Martinoli et al. (2001) found that the average length of these foraging bouts was around two hours. The remains of larger prey were brought to the den and eaten. In Sweden, Stoats studied by Sam Erlinge and Mikeal Sandell across several years established spaced-out home ranges in early autumn, around the same time that young of the year are dispersing in search of their own territories for the winter. Both males and females excluded other members of the same sex from their home range, and there was little contact between males and females over winter. Adult females hunted more in small rodent tunnels than males and would sometimes use a restricted part of an adult male's home range but they still stayed away from the male as much as possible (Erlinge 1977b). Juvenile Stoats of both sexes also avoided adult males. Young female Stoats usually settled and remained in their natal area throughout their lifetime, whereas juvenile males generally stayed there for their first autumn and winter but would then disperse more widely in spring. Female home ranges were usually well spaced out but, where they did share a common boundary, they would try to avoid any direct contact. Radio tracking showed that resident animals would regularly 'patrol' their ranges, with some areas being visited almost daily. All the resident Stoats in Erlinge's (1977) study used core parts of their home ranges more than others, spending more time hunting in areas with the highest abundance of prey. Hunting was usually in short bursts (10–45 minutes) of activity separated by longer periods of rest (3–5 hours). Not unexpectedly, the Stoats spent a longer time hunting and over larger areas when rodent numbers were low. Stoats are active both day and night but tended to be more nocturnal in winter and diurnal in summer. In spring, the established home ranges gradually broke down. Some male Stoats made long excursions of 5–6 km, whereas others stayed put but moved about more within their ranges. Some disappeared altogether (Sandell 1986). The difference in behaviour was correlated with age, body size and therefore social dominance, which has been shown to be higher in older males (Erlinge 1977a). Sandell and Liberg (1992) called these large, assertive males wandering widely in spring 'roamers', whereas the smaller, low-ranking males that stayed put were labelled as 'stayers'. Erlinge and Sandell (1986) suggested that this marked seasonal change in behaviour is because there is a shift in decisive resources at this time of year. In winter, the limiting resource is food, which is predictable and concentrated within areas sufficiently small to be defensible by a territory holder. However, in spring and early summer, the most important resource for males is receptive females with which to mate. These are widely dispersed and also unpredictable, as adult females are only in oestrus for a short period which occurs sometime between late April to the end of May. Under these circumstances where it would be impossible for a male to defend more than one female at a time, the most matings can be achieved by a dominant male if he roams in search of females over a large (but not too large) area. There is a limit to the number of females that a dominant male can keep track of. If he is overambitious and tries to find too many females, he might mistime his visits to any (or all) of them and miss potential mating opportunities. At the same time, a low-ranking 'stayer' who would probably lose possession of the resident female to a dominant roamer, should one arrive, might just be in the right place at the right time if he stays close.

These seasonal changes in the activity patterns of male and female Stoats (which are also evident in strong seasonal differences in the sex ratio of Stoats caught in traps) impact on estimates of density and home range size. It might be assumed that the huge home range sizes of New Zealand and Scottish male Stoats shown in Table 16 might be because these animals are relatively large compared with those of the same species in different regions. However, even the small male Stoats in Alberta, studied by Lisgo (1999), also maintained similarly large home ranges with an average size of 150 ha. That being said, the small Stoats found in Ontario and Switzerland do have proportionately smaller home ranges, more similar in size to those of Weasels. Nonetheless, similar patterns emerge to those seen with the other small mustelids, namely that Stoats of all sizes need larger

TABLE 16. Home range sizes of Stoats *Mustela erminea* reported in the literature.

Country/province (years)	Habitat	Males (ha)	Females (ha)
[1]Scotland (1977–79)	Farmland	254	114
[2]Sweden (1973–82)	Pasture and marshes	8–13 (winter)	2–7
[3]Switzerland (1977–80)	Alpine	8–40	2–7
[4]Finland (1952–58)	Mixed	29–40	4–17
[5]Finland (1998–99)	Subarctic birch forest, tundra	121–207 (summer)	35–66 (summer)
[6]Russia (1970–71)	Meadows / Scrub forest	11–69* / 120–124*	11–69* / 120–124*
[7]New Zealand (1990–91)	Southern beech forest, mice abundant	93 (summer/autumn)	69 (summer/autumn)
[7]New Zealand (1991–92)	Southern beech forest, mice scarce	206 (summer/autumn)	124 (summer/autumn)
[8]New Zealand (1996)	Southern beech forest, mice scarce	223 (spring)	94 (spring)
[9]New Zealand (1997–98)	Podocarp forest	256 (spring) / 145 (autumn)	44–123
[10]New Zealand (1992, 1995)	Rough grassland	66–215	32–135
[11]New Zealand (2001–2)	Braided riverbed	315 (spring) / 185 (autumn)	127 (spring) / 116 (autumn)
[12]Alberta (1995, 1996)	Mixed wood boreal forest	123–205	66–95

Data from [1]Pounds (1981), [2]Erlinge (1977b), [3]Debrot & Mermod (1983), [4]Nyholm (1959), [5]Hellstedt & Henttonen (2006), [6]Vaisfeld (1972), [7]Murphy & Dowding (1995), [8]Alterio (1998), [9]Miller *et al.* (2001), [10]Moller & Alterio (1999), [11]Dowding & Elliot (2003) & [12]Lisgo (1999).
*Both sexes combined.

home ranges when there is a scarcity of prey, the home ranges of male Stoats are larger than those of females, and neither will share their space with another of the same sex.

The total population of Stoats in Britain was estimated by Harris *et al.* (1995) to be 462,000 (comprised of 245,000 in England, 180,000 in Scotland and 37,000 in Wales). These figures were derived from expert opinion of density estimates for all habitat types combined with habitat availability. By adjusting results to reflect temporal changes in the British landscape, Mathews *et al.* (2018) produced a new estimate of 399,000 but acknowledged that this had low reliability. The density estimates were taken from Harris *et al.* (1995) but were not based on any published data for Britain because there was no literature that reported pre-breeding population, density estimates, trends, occupancy or the effect of environmental variables on relative density. A decline in specialist prey species was identified as a potential driver of population change for Stoats between 1995 and 2018. However, it is not possible to assess the significance of the apparent reduction. Further research is needed to improve the reliability of population size estimates for this species. The density estimates used were more than 10 years old and based on opinion. They did not reflect variation within habitats; for example, Stoats occur in arable land and grassland but are more likely to use boundary features such as hedgerows. It was unclear if this was accounted for in the density estimates for each habitat type. Mathews *et al.* (2018) found that there was a similar lack of published occupancy or density data on which to base population estimates for Weasels. The review by Harris *et al.* (1995) suggested a total British Weasel population of c.450,000 (with c.308,000 in England, c.106,000 in Scotland and c.36,000 in Wales) based on a ratio of Weasels to Stoats of 1:1 overall. This is thought to be extremely unreliable given that the Weasel population shows more regional variation than that of Stoats.

POLECAT POPULATION TRENDS

For European Polecat, there were four published studies that reported relevant data for Britain. Polecats can occur in most habitats, so population density estimates were area-specific, rather than habitat-specific (1.63 per km^2) (Birks & Kitchener 1999). Nevertheless, polecats are unlikely to be found in urban areas (Birks 2015), so these were removed from the total area within the distributional range for the purposes of estimating population size. This resulted in a polecat population of a little shy of 83,000 in Britain, of which 16,800 are in Wales and 66,000 are in England (Mathews *et al.* 2018). Records of true polecats in Scotland are very rare at present. Although there is considerable uncertainty about polecat population size in Britain, the lower plausible population estimate is well above the threshold for IUCN Red List criteria of population size estimated to number fewer than 250 mature individuals, and the geographical range is not highly restricted in Britain, particularly in Wales or southern and central England. However, elsewhere in Europe, declines of polecat populations are reported from several countries, although a lack of data means that in most cases trends are poorly understood. Croose *et al.* (2018) collated information and opinion from across the polecat's European range. Caution is needed when interpreting population trend assessments based on information collected over short time periods, but where trends could be identified, the polecat was known or thought to be declining in 20 out of 34 countries.

The European Polecat is listed on Appendix III of the Bern Convention on the conservation of European wildlife and natural habitats (Council of Europe 2016), which allows signatory countries the discretion to provide protection through 'appropriate and necessary legislative and administrative measures'. It is also an Annex V species under the EU Habitats Directive on the conservation of natural habitats and of wild fauna and flora, meaning that member states must ensure that their exploitation and taking in the wild is compatible with maintaining them in a favourable conservation status. Consequently, while polecats have some level of limited legal protection in countries such as Britain, elsewhere they can still be legally hunted for fur or trapped as a perceived pest species. In the IUCN Red List of species, the polecat is classed as Least Concern, but in 2014–16 this category was reviewed. Researchers from several countries in the polecat's range reported that populations were declining or suspected to be in decline, while some had insufficient data to draw conclusions about population trends. Nevertheless, the review concluded that the overall global population trend was unlikely to reach the rate of decline that would justify re-classification to Near Threatened. This would have required an overall loss of 20 per cent or more of the global population over either the previous 13 years or 3 generations, or an anticipated loss of that magnitude over the following 13 years. For now, the polecat's Red List categorisation remains as Least Concern (IUCN 2016). Thresholds have to be applied to any criteria, but this means that the polecat could decline at a globally averaged rate of 19 per cent per 13 years and would still correctly be classified as Least Concern, although most people would consider this of very great concern. During the Red List assessment, it was acknowledged that confidence in the categorisation is low, given the lack of recent and accurate information on population trends across most of the polecat's range. Widespread uncertainty of the status of polecats in many countries highlights that further action is needed. It should be noted that for some predator species, their abundance fluctuates in response to that of their small mammal prey. Therefore, in countries or regions where small mammals constitute a high proportion of polecat diet, polecat abundance may also fluctuate accordingly. Consequently, long-term data series are needed to detect true population trends. Unfortunately, there is a lack of robust monitoring methods for this species, as for other small mustelids. The national polecat surveys of Britain are carried out periodically (every ten years) and use a wide-scale, citizen scientist approach. Naturalists, members of the public and other researchers are asked to send in georeferenced records of polecat sightings and roadkill carcasses throughout the course of the year. These data are invaluable for monitoring polecat distribution as the species recolonises its former range; however, they do not provide any information on population density. The main drivers of polecat decline are likely to differ across the species' range, but the loss, degradation and fragmentation of habitat are likely to be the principal causes in many countries. This includes drainage and loss of wetlands which is thought to have led to a decline in Germany, France and, historically, in Switzerland. Hedgerows are another important habitat feature used by polecats and other mustelids. Hedgerow removal, resulting from changes in agricultural practice, is likely to have reduced landscape connectivity for polecats and many other species. Loss and degradation of habitat may also have impacted prey species with negative effects on polecats. Declines of amphibians and rabbits, both of which can make up a significant proportion of polecat diet, have been implicated in polecat population declines in some countries. Competition with invasive non-native species such as American Mink may also pose a threat to polecats. As discussed in Chapter 9, polecats can be affected by secondary exposure to anticoagulant rodenticides through eating poisoned rodent prey.

FIG 177. European Polecat *Mustela putorius* in farmland.

Polecats living in agricultural habitats prey heavily on rats around farm buildings, especially in autumn and winter, making them particularly vulnerable to rodenticide exposure (Shore *et al.* 2003). Although the polecat is still legally harvested in parts of its range, the price of polecat pelts has always been much lower than that of other furbearers such as martens and mink, so numbers taken are relatively small. The polecat is still seen by many as a pest or vermin, particularly where conflict arises through actual or perceived threats of polecat predation on poultry or game birds (Packer & Birks 1999). Even in countries where polecats have some level of legal protection, they are still vulnerable to accidental mortality in traps set for other species such as American Mink, Stoat, Weasel or rat.

Road traffic collisions are another cause of mortality in polecats. Recent analysis showed that polecat casualties were associated with road density, traffic volume, presence of rabbits, habitat patchiness and the proximity of improved grassland (Barg *et al.* 2022). Increases in road density and traffic flow across Europe are therefore likely to impact polecat populations. Hybridisation is another potential threat. As we have seen, polecats in Britain hybridise with domestic and feral Ferrets to produce Polecat-Ferret hybrids, particularly where polecats are at low density. The longer-term impact of widespread introgression of Ferret DNA into the polecat population is not yet known (Costa *et al.* 2013).

CURRENT AND FUTURE THREATS

A recent systematic review used published literature on the mustelids to investigate the conservation status, population trends and threats to all of the mustelids (Wright *et al.* 2022). This found that the majority of studies focussed on very few species and were mostly from North America or Europe. A large number of these were of *Martes* species (136 papers), many of which are or were highly valuable furbearers and therefore subjected to intensive trapping. Overharvesting combined with their relatively low reproductive rates resulted in extensive historical declines for many species. Some populations are now recovering, although most have yet to achieve their former numbers or historical range. Translocations and reintroductions have been needed for the return of some species. This has raised the potential risk of overharvesting of donor populations for translocations where these are seen as the solution to historical local extinctions. Another potential threat arising from successful species restoration programmes is the resumption of trapping for predator control if legal protection is removed too quickly. Some have been calling for this in recent years for Pine

FIG 178. Wolverines *Gulo gulo* are one of the mustelids likely to be impacted by future climate change.

Martens in Scotland and Ireland, in response to the perception that the species has sufficiently recovered. A further threat for martens includes the loss and alteration of woodland through forestry practices such as logging. Being solitary, with large home ranges and mainly dependent on forested habitats, makes the Pine Marten particularly vulnerable to forest fragmentation and degradation. This is also likely to arise from future development and infrastructure projects as well as increasing recreational and multifunction use of forests.

The impact of future climate change has been intensively studied for some mustelids such as the Wolverine. Female Wolverines require deep, persistent snowpack to raise their young from February to May, making this species particularly vulnerable to a warming climate and declining snowpack (Peacock 2011). Models predict that by 2050, Wolverines may lose a third of their existing range in North America and that loss of existing range could be as much as two-thirds by the end of the century. Milder winters may have less dramatic but no less significant impacts on other mustelid species, such as increasing the niche overlap between Fisher and American Marten where both species occur (Suffice *et al.* 2017). Future climate scenarios predict conditions that favour the Fisher, although American Martens may be capable of shifting their niche to warmer and less snowy habitats (Zielinski *et al.* 2017). For some species, such as the polecat, a reduction in winter snow cover may actually increase the area of suitable habitat. A study in Sweden showed that reduced snow cover will likely result in a northward range expansion of polecat distribution in Sweden and therefore an increase in the population of polecats in Scandinavia (Osinga *et al.* 2022). This may also be true as polecats recover in northern Britain.

FIG 179. Stoat *Mustela erminea* in its pied winter coat.

WINTER WHITENING IN WEASELS AND STOATS

Species such as the Stoat, however, with its adaptation for winter whitening in cold regions, may find that its winter camouflage becomes out of step with its environment and might not keep up with a rapidly warming climate. Rather than being a hard dividing line where animals on one side turn white in winter but those on the other do not, the southern limit of regular winter whitening in Stoats is a zone around 350 km wide, where white, brown and pied animals occur together in various combinations. This 'transition zone' corresponds approximately to the southern limit of the region where a decent cover of lying snow can be relied upon for around 50 days each year. In Britain, Stoats with white winter coats are common in Scotland as well as in the far north of England and Wales, whereas they stay brown all year round in much of England (Fig. 180). Mottled or piebald Stoats are often seen in the intermediate zones.

FIG 180. In the winter of 1983–84, viewers of the BBC programme *Wildtrack* were asked to send in details of their observations of Stoats *Mustela erminea* in their winter coats from which Carolyn King calculated the boundary, at that time, between white and brown winter coloration. Redrawn from King (1989).

FIG 181. Stoats *Mustela erminea* in their white winter coats in conditions with and without snow cover show how coat colour affects their camouflage.

Given the relatively warm, maritime climate of Britain, few areas experience lying snow for sustained periods. Even in Scotland, the cold weather events when snowfall occurs are usually interspersed by warmer periods when much of the lying snow can melt, although some isolated and sheltered snow patches can remain right through from one winter to the next. Nonetheless, British Stoats seem to whiten even in the relatively mild winters. That being said, the data used for Figure 180 are now 40 years old. Recent analyses have shown significant decreases in the number of days of lying snow between 1961 and 1962 and between 2004 and 2005 in all regions of Britain (Perry 2006). Across Wales, snow cover has decreased from a maximum of 61–70 days' cover (1997–81) to less than half, at approximately 21–30 days (2001–5). The area of snow cover has also declined from widespread coverage to small clusters along the spine of the

Cambrian Mountains (Biggs & Atkinson 2011). It is not known if this has altered the pattern of winter whitening in these areas and it would be timely to repeat the 1983 study.

Weasels in Britain do not change colour in winter as they do elsewhere in Europe. They are the same species, *Mustela nivalis*, across their wide Holarctic distribution but were previously characterised as two separate subspecies: Least Weasel (*M. nivalis nivalis*) and Common Weasel (*M. nivalis vulgaris*) based on coat characteristics, winter whitening and size variation. However, recent genetic analyses confirm that they are the same species, namely the Least Weasel, but represent different morphotypes (morphs) reflecting differences in their phylogenetic history. The *nivalis* morph is found at higher latitudes and altitudes

FIG 182. White and brown winter colour morphs of the Least Weasel *Mustela nivalis nivalis* in snow.

with predictable winter snow cover and is characterised in summer pelage by a straight demarcation line separating the brown dorsal fur from the creamy white fur on the underside. In winter, the coat is typically all white. The *vulgaris* morph is generally larger and stays brown all year long, with a ragged division between back and belly fur. Interestingly, when the *nivalis* and *vulgaris* morphs were experimentally crossbred, the two coloration traits (winter whitening and straight/ragged dorsoventral line) did not separate, suggesting that they are determined by the same or closely linked genes, or segments of DNA (Frank 1985). This also showed that the morphs follow a Mendelian pattern of inheritance, with *vulgaris* being the dominant trait and *nivalis* the recessive. Recent work by Miranda *et al.* (2022) investigated the genetic basis and evolution of winter coat coloration morphs in Weasels from samples collected across two separate transition zones between the *nivalis* and morphs in Europe. Using genotyping of wild populations, they were able to narrow down the association to a specific pigmentation gene (MC1R).

The southernmost limit of regular winter whitening in the Weasel occurs at a much higher latitude than for Stoats. In Sweden, Weasels stay brown all

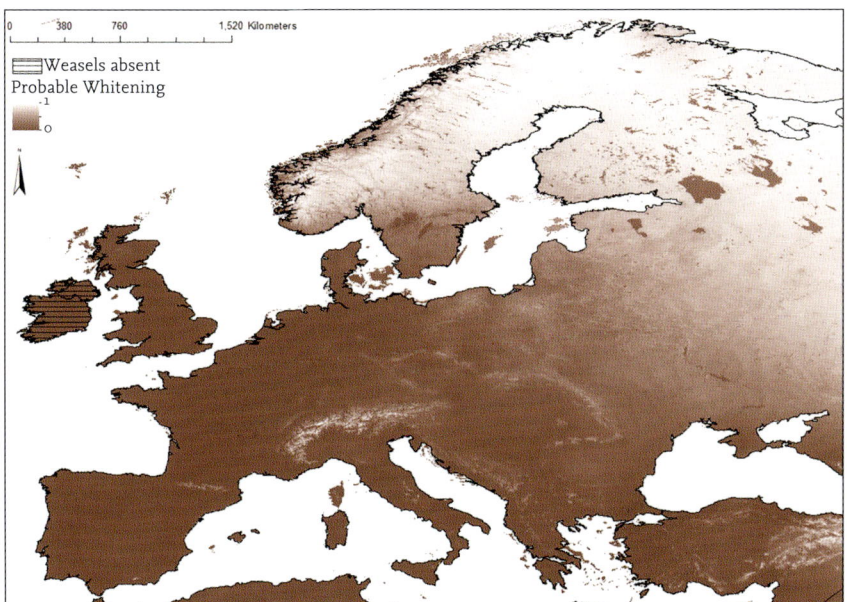

FIG 183. Distribution of winter coat colour morphs shown as the probability of winter white coats across the European range of Weasel *Mustela nivalis*. Data from Mills *et al.* (2018).

winter as far north as about 59–60°, the same latitude as the Orkney Islands off the far north coast of Britain. Above this, there is a narrow boundary zone of about 100 km beyond which the Weasels all turn white. Within the boundary or transition zone, white or brown animals may be found in different proportions, apparently varying with location and temperature.

Salomonsen (1939) looked at the geographical distribution of white and brown winter Weasel skins from Greenland, Scandinavia and Britain and concluded that there was a critical minimum temperature associated with whitening in Common/Least Weasels, at least five degrees lower than that for Stoats. The likelihood of a regular covering of snow every winter increases with increasing latitude and elevation. In cold climates, the temperature drops quickly in autumn, and all northern Stoats and Weasels turn completely white, with the exception of the Stoat's tail tip, which remains black throughout the year. The autumn moult is initiated by shortening day length, and Stoats and Weasels everywhere grow a new coat for the winter. However, the mechanism controlling the growth of the new hair is different in those that exhibit seasonal colour change compared with those that do not. The hair follicles need melanin to produce brown hair. In Stoats and Weasels living in milder climates, melanocyte-stimulating hormone (MSH) is produced on cue by the pituitary gland, stimulating melanocytes (specialised skin cells) to produce melanin, and the new growth of winter fur is brown. If an inhibitor is present that prevents the pituitary from producing MSH or prevents the follicles from responding to it, the MSH is 'switched off' and new hair growth is white (Rust & Meyer 1969). If temperature or another environmental condition was the only controlling factor, then Stoats or Weasels, transferred from a region where all individuals whiten in winter to an area where they do not, would adjust their winter coat colour accordingly, but experimental translocation of these species shows that this is not the case. In 1953, Rothschild and Lane (1957) brought a young Stoat caught in the Swiss Alps back with them to England, about five degrees of latitude further north. In autumn, despite the milder climate of its new home, the transplanted animal turned pure white, in contrast to the locally resident Stoats, which stayed brown. In a series of more controlled experiments, Feder (1990) kept eight captive Stoats under the same short-day photoperiod conditions, but two Stoats from Alaska and two from Oregon were kept at a warm 18–20 °C, while two from each region were kept at below 0 °C. Regardless of the temperatures in which they moulted, all of the Alaskan Stoats changed into white winter coats and all four from Oregon moulted to brown. Therefore, heredity must be involved in the control of winter whitening.

More recent research has confirmed that the fact that Weasels do not turn white in Britain is a result of genetics and evolutionary history rather than the effect of climate. The transition zone where some individuals of the same species of Weasel turn white or not marks the meeting of two distinct clades, or lineages, of *nivalis*, which have different histories and genetic makeup and which are distinctly different in their summer coats, as well as in their winter coats. In Europe, the distribution of the two morphs is correlated with snow cover variables (Mills *et al.* 2018). The *nivalis* morph occurs at higher latitudes and altitudes where winter snow cover is more persistent, whereas the *vulgaris* morph is found in southern and coastal areas. This pattern of distribution suggests that the distribution of morphotypes is maintained by environmental selection for camouflage. Current belief is that the *vulgaris* morph evolved and recolonised Europe from a more southerly glacial refugium than the *nivalis* morph, which may have had a selective advantage in colder regions (McDevitt *et al.* 2012). Consequently, in the *vulgaris* group, the allele (matching pair of genes) for winter whitening has become rare, whereas in the *nivalis* group it is still widespread or even fixed. In the far north and far south, where winter snow cover (or lack of it) is fairly predictable, only the genes for either white or brown winter coats would be expressed. However, in the transition zones where winter conditions are more variable, there could be different proportions of individuals with different genetic origins, including hybrids between the two morphs. Postglacial recolonisation is thought to have brought populations of winter white Weasels from refugia in the colder north into contact with those of winter brown weasels from milder southern refugia (McDevitt *et al.* 2012). In the contact, or 'suture', zone where the populations meet, extensive hybridisation between the two combined with local adaptation to the variable winters would result in the patterns we can see today.

Similarly, for Stoats, it is reasonable to assume that those living in the far north of Britain would retain their adaptation to cold winters, including the expression of genes for winter whitening, whereas in the milder climate in England and parts of Wales, standing out as a snowy white animal in winters where snow is infrequent would actually be a disadvantage. In northeast Scotland, Hewson and Watson (1979) found that the vast majority (more than 90 per cent) of Stoats turned white in winter, whereas in southwest Scotland, most were pied, and in northeast England, few changed colour. These differences were associated with regional differences in the number of days of snowfall and snow cover and in the monthly minimum temperature, but not with mean temperature or exposure. Conversely, most Stoats in England stay brown all year round. The results of a questionnaire survey carried out by Flintoff (1933) in Yorkshire from 1931 to 1932 showed that out of 2,930 Stoats killed in winter, only 21 were completely in ermine and a further 175 were pied. Pied Stoats may look as if they have been caught halfway through moulting, but the condition is probably a

FIG 184. Stoat *Mustela erminea* in winter coat showing pied coloration.

result of them being heterozygous for (i.e. having two different versions of) the alleles that control winter whitening.

What of the ancestors of the Stoats that were shipped to New Zealand centuries ago? How, or indeed, have they adapted to winters in the southern hemisphere? King and Moody (1982) collected Stoat carcasses over a period of four years in New Zealand. In Fiordland, in the cool, far south, very few Stoats collected between the end of the autumn moult in June and the beginning of the spring moult in November were white. In one region, encompassing six collection areas between 44 °S and 45 °S, only 71 per cent of the 34 females collected and 47 per cent of the 62 males collected had any white hairs at all. Out of a total of 429 Stoats collected from the whole country during the winter months, only eight were almost fully white and even they still had some brown hairs remaining around the eyes. The majority were pied and many had just a few white hairs at the start of the black tail tip. Variation in the proportions of white or pied Stoats in samples collected in the southern winter and spring months (June–November) was significantly correlated, both with the mean daily minimum temperature in July (the coldest month) and with the number of days of ground frost per year. Although the vast majority of Stoats stayed completely (or mostly) brown, white and pied animals were more common at higher elevations and more southerly latitudes, consistent with the idea of local adaptation linked to temperature (King & Moody 1982b).

The conclusion from these studies is that Stoats from the cool, far north of Britain turn white more consistently than those from the cool end of New Zealand (the far south), despite not being exposed to a climate that is significantly

FIG 185. Ptarmigan *Lagopus muta* (a) and Mountain Hare *Lepus timidus* (b), the only two species other than Stoat *Mustela erminea* that change colour in winter in Britain.

colder or snowier than those sampled by King and Moody (1982b). Both regions have minimum winter temperatures that are usually between minus and plus 5 °C and lying snow for fewer than 20 days in most winters. It seems that the New Zealand Stoats are more like those found in England than in Scotland. This should come as no surprise, given that all New Zealand Stoats are probably descended from English (non-whitening) stock, and they have only lived in their new homeland for about 120 generations. Therefore, even for those populations that now occupy the coldest regions of the country, natural selection may not yet have had time to retrieve the lost or latent genes for whitening. This all makes sense if it is heredity rather than temperature that determines winter coat coloration.

What might be the advantage of going to all the trouble of changing colour once a year? Only a handful of animals do it (just Mountain Hares, Stoats and Ptarmigan *Lagopus muta* in Britain), so there are obviously costs to be weighed up against the potential benefits. Two of the suggested advantages of seasonal coat colour change are related to heat loss in the colder temperatures of winter. The first is that a coat of white fur is warmer because it loses less heat by radiation than darker colours. However, this is not true for animals because they radiate heat from their bodies in the far infrared range, regardless of coat colour. So white- and dark-coated animals all experience heat loss by radiation at equal rates (Hammel 1956). It is the density and length of the fur rather than its colour that determine how well it conserves heat. Because of their structure, white guard hairs contain more air as opposed to pigment than brown hairs, increasing their insulative properties. In some species, the lower part of the guard hair is thinner, providing space for more down hairs at the skin or by increasing the volume of air trapped in the bottom layer of fur. This is not the case for Weasels and Stoats, which have short fur all year round. Therefore, their summer and winter coats have the same thermal properties, regardless of colour.

FIG 186. Stoat *Mustela erminea* in ermine stands out against the landscape when there is no snow cover.

The other potential benefits of a white coat in winter relate to crypsis, or camouflage, which could be useful to Stoats and Weasels both as predator and prey species. Ambush predators (i.e. those that lie in wait for their prey or creep up on them quietly from a distance) rely on both camouflage and stealthy behaviour to remain hidden from their target prey for as long as possible. However, Stoats and Weasels do not usually hunt in this way. Their method is to move constantly, sometimes up to 2 km in a single hunting expedition of a few hours, exploring every likely tunnel and runway for their small mammal prey. Even if a white coat against a snowy background was perfect camouflage, which it is not, the rapid movements of a Stoat or Weasel would soon give it away. These are animals that are hardly ever still except when they are asleep. The most likely advantage of winter whitening for Stoats and Weasels is as camouflage that helps them to avoid detection by larger predators such as raptors, owls and foxes. Field experiments by Atmeh et al. (2018) used brown and white Weasel models and placed them against backgrounds of different colours. Their results confirmed that contrasting models faced significantly higher detection by predators. As camouflage from predators seems to be the primary advantage of seasonal colour change for Stoats and (in parts of their range) Weasels, the fitness cost of coat colour mismatches to their environment could be particularly harmful to winter white populations facing decreased snow cover duration driven by global climate change.

LOOKING AHEAD

It is very many years since we have had a full suite of small mustelids across much of Britain. This will inevitably lead to some changes as species readjust to living alongside each other again. We don't expect there to be much interaction or competition between Pine Martens and European Polecats, as they occupy sufficiently different niches, with polecats being able to exploit more open and agricultural habitats, preying on rats, rabbits and amphibians, whereas martens are more tied to wooded areas, where they eat mainly voles, beetles and berries. However, polecats are often seen on camera traps in woodland so it will be interesting to observe what happens. Other native carnivores such as the Wildcat are also a focus of conservation reintroductions in some parts of the country, which will add to the number of predator species competing for some prey. They can't all eat the same rabbit, rat or vole, and so there will inevitably be winners and losers. Prey species, too, are not immune to the potential impacts of future climate change, such as increased rainfall. Rabbits and voles do not burrow in waterlogged soils. Rabbits have also undergone significant changes to their populations in previous years as a result of diseases such as myxomatosis and, more recently, rabbit haemorrhagic disease.

New emerging diseases are a threat to mustelids, too. During the COVID-19 pandemic, the virus was detected in American Mink at 400 fur farms in 8 European countries. Huge numbers of animals were culled in response. Research found that American Mink and Ferrets were highly susceptible to the virus, which was most likely transmitted via direct and indirect contact from asymptomatic humans (Boklund *et al.* 2021). Affected animals showed symptoms including laboured breathing and nasal discharge. The disease was fatal to those with moderate to severe symptoms. By 2021, antibodies to the virus had also been detected in a small number of wild mustelids in France, including three Pine Martens. A disease risk assessment concluded that there was a high likelihood of disease and death in Pine Martens exposed to the virus and, for martens in Britain or Ireland, exposure was most likely through contact with people in the course of conservation fieldwork activities. The recommendation was that a stringent set of biosecurity measures be put in place for anyone carrying out fieldwork with Pine Martens (Carraro *et al.* 2021). As well as direct contact (e.g. during trapping and handling for translocations), this included indirect contact, such as putting out food or bait at camera traps or checking den boxes for signs of occupancy. Thankfully, to date, there has been no evidence of transmission to wild Pine Martens in Britain, but disease surveillance is an important aspect of monitoring wildlife populations, particularly those that are still recovering from past declines, all of which underlines the fact that we need to cherish all of our biodiversity, monitor and appreciate our mustelids and take care that we do not lose them in future as, with some of them, we so very nearly did in the past.

FIG 187. The author enjoying watching a Pine Marten *Martes martes*, a regular visitor to a house in Scotland.

References

Abbott, G.F. (1903). *Macedonian Folklore*. Cambridge University Press.

Aebischer, N. (2019). Fifty-year trends in UK hunting bags of birds and mammals, and calibrated estimation of national bag size, using GWCT's National Gamebag Census. *European Journal of Wildlife Research* 65, 64.

Aebischer, N., Davey, P. & Kingdon, N. (2011). National gamebag census: mammal trends to 2009. Game and Wildlife Conservation Trust, Fordingbridge.

Aelian *On the Nature of Animals* (translation from Aelian, *On the Characteristics of Animals*, trans. A. F. Scholfield, Cambridge, MA, Harvard University Press, 1958).

Aliev, F. & Sanderson, G.C. (1970). The american mink, *Mustela vison* (Schreber 777), in the USSR. *Saugetierkundliche Mitteilungen* 18, 122–127.

Alston, E.R. (1879). On the specific identity of the British martens. *Journal of Zoology* 47, 468–474.

Alterio, N. (1998). Spring home range, spatial organisation and activity of stoats *Mustela erminea* in a South Island *Nothofagus* forest, New Zealand. *Ecography* 21, 18–24.

Amar, A., Hewson, C.M., Thewlis, R.M. *et al.* (2008). *What's Happening to Our Woodland Birds?* RSPB, Sandy, and British Trust for Ornithology, Thetford.

Armstrong, D.P. & Seddon, P.J. (2008). Directions in reintroduction biology. *Trends in Ecology & Evolution* 23, 20–25.

Auld, M., Ayling, B., Bambini, L. *et al.* (2019). Safeguarding Orkney's native wildlife from non-native invasive stoats. In C.R. Veitch, M.N. Clout & D.R. Towns (eds.), *Island invasives: scaling up to meet the challenge*. IUCN, Gland, Switzerland, pp. 244–248.

Auster, R.E., Barr, S.W. & Brazier, R.E. (2020). Improving engagement in managing reintroduction conflicts: learning from beaver reintroduction. *Journal of Environmental Planning and Management*, 1–22.

Baillie, S.R., Marchant, J., Leech, D.I. *et al.* (2009). *Breeding Birds in the Wider Countryside: Their Conservation Status 2008*. BTO Research Report 516. British Trust for Ornithology, Thetford.

Baines, D. (1996). The implications of grazing and predator management on the habitats and breeding success of black grouse *Tetrao tetrix*. *Journal of Applied Ecology* 33, 54–62.

Baines, D., Moss, R. & Dugan, D. (2004). Capercaillie breeding success in relation to forest habitat and predator abundance. *Journal of Applied Ecology* 41, 59–71.

Balestrieri, A., Bogliani, G., Boano, G. *et al.* (2016). Modelling the distribution of forest-dependent species in human-dominated landscapes: patterns for the pine marten in intensively cultivated lowlands. *PLoS One* 11, e0158203.

Balestrieri, A., Remonti, L., Ruiz-González, A. *et al.* (2015). Distribution and habitat use by *Martes martes* in a riparian corridor crossing intensively cultivated lowlands. *Ecological Research* 30, 153–162.

Balharry, D. (1993). Factors affecting the distribution and population density of pine martens (*Martes martes* L.) in Scotland. PhD Thesis, University of Aberdeen.

Balharry, E.A. (1998). A non-lethal approach to the exclusion of a protected carnivore, the pine marten, from game and poultry pens. In T.V.W. Trust (ed.). Vincent Wildlife Trust, London.

Balharry, E.A., McGowan, G.M., Kruuk, H. & Halliwell, E. (1996). Distribution of pine martens in Scotland as determined by field survey and questionnaire. Scottish Natural Heritage Survey & Monitoring Report No. 48.

Barg, A., MacPherson, J. & Caravaggi, A. (2022). Spatial and temporal trends in western polecat road mortality in Wales. *PeerJ* 10, e14291.

Barnouw, V. (1977). *Wisconsin Chippewa Myths & Tales and Their Relation to Chippewa Life* (based on folktales collected by Victor Barnouw, Joseph B. Casagrande, Ernestine Friedl, and Robert E. Ritzenthaler). University of Wisconsin Press.

Bartolommei, P., Manzo, E. & Cozzolino, R. (2012). Evaluation of three indirect methods for surveying European pine marten in a forested area of central Italy. *Hystrix* 23, 91.

Bateman, J.A. (1988). *Animal Traps and Trapping*. David & Charles, London.

Bavin, D. (2021). Social and behavioural aspects of a pine marten translocation. MPhil Thesis, University of Exeter.

Beck, J. (1993). The dream of Leonardo da Vinci. *Artibus et Historiae* 14, 185–198.

Bekker, J.P. (1988). Daubenton's bat *Myotis daubentonii* as a prey of beech marten *Martes foina* in subterranean marl pits. *Lutra* 31, 82–85.

Bellamy, P.E., Rothery, P., Hinsley, S.A. & Newton, I. (2000). Variation in the relationship between numbers of breeding pairs and woodland area for passerines in fragmented habitat. *Ecography* 23, 130–138.

Beltran, R.S., Sadou, M.C., Condit, R. *et al.* (2015). Fine-scale whisker growth measurements can reveal temporal foraging patterns from stable isotope signatures. *Marine Ecology Progress Series* 523, 243–253.

Berger-Tal, O., Blumstein, D. & Swaisgood, R.R. (2020). Conservation translocations: a review of common difficulties and promising directions. *Animal Conservation* 23, 121–131.

Biggs, E.M. & Atkinson, P.M. (2011). A characterisation of climate variability and trends in hydrological extremes in the Severn Uplands. *International Journal of Climatology* 31, 1634–1652.

Birks, J.D.S. (1990). Feral mink and nature conservation. *British Wildlife* 1, 313–323.

Birks, J.D.S. (1998). Secondary rodenticide poisoning risk arising from winter farmyard use by the European polecat *Mustela putorius*. *Biological Conservation* 85, 233–240.

Birks, J.D.S. (2008). The Polecat Survey of Britain 2004–2006. Vincent Wildlife Trust, Ledbury.

Birks, J.D.S. (2015). *Polecats*. Whittet Books, Stansted.

Birks, J.D.S. (2017). *The Pine Marten*. Whittet Books, Stansted,

Birks, J.D.S. & Kitchener, A. (1999). The distribution and status of the polecat *Mustela putorius* in Britain in the 1990s. Vincent Wildlife Trust, London.

Birks, J.D.S. & Messenger, J. (2010). Evidence of pine martens in England and Wales 1996–2007. Vincent Wildlife Trust, Ledbury.

Bjornsson, T. & Hersteinsson, P. (1991). Mink in southern Breidafjordur Bay. *Wildlife Management News (Iceland)* 7, 3–12.

Blackburn, T.M., Lockwood, J.L. & Cassey, P. (2015). The influence of numbers on invasion success. *Molecular Ecology* 24, 1942–1953.

Blandford, P.R.S. (1986). Behavioural ecology of the polecat *Mustela putorius* in Wales. PhD Thesis, University of Exeter.

Blandford, P.R.S. (1987). Biology of the polecat *Mustela putorius*: a literature review. *Mammal Review* 17, 155–198.

Bodey, T.W., Bearhop, S. & McDonald, R.A. (2011a). The diet of an invasive nonnative predator, the feral ferret *Mustela furo*, and implications for the conservation of ground-nesting birds. *European Journal of Wildlife Research* 57, 107–117.

Bodey, T.W., Bearhop, S. & McDonald, R.A. (2011b). Localised control of an introduced predator: creating problems for the future? *Biological Invasions* 13, 2817–2828.

Bodey, T.W., McDonald, R.A., Sheldon, R.D. & Bearhop, S. (2011c). Absence of effects of predator control on nesting success of northern lapwings *Vanellus vanellus*: implications for conservation. *Ibis* 153, 543–555.

Bogdziewicz, M. & Zwolak, R. (2014). Responses of small mammals to clear-cutting in temperate and boreal forests of Europe: a meta-analysis and review. *European Journal of Forest Research* 133, 1–11.

Boklund, A., Gortázar, C., Pasquali, P. *et al.* (2021). Monitoring of SARS-CoV-2 infection in mustelids. *EFSA Journal*. European Food Safety Authority, European Centre for Disease Prevention Control.

Bonesi, L. & Palazon, S. (2007). The American mink in Europe: status, impacts, and control. *Biological Conservation* 134, 470–483.

Boonstra, R. & Krebs, C.J. (1979). Viability of large- and small-sized adults in fluctuating vole populations. *Ecology* 60, 567–573.

Brainerd, S.M. (1990). The pine marten and forest fragmentation: a review and general hypothesis. *Transaction International Congress of Game Biologists*, 421–434.

Brainerd, S.M. & Rolstad, J. (2002). Habitat selection by Eurasian pine martens *Martes martes* in managed forests of southern boreal Scandinavia. *Wildlife Biology* 8, 289–297.

Breitenmoser, U. (1998). Large predators in the Alps: the fall and rise of man's competitors. *Biological Conservation* 83, 279–289.

Bright, P.W. (2000). Lessons from lean beasts: conservation biology of the mustelids. *Mammal Review* 30, 217–226.

Bright, P.W. & Halliwell, E. (1999). Species Recovery Programme for the Pine Marten in England: 1996–1998. English Nature Research Report No. 306. English Nature, Peterborough.

Bright, P.W. & Smithson, T.J. (1997). Species Recovery Programme for the Pine Marten in England: 1995–1996. English Nature Research Report No. 240. English Nature, Peterborough.

Brown, S. (2001). The behavioural responses of stoats (*Mustela erminea*) to trapping tunnels. MSc Thesis, Lincoln University.

Bubac, C.M., Johnson, A.C., Fox, J.A. & Cullingham, C.I. (2019). Conservation translocations and post-release monitoring: Identifying trends in failures, biases, and challenges from around the world. *Biological Conservation* 238, 108239.

Buckley, D.J. & Lundy, M. (2013). The current distribution and potential for future range expansion of feral ferret *Mustela putorius furo* in Ireland. *European Journal of Wildlife Research* 59, 323–330.

Buckley, D.J., Sleeman, D. & Murphy, J. (2007). Feral ferrets *Mustela putorius furo* L. in Ireland. *The Irish Naturalists' Journal* 28, 356–360.

Buckley, K.P., Byrne, É.B. & Sleeman, D.P. (2015). Diet of Irish stoats (*Mustela erminea hibernica*) in two habitats. *Irish Naturalists' Journal* 34, 8–12.

Bull, E.L. & Heater, T.W. (2001). Survival, causes of mortality, and reproduction in the American marten in northeastern Oregon. *Northwestern Naturalist* 82, 1–6.

Burian, A., Mauvisseau, Q., Bulling, M. *et al.* (2021). Improving the reliability of eDNA data interpretation. *Molecular Ecology Resources* 21, 1422–1433.

Burton, J.A., Jefferies, D. & Birks, J. (2018). Was the stone marten once established in the wild in the British Isles? *British Wildlife* 30, 99–106.

Byrnes, E. & Little, D. (2007). A history of woodland management in Ireland: an overview. Forest Service Native Woodland Scheme Information Note No. 2.

Calvete, C. (2006). Modeling the effect of population dynamics on the impact of

rabbit hemorrhagic disease. *Conservation Biology* 20, 1232–1241.
Carey, M., Hamilton, G., Poole, A. & Lawton, C. (2007). The Irish Squirrel Survey 2007. COFORD, Dublin.
Carraro, C., Common, S. & Sainsbury, A.W. (2021). Risk from SARS-CoV-2 to the reintroduced (*Martes martes*) population in England and mitigation recommendations for conservation fieldworkers. Zoological Society of London.
Caryl, F.M. (2008). Pine marten diet and habitat use within a managed coniferous forest. PhD Thesis, University of Stirling.
Caryl, F.M., Quine, C.P. & Park, K.J. (2012a). Martens in the matrix: the importance of nonforested habitats for forest carnivores in fragmented landscapes. *Journal of Mammalogy* 93, 464–474.
Caryl, F.M., Raynor, R., Quine, C.P. & Park, K.J. (2012b). The seasonal diet of British pine marten determined from genetically identified scats. *Journal of Zoology* 288, 252–259.
Chamberlain, D.E., Glue, D.E. & Toms, M.P. (2009). Sparrowhawk *Accipiter nisus* presence and winter bird abundance. *Journal of Ornithology* 150, 247–254.
Chambert, T., Miller, D.A. & Nichols, J.D. (2015). Modeling false positive detections in species occurrence data under different study designs. *Ecology* 96, 332–339.
Charles, W.N. (1981). Abundance of field voles (*Microtus agrestis*) in conifer plantations. In F.T. Last & A.S. Gardiner (eds.), *Forest and woodland ecology: an account of research being done in ITE*. NERC/Institute of Terrestrial Ecology, Cambridge, pp. 135–137.
Chauvenet, A.L., Canessa, S. & Ewen, J.G. (2016). Setting objectives and defining the success of reintroductions. In D.S. Jachowski, J.J. Millspaugh, P.L. Angermeier & R. Slotow (eds.), *Reintroduction of Fish and Wildlife Populations*. University of California Press, pp. 105–122.
Choquenot, D., Ruscoe, W.A. & Murphy, E. (2001). Colonisation of new areas by stoats: time to establishment and requirements for detection. *New Zealand Journal of Ecology* 25, 83–88.
Clayton, R., Byrom, A., Anderson, D. et al. (2011). Density estimates and detection models inform stoat (*Mustela erminea*) eradication on Resolution Island, New Zealand. In C.R. Veitch, M.N. Clout & D.R. Towns (eds.), *Island invasives: eradication and management*. IUCN, Gland, Switzerland, pp. 413–417.
Clevenger, A. (1993). Pine marten (*Martes martes*) comparative feeding ecology in an island and mainland population of Spain. *Zeitschrift für Saugetierkunde* 58, 212–224.
Clout, M.N. & Russell, J. (2006). The eradication of mammals from New Zealand islands. In F. Koike, M.N. Clout, M. Kawamichi, M. De Poorter & K. Iwatsuki (eds.), *Assessment and Control of Biological Invasion Risks*. IUCN, Gland, Switzerland, pp. 127–141.
Coope, R. (2007). A preliminary investigation of the food and feeding behaviour of pine martens in productive forestry from an analysis of the contents of their scats collected in Inchnacardoch Forest, Fort Augustus. *Scottish Forestry* 61, 3.
Corbet, G.B. (1966). *The Terrestrial Mammals of Western Europe*. Foulis & Co., London.
Corbet, G.B. & Harris, S. (1991). *Handbook of British Mammals*. Published for the Mammal Society by Blackwell Scientific Publications, Oxford.
Costa, M. (2014). Phylogeography and hybridisation in the European polecat (*Mustela putorius*). PhD Thesis, University of Lisbon.
Costa, M., Fernandes, C., Birks, J.D.S. et al. (2013). The genetic legacy of the 19th-century decline of the British polecat: Evidence for extensive introgression from feral ferrets. *Molecular Ecology* 22, 5130–5147.
Costa, M., Fernandes, C., Rodrigues, M., Santos-Reis, M. & Bruford, M.W. (2012). A panel of microsatellite markers for genetic studies of European polecats (*Mustela putorius*) and ferrets (*Mustela furo*). *European Journal of Wildlife Research* 58, 629–633.

Côté, I.M. & Sutherland, W.J. (1997). The effectiveness of removing predators to protect bird populations. *Conservation Biology* 11, 395–405.

Coughlin, C.E. & Van Heezik, Y. (2015). Weighed down by science: do collar-mounted devices affect domestic cat behaviour and movement? *Wildlife Research* 41, 606–614.

Craik, C. (1998). Recent mink related declines in gulls and terns in west Scotland and the beneficial effects of mink control. *Argyll Bird Report* 14, 99–110.

Craik, C. (2000). Breeding success of common gulls *Larus canus* in west Scotland II. Comparison between colonies. *Atlantic Seabirds* 2, 1–12.

Cresswell, W. (2011). Predation in bird populations. *Journal of Ornithology* 152, 251–263.

Croose, E. (2016). The distribution and status of the polecat (*Mustela putorius*) in Britain 2014–2015. Vincent Wildlife Trust, London.

Croose, E., Birks, J.D.S, Martin, J. et al. (2019). Comparing the efficacy and cost-effectiveness of sampling methods for estimating population abundance and density of a recovering carnivore: the European pine marten (*Martes martes*). *European Journal of Wildlife Research* 65, 37.

Croose, E., Birks, J.D.S, O'Reilly, C. et al. (2016). Sample diversity adds value to non-invasive genetic assessment of a pine marten (*Martes martes*) population in Galloway Forest, southwest Scotland. *Mammal Research* 61, 131–139.

Croose, E., Birks, J.D.S. & Schofield, H.W. (2013). Expansion Zone Survey of Pine Marten (*Martes martes*) Distribution in Scotland. NatureScot Commissioned Report No. 520.

Croose, E., Birks, J.D.S., Schofield, H.W. & O'Reilly, C. (2014). Distribution of the pine marten (*Martes martes*) in southern Scotland in 2013. NatureScot Commissioned Report No. 740.

Croose, E. & Carter, S.P. (2019). A pilot study of a novel method to monitor weasels (*Mustela nivalis*) and stoats (*M. erminea*) in Britain. *Mammal Communications* 5, 6–12.

Croose, E., Duckworth, J., Ruette, S. et al. (2018). A review of the status of the Western polecat *Mustela putorius*: a neglected and declining species? *Mammalia* 82, 550–564.

Croose, E., Hanniffy, R., Hughes, B. et al. (2022). Assessing the detectability of the Irish stoat *Mustela erminea hibernica* using two camera trap-based survey methods. *Mammal Research* 67, 1–8.

Cuthbert, J.H. (1973). The origin and distribution of feral mink in Scotland. *Mammal Review* 3, 97–103.

Cuthbert, R., Bartlett, A., Turbelin, A. et al. (2021). Economic costs of biological invasions in the United Kingdom. *NeoBiota* 67, 299–328.

Davis, M.H. (1983). Post-release movements of introduced marten. *Journal of Wildlife Management* 47, 59–66.

Davison, A., Birks, J.D.S, Brookes, R.C., Braithwaite, T.C. & Messenger, J.E. (2002). On the origin of faeces: morphological versus molecular methods for surveying rare carnivores from their scats. *Journal of Zoology* 257, 141–143.

Davison, A., Birks, J.D.S., Griffiths, H. et al. (1999). Hybridization and the phylogenetic relationship between polecats and domestic ferrets in Britain. *Biological Conservation* 87, 155–161.

Day, M. (1966). Identification of hair and feather remains in the gut and faeces of stoats and weasels. *Journal of Zoology* 148, 201–217.

Day, M. (1968). Food habits of British stoats (*Mustela erminea*) and weasels (*Mustela nivalis*). *Journal of Zoology* 155, 485–497.

Dayan, T. & Simberloff, D. (1994). Character displacement, sexual dimorphism, and morphological variation among British and Irish mustelids. *Ecology* 75, 1063–1073.

Deane, C.D. & O'Gorman, F. (1969). The spread of feral mink in Ireland. *The Irish Naturalists' Journal* 16, 198–202.

De Azevedo, C.S. & Young, R.J. (2021). Animal personality and conservation: basics for inspiring new research. *Animals* 11, 1019.

Debrot, S. & Mermod, C. (1983). The spatial and temporal distribution pattern of the

stoat (*Mustela erminea* L.). *Oecologia* 59, 69–73.
DEFRA (2021). Reintroductions and other conservation translocations: code and guidance for England. DEFRA, London.
Department of Agriculture, Food and the Marine (2023). Forest Statistics Ireland 2023.
Dickens, M.J., Delehanty, D.J. & Romero, L.M. (2010). Stress: an inevitable component of animal translocation. *Biological Conservation* 143, 1329–1341.
Dilks, P. & Lawrence, B. (2000). The use of poison eggs for the control of stoats. *New Zealand Journal of Zoology* 27, 173–182.
Doi, H., Fukaya, K., Oka, S.-I. *et al.* (2019). Evaluation of detection probabilities at the water-filtering and initial PCR steps in environmental DNA metabarcoding using a multispecies site occupancy model. *Scientific Reports* 9, 3581.
Donadio, E. & Buskirk, S.W. (2006). Diet, morphology, and interspecific killing in Carnivora. *The American Naturalist* 167, 524–536.
Dowding, J.E. & Elliot, M.J. (2003). Ecology of stoats in a South Island braided river valley. New Zealand Department of Conservation, Wellington.
Dowding, J.E. & Murphy, E.C. (1996). Predation of Northern New Zealand dotterels (*Charadrius obscurus aquilinius*) by stoats. *Notornis* 43, 144–146.
Dunn, E. (1977). Predation by weasels (*Mustela nivalis*) on breeding tits (*Parus* spp.) in relation to the density of tits and rodents. *Journal of Animal Ecology* 46, 633–652.
Dunstone, N. (1993). *The Mink*. T. & A.D. Poyser, London.
Dyczkowski, J. & Yalden, D. (1998). An estimate of the impact of predators on the British field vole *Microtus agrestis* population. *Mammal Review* 28, 165–184.
Edge, K., Crouchley, D., McMurtrie, P., Willans, M. & Byrom, A. (2011). Eradicating stoats (*Mustela erminea*) and red deer (*Cervus elaphus*) off islands in Fiordland. In C.R. Veitch, M.N. Clout & D.R. Towns (eds.), *Island invasives: eradication and management*. IUCN, Gland, Switzerland, pp. 166–171.

Elliott, G., Willans, M., Edmonds, H. & Crouchley, D. (2010). Stoat invasion, eradication and re-invasion of islands in Fiordland. *New Zealand Journal of Zoology* 37, 1–12.
Elmeros, M., Christensen, T.K. & Lassen, P. (2011). Concentrations of anticoagulant rodenticides in stoats *Mustela erminea* and weasels *Mustela nivalis* from Denmark. *Science of the Total Environment* 409, 2373–2378.
Erlinge, S. (1977a). Agonistic behaviour and dominance in stoats (*Mustela erminea* L.). *Zeitschrift für Tierpsychologie* 44, 375–388.
Erlinge, S. (1977b). Spacing strategy in stoat *Mustela erminea*. *Oikos* 28, 32–42.
Erlinge, S. (1983). Demography and dynamics of a stoat *Mustela erminea* population in a diverse community of vertebrates. *Journal of Animal Ecology* 52, 705–726.
Erlinge, S. (1987). Why do European stoats *Mustela erminea* not follow Bergmann's rule? *Ecography* 10, 33–39.
Erlinge, S. & Sandell, M. (1986). Seasonal changes in the social organization of male stoats, *Mustela erminea*: an effect of shifts between two decisive resources. *Oikos* 47, 57–62.
Ernest, H.B., Rubin, E.S. & Boyce, W.M. (2002). Fecal DNA analysis and risk assessment of mountain lion predation of bighorn sheep. *Journal of Wildlife Management* 66, 75–85.
Evans, K.L. (2004). The potential for interactions between predation and habitat change to cause population declines of farmland birds. *Ibis* 146, 1–13.
Facka, A.N., Lewis, J.C., Happe, P. *et al.* (2016). Timing of translocation influences birth rate and population dynamics in a forest carnivore. *Ecosphere* 7, e01223.
Fairley, J.S. (1971). New data on the Irish stoat. *The Irish Naturalists' Journal* 71, 49–57.
Fairley, J.S. (2001). *A Basket of Weasels: The Weasel Family of Ireland and Other Furred Irish Beasts: Bats, the Rabbit, Hares and Rodents*. James Fairley, Belfast.
Fairlie, S. (2009). A short history of enclosure in Britain. *The Land* 7, 12.

Fairnell, E.H. & Barrett, J.H. (2007). Fur-bearing species and Scottish islands. *Journal of Archaeological Science* 34, 463–484.

Fediajevaite, J., Priestley, V., Arnold, R. & Savolainen, V. (2021). Meta-analysis shows that environmental DNA outperforms traditional surveys, but warrants better reporting standards. *Ecology and Evolution* 11, 4803–4815.

Fischer, J. & Lindenmayer, D.B. (2000). An assessment of the published results of animal relocations. *Biological Conservation* 96, 1–11.

Forest Research (2023). Woodland Statistics dataset. https://www.forestresearch.gov.uk/tools-and-resources/statistics/statistics-by-topic/woodland-statistics/ (accessed December 2023)

Forrester, N., Boag, B., Buckley, A., Moureau, G. & Gould, E. (2009). Co-circulation of widely disparate strains of rabbit haemorrhagic disease virus could explain localised epidemicity in the United Kingdom. *Virology* 393, 42–48.

Franzen, M., Kloetzer, L., Ponti, M., Trojan, J. & Vicens, J. (2021). Machine learning in citizen science: promises and implications. In K. Vohland, A. Land-Zandstra, L. Ceccaroni *et al.* (eds.), *The Science of Citizen Science*. Springer, New York, pp. 183–198.

Fuller, R.J., Noble, D.G., Smith, K.W. & Vanhinsbergh, D. (2005). Recent declines in populations of woodland birds in Britain. *British Birds* 98, 116–143.

Gamberg, M. & Atkinson, J.L. (1988). Prey hair and bone recovery in ermine scats. *Journal of Wildlife Management* 52, 657–660.

Garthwaite, D., Hudson, S., Barker, I. *et al.* (2012). Pesticide Usage Survey Report 250: Arable Crops in the United Kingdom. DEFRA, London.

Garvey, P.M., Glen, A.S. & Pech, R.P. (2015). Foraging ermine avoid risk: behavioural responses of a mesopredator to its interspecific competitors in a mammalian guild. *Biological Invasions* 17, 1771–1783.

Gerell, R. (1967). Dispersal and acclimatisation of the mink (*Mustela vison* Schreber) in Sweden. *Viltrevy* 4, 1–38.

Gillingham, B.J. (1984). Meal size and feeding rate in the least weasel (*Mustela nivalis*). *Journal of Mammalogy* 65, 517–519.

Gleeson, D.M., Byrom, A.E. & Howitt, R.L. (2010). Non-invasive methods for genotyping of stoats (*Mustela erminea*) in New Zealand: potential for field applications. *New Zealand Journal of Ecology* 34, 356–359.

Goldberg, C.S., Strickler, K.M. & Fremier, A.K. (2018). Degradation and dispersion limit environmental DNA detection of rare amphibians in wetlands: Increasing efficacy of sampling designs. *Science of the Total Environment* 633, 695–703.

Goldstein, E.A. (2014). Ecology of frontier populations of the invasive grey squirrel (*Sciurus carolinensis*) in Ireland. PhD Thesis, University College Cork.

Golley, F.B. (1960). Energy dynamics of a food chain of an old-field community. *Ecological Monographs* 30, 187–206.

Gooch, S., Baillie, S. & Birkhead, T. (1991). Magpie *Pica pica* and songbird populations. Retrospective investigation of trends in population density and breeding success. *Journal of Applied Ecology* 28, 1068–1086.

Goszczynski, J. (1986). Diet of foxes and martens in central Poland. *Acta Theriologica* 31, 491–506.

Graham, I.M. & Lambin, X. (2002). The impact of weasel predation on cyclic field-vole survival: the specialist predator hypothesis contradicted. *Journal of Animal Ecology* 71, 946–956.

Grayson, D.K. (2001). The archaeological record of human impacts on animal populations. *Journal of World Prehistory* 15, 1–68.

Gregory, R.D., Vorisek, P., Van Strien, A. *et al.* (2007). Population trends of widespread woodland birds in Europe. *Ibis* 149, 78–97.

Griffith, B., Scott, J.M., Carpenter, J.W. & Reed, C. (1989). Translocation as a species conservation tool: status and strategy. *Science* 245, 477–480.

Gurnell, J. (1987). *The Natural History of Squirrels*. Christopher Helm, London.

Halliwell, E.C. (1997). The ecology of red squirrels in Scotland in relation to pine marten predation. PhD Thesis, University of Aberdeen.

Hammershøj, M., Thomsen, E.A. & Madsen, A.B. (2004). Diet of free-ranging American mink and European polecat in Denmark. *Acta Theriologica* 49, 337–347.

Hansson, L. (1983). Competition between rodents in successional stages of taiga forests: *Microtus agrestis* vs. *Clethrionomys glareolus*. *Oikos* 40, 258–266.

Harrington, L.A., Harrington, A.L. & Macdonald, D.W. (2008). Estimating the relative abundance of American mink *Mustela vison* on lowland rivers: evaluation and comparison of two techniques. *European Journal of Wildlife Research* 54, 79–87.

Harris, S., Massimino, D., Newson, S. *et al.* (2015). The Breeding Bird Survey 2014. BTO Research Report No. 673. British Trust for Ornithology, Thetford.

Harris, S., Morris, P., Wray, S. & Yalden, D. (1995). *A Review of British Mammals: Population Estimates and Conservation Status of British Mammals Other Than Cetaceans.* Joint Nature Conservation Committee, Peterborough.

Harris, S. & Yalden, D.W. (2008). *Mammals of the British Isles: Handbook.* Mammal Society, Southampton.

Heinsohn, T. (2001). Human influences on vertebrate zoogeography: animal translocation and biological invasions across and to the east of Wallace's Line. In I. Metcalfe, J.M.B. Smith, M. Morwood & I. Davidson (eds.), *Faunal and Floral Migration and Evolution in SE Asia–Australasia.* A.A. Balkema, Lisse, pp. 153–170.

Hellstedt, P. & Henttonen, H. (2006). Home range, habitat choice and activity of stoats (*Mustela erminea*) in a subarctic area. *Journal of Zoology* 269, 205–212.

Hewson, C.M. & Noble, D.G. (2009). Population trends of breeding birds in British woodlands over a 32-year period: relationships with food, habitat use and migratory behaviour. *Ibis* 151, 464–486.

Hewson, R. & Watson, A. (1979). Winter whitening of stoats (*Mustela erminea*) in Scotland and north-east England. *Journal of Zoology* 187, 55–64.

Heydon, M.J. & Reynolds, J.C. (2000). Demography of rural foxes (*Vulpes vulpes*) in relation to cull intensity in three contrasting regions of Britain. *Journal of Zoology* 251, 265–276.

Hodgman, T.P., Harrison, D.J., Katnik, D.D. & Elowe, K.D. (1994). Survival in an intensively trapped marten population in Maine. *Journal of Wildlife Management* 58, 593–600.

Holling, C.S. (1959a). The components of predation as revealed by a study of small-mammal predation of the European pine sawfly. *Canadian Entomologist* 91, 293–320.

Holling, C.S. (1959b). Some characteristics of simple types of predation and parasitism. *Canadian Entomologist* 91, 385–398.

Holt, A.R., Davies, Z.G., Tyler, C. & Staddon, S. (2008). Meta-analysis of the effects of predation on animal prey abundance: evidence from UK vertebrates. *PLoS One* 3, e2400.

Hoskins, W.G. (1964). Harvest fluctuations and English economic history 1480–1619. *Agricultural History Review* 12, 28–46.

IUCN (2013). Guidelines for Reintroductions and Other Conservation Translocations. IUCN Species Survival Commission, Gland, Switzerland.

Jędrzejewski, W. & Jędrzejewska, B. (1996). Rodent cycles in relation to biomass and productivity of ground vegetation and predation in the Palearctic. *Acta Theriologica* 41, 1–34.

Jędrzejewski, W., Jędrzejewska, B. & Szymura, L. (1995). Weasel population response, home range, and predation on rodents in a deciduous forest in Poland. *Ecology* 76, 179–195.

Jefferies, D. & Pendlebury, J. (1968). Population fluctuations of stoats, weasels and hedgehogs in recent years. *Journal of Zoology* 156, 513–517.

Jeffery, K.J., Abernethy, K.A., Tutin, C.E. & Bruford, M.W. (2007). Biological and environmental degradation of gorilla hair and microsatellite amplification success. *Biological Journal of the Linnean Society* 91, 281–294.

Jones, C., Pech, R., Forrester, G., King, C.M. & Murphy, E.C. (2011). Functional responses of an invasive top predator *Mustela erminea* to invasive meso-predators *Rattus rattus* and *Mus musculus*, in New Zealand forests. *Wildlife Research* 38, 131–140.

Jordan, N.R. (2011). A strategy for restoring the pine marten to England and Wales. Vincent Wildlife Trust, Ledbury.

Jordan, N.R., Messenger, J., Turner, P. et al. (2012). Molecular comparison of historical and contemporary pine marten (*Martes martes*) populations in the British Isles: evidence of differing origins and fates, and implications for conservation management. *Conservation Genetics* 13, 1195–1212.

Jule, K.R., Leaver, L.A. & Lea, S.E.G. (2008). The effects of captive experience on reintroduction survival in carnivores: a review and analysis. *Biological Conservation* 141, 355–363.

Kastdalen, L. & Wegge, P. (1985). Animal food in capercaillie and black grouse chicks in south east Norway – a preliminary report. *Proceeding of the International Grouse Symposium* 3, 499–513.

Kauhala, K., Helle, P. & Helle, E. (2000). Predator control and the density and reproductive success of grouse populations in Finland. *Ecography* 23, 161–168.

Kilpi, M. (1995). Breeding success, predation and local dynamics of colonial common gulls *Larus canus*. *Annales Zoologici Fennici* 32, 175–182.

King, C.M. (1975). The home range of the weasel (*Mustela nivalis*) in an English woodland. *Journal of Animal Ecology* 44, 639–668.

King, C.M. (1980). The weasel *Mustela nivalis* and its prey in an English woodland. *Journal of Animal Ecology* 49, 127–159.

King, C.M. (1983). The relationships between beech (*Nothofagus* sp.) seedfall and populations of mice (*Mus musculus*), and the demographic and dietary responses of stoats (*Mustela erminea*), in three New Zealand forests. *Journal of Animal Ecology* 52, 141–166.

King, C.M. (1989). *The Natural History of Weasels and Stoats*. Christopher Helm, London.

King, C.M. (2017a). The chronology of a sad historical misjudgement: the introductions of rabbits and ferrets in nineteenth-century New Zealand. *International Review of Environmental History* 3, 139–173.

King, C.M. (2017b). The history of transportations of stoats (*Mustela Erminea*) and weasels (*M. Nivalis*) to New Zealand, 1883–92. *International Review of Environmental History* 3, 51–87.

King, C.M. & Moody, J.E. (1982a). The biology of the stoat (*Mustela erminea*) in the National Parks of New Zealand: II. Food habits. *New Zealand Journal of Zoology* 9, 57–80.

King, C.M. & Moody, J.E. (1982b). The biology of the stoat (*Mustela erminea*) in the National Parks of New Zealand: V. Moult and colour change. *New Zealand Journal of Zoology* 9, 119–130.

King, C.M. & Moors, P.J. (1979a). On co-existence, foraging strategy and the biogeography of weasels and stoats (*Mustela nivalis* and *M. erminea*) in Britain. *Oecologia* 39, 129–150.

King, C.M. & Moors, P.J. (1979b). The life-history tactics of mustelids, and their significance for predator control and conservation in New Zealand. *New Zealand Journal of Zoology* 6, 619–622.

King, C.M., Norbury, G. & Veale, A. (2017). Small mustelids in New Zealand: invasion ecology in a different world. In D.W. Macdonald, C. Newman & L.A. Harrington (eds.), *The Biology and Conservation of Musteloids*. Oxford University Press, pp. 257–277.

King, C.M. & Powell, R.A. (2007). *The Natural History of Weasels and Stoats: Ecology, Behavior, and Management*, 2nd edition. Oxford University Press.

King, C.M., Veale, A., Patty, B. & Hayward, L. (2014). Swimming capabilities of stoats and the threat to inshore sanctuaries. *Biological Invasions* 16, 987–995.

Kitchener, A.C. & Birks, J. (2008). Ferret. In S. Harris & D.W. Yalden (eds.), *Mammals of the British Isles: Handbook*. Mammal Society, Southampton, pp. 485–487.

Klare, U., Kamler, J.F. & Macdonald, D.W. (2011). A comparison and critique of different scat-analysis methods for determining carnivore diet. *Mammal Review* 41, 294–312.

Knapp, J.L. (1829). *The Journal of a Naturalist*. John Murray, London.

Koepfli, K.-P., Deere, K.A., Slater, G.J. et al. (2008). Multigene phylogeny of the Mustelidae: resolving relationships, tempo and biogeographic history of a mammalian adaptive radiation. *BMC Biology* 6, 10.

Kohl, K.D., Coogan, S.C. & Raubenheimer, D. (2015). Do wild carnivores forage for prey or for nutrients? Evidence for nutrient-specific foraging in vertebrate predators. *BioEssays* 37, 701–709.

Koivisto, E., Koivisto, P., Hanski, I.K. et al. (2016). Prevalence of anticoagulant rodenticides in non-target predators and scavengers in Finland. *Report of the Finnish Safety and Chemicals Agency (Tukes)*, 40.

Korpimäki, E. & Norrdahl, K. (1989). Avian predation on mustelids in Europe 1: occurrence and effects on body size variation and life traits. *Oikos* 55, 205–215.

Korpimäki, E. & Norrdahl, K. (1998). Experimental reduction of predators reverses the crash phase of small-rodent cycles. *Ecology* 79, 2448–2455.

Korpimäki, E., Norrdahl, K., Klemola, T., Pettersen, T. & Stenseth, N.C. (2002). Dynamic effects of predators on cyclic voles: field experimentation and model extrapolation. *Proceedings of the Royal Society of London B* 269, 991–997.

Korpimäki, E., Oksanen, L., Oksanen, T. et al. (2005). Vole cycles and predation in temperate and boreal zones of Europe. *Journal of Animal Ecology* 74, 1150–1159.

Krebs, C.J. (2013). *Population Fluctuations in Rodents*. University of Chicago Press.

Křivan, V. (1996). Optimal foraging and predator–prey dynamics. *Theoretical Population Biology* 49, 265–290.

Kubasiewicz, L.M. (2014). Monitoring European pine martens (*Martes martes*) in Scottish forested landscapes. PhD Thesis, Stirling University.

Lahoz-Monfort, J.J., Guillera-Arroita, G. & Tingley, R. (2016). Statistical approaches to account for false-positive errors in environmental DNA samples. *Molecular Ecology Resources* 16, 673–685.

Lambin, X. (2017). The population dynamics of bite-sized predators: prey dependence, territoriality, and mobility. In D.W. Macdonald, C. Newman & L.A. Harrington (eds.), *Biology and Conservation of Musteloids*. Oxford University Press, pp. 129–148.

Lambin, X., Elston, D.A., Petty, S.J. & MacKinnon, J.L. (1998). Spatial asynchrony and periodic travelling waves in cyclic populations of field voles. *Proceedings of the Royal Society of London B* 265, 1491–1496.

Lambin, X. & Graham, I.M. (2003). Testing the specialist predator hypothesis for vole cycles. *Trends in Ecology & Evolution* 18, 493.

Lambin, X., Horrill, J. & Raynor, R. (2019). Achieving large scale, long-term invasive American mink control in northern Scotland despite short-term funding. In C.R. Veitch, M.N. Clout & D.R. Towns (eds.), *Island invasives: scaling up to meet the challenge*. IUCN, Gland, Switzerland, pp. 651–657.

Lambin, X., Petty, S.J. & MacKinnon, J.L. (2000). Cyclic dynamics in field vole populations and generalist predation. *Journal of Animal Ecology* 69, 106–119.

Langley, P.J.W. & Yalden, D.W. (1977). The decline of the rarer carnivores in Great Britain during the nineteenth century. *Mammal Review* 7, 95–116.

Lanszki, J. (2003). Feeding habits of stone martens in a Hungarian village and its surroundings. *Folia Zoologica* 52, 367–377.

Lanszki, J., Sardi, B. & Széles, L.G. (2009). Feeding habits of the stone marten (*Martes foina*) in villages and farms in Hungary. *Natura Somogyiensis* 15, 231–246.

Laundré, J.W., Hernández, L. & Altendorf, K.B. (2001). Wolves, elk, and bison: reestablishing the "landscape of fear" in Yellowstone National Park, USA. *Canadian Journal of Zoology* 79, 1401–1409.

Lawton, C., Flaherty, M., Goldstein, E.A. et al. (2015). Irish Squirrel Survey 2012. *Irish Wildlife Manuals* No. 89. National Parks and Wildlife Service, Department of the Environment, Heritage and the Gaeltacht, Ireland.

Lawton, C., Hanniffy, R., Molloy, V. et al. (2020). All-Ireland Squirrel and Pine Marten Survey 2019. *Irish Wildlife Manuals* No. 121. National Parks and Wildlife Service, Department of the Environment, Heritage and the Gaeltacht, Ireland.

Leempoel, K., Hebert, T. & Hadly, E.A. (2020). A comparison of eDNA to camera trapping for assessment of terrestrial mammal diversity. *Proceedings of the Royal Society of London B* 287, 20192353.

Lestrade, M., Vergne, T., Guinat, C. et al. (2021). Risk of anticoagulant rodenticide exposure for mammals and birds in Parc National des Pyrénées, France. *Journal of Wildlife Diseases* 57, 637–642.

Lightfoot, K.G., Panich, L.M., Schneider, T.D. & Gonzalez, S.L. (2013). European colonialism and the Anthropocene: A view from the Pacific Coast of North America. *Anthropocene* 4, 101–115.

Lima, S.L. (2009). Predators and the breeding bird: behavioral and reproductive flexibility under the risk of predation. *Biological Reviews* 84, 485–513.

Lima, S.L. & Dill, L.M. (1990). Behavioral decisions made under the risk of predation: a review and prospectus. *Canadian Journal of Zoology* 68, 619–640.

Lindahl, T. (1993). Instability and decay of the primary structure of DNA. *Nature* 362, 709–715.

Lindström, E.R., Brainerd, S.M., Helldin, J. & Overskaug, K. (1995). Pine marten–red fox interactions: a case of intraguild predation? *Annales Zoologici Fennici*, 123–130.

Lindström, J. (1994). Tetraonid population studies – state of the art. *Annales Zoologici Fennici* 31, 347–364.

Lisgo, K.A. (1999). Ecology of the short-tailed weasel (*Mustela erminea*) in the mixedwood boreal forest of Alberta. BSc Thesis, University of British Columbia.

Lockie, J.D. (1961). The food of the pine marten, *Martes martes*, in West Ross-shire, Scotland. *Proceedings of the Zoological Society of London*. Wiley Online Library, 187–195.

Lockie, J.D. (1964). Distribution and fluctuations of the pine marten, *Martes martes* (L.), in Scotland. *Journal of Animal Ecology* 33, 349–356.

Lockie, J.D. (1966). Territory in small carnivores. In P. Jewell & C. Loizos (eds.), *Play, Exploration and Territory in Mammals*. Symposium of the Zoological Society of London 18. Academic Press, London and New York, pp. 143–165.

Lodé, T. (1997). Trophic status and feeding habits of the European polecat *Mustela putorius* L. 1758. *Mammal Review* 27, 177–184.

Lovegrove, R. (2007). *Silent Fields: The Long Decline of a Nation's Wildlife*. Oxford University Press.

Lyet, A., Pellissier, L., Valentini, A. et al. (2021). eDNA sampled from stream networks correlates with camera trap detection rates of terrestrial mammals. *Scientific Reports* 11, 1–14.

Lynch, Á.B. & Mccann, Y. (2007). The diet of the pine marten (*Martes martes*) in Killarney National Park. *Biology & Environment: Proceedings of the Royal Irish Academy, 2007*. The Royal Irish Academy, 67–76.

Mac Coitir, N. (2015). *Ireland's Animals: Myths, Legends and Folklore*. Gill & Macmillan Ltd, Wilton, Cork.

Macdonald, D.W. (1978). Radio-tracking: some applications and limitations. In B. Stonehouse (ed.), *Animal Marking*. Springer, New York, pp. 192–204.

Macdonald, D.W. (2009). *The Encyclopedia of Mammals*. Oxford University Press.

Macdonald, D.W., Sidorovich, N.V., Maran, T. & Kruuk, H. (2002). *European mink, Mustela lutreola: analyses for conservation*. Wildlife Conservation Research Unit, Oxford.

Macdonald, D.W., Tew, T.E. & Todd, I.A. (2004). The ecology of weasels (*Mustela nivalis*) on mixed farmland in southern England. *Biologia* 59, 235–241.

Macleod, I., Maclennan, D., Raynor, R., Thompson, D. & Whitaker, S. (2019). Large scale eradication of non-native invasive American Mink (*Neovison vison*) from the Outer Hebrides of Scotland. In C.R. Veitch, M.N. Clout & D.R. Towns (eds.), *Island invasives: scaling up to meet the challenge*. IUCN, Gland, Switzerland, pp. 261–266.

Macpherson, H.A. (1892). *A Vertebrate Fauna of Lakeland: Including Cumberland and Westmorland with Lancashire North of the Sands*. David Douglas, Edinburgh.

MacPherson, J.L. (2017). Pine marten translocations: the road to recovery and beyond. *In Practice: Bulletin of the Chartered Institute of Ecology and Environmental Management* 95, 32–36.

MacPherson, J.L. & Bright, P.W. (2010). A preliminary investigation into whether grazing marsh is an effective refuge for water voles from predation. *Lutra* 53, 21–28.

MacPherson, J.L., Croose, E., Bavin, D. et al. (2014). Feasibility assessment for reinforcing pine marten numbers in England and Wales. Vincent Wildlife Trust, Ledbury.

MacPherson, J.L., Denman, H., Tosh, D.G., McNicol, C.M. & Halliwell, E. (2016). A review of the curent evidence for impacts of the pine marten (*Martes martes*) on non-native and native squirrel populations. In C. Shuttleworth, P.W.W. Lurz & J. Gurnell (eds.), *The Grey Squirrel: Ecology and Management of an Invasive Species in Europe*. European Squirrel Initiative, Stoneleigh Park, pp. 289–304.

MacPherson, J.L. & Wright, P. (2021). A long-term strategic recovery plan for pine marten *Martes martes* in Britain. Vincent Wildlife Trust, Ledbury.

Manzo, E., Bartolommei, P., Rowcliffe, J.M. & Cozzolino, R. (2012). Estimation of population density of European pine marten in central Italy using camera trapping. *Acta Theriologica* 57, 165–172.

Marchesi, P. (1989). Ecologie et comportement de la martre (*Martes martes* L.) dans le Jura Suisse. PhD Thesis, Université de Neuchatel.

Marchesi, P., Lachat, N., Lienhard, R., Debieve, P. & Mermod, C. (1989). Comparaison des régimes alimentaires de la fouine (*Martes foina*) et de la martre (*Martes martes*) dans une région du Jura suisse. *Revue suisse de Zoologie* 96, 281–296.

Marcstrom, V., Kenward, R.E. & Engren, E. (1988). The impact of predation on boreal tetraonids during vole cycles: an experimental study. *Journal of Animal Ecology* 57, 859–872.

Marnell, F., Kingston, N. & Looney, D. (2009). Ireland Red List No. 3: Terrestrial Mammals. National Parks & Wildlife Service, Dublin, Ireland. Available online: https://www.npws.ie/content/publications/irish-red-list-no-3-terrestrial-mammals.

Maroo, S. & Yalden, D. (2000). The mesolithic mammal fauna of Great Britain. *Mammal Review* 30, 243–248.

Martin, A.R. & Lea, V.J. (2020). A mink-free GB: perspectives on eradicating American mink *Neovison vison* from Great Britain and its islands. *Mammal Review* 50, 170–179.

Martinoli, A., Preatoni, D.G., Chiarenzi, B., Wauters, L.A. & Tosi, G. (2001). Diet of stoats (*Mustela erminea*) in an Alpine habitat: the importance of fruit consumption in summer. *Acta Oecologica* 22, 45–53.

Mathews, F. & Harrower, C. (2020). IUCN-compliant Red List for Britain's Terrestrial Mammals. Assessment by the Mammal Society under contract to Natural England, Natural Resources Wales and Scottish Natural Heritage.

Mathews, F., Kubasiewicz, L., Gurnell, J. et al. (2018). A review of the population and conservation status of British mammals: technical summary. A report by the Mammal Society under contract to Natural England, Natural Resources Wales and Scottish Natural Heritage. Natural England, Peterborough.

McAney, K. (2010). A pilot study to test the use of hair tubes to detect the Irish stoat along hedgerows in County Galway. Heritage Council, Ireland.

McDevitt, A.D., Zub, K., Kawałko, A. et al. (2012). Climate and refugial origin influence the mitochondrial lineage distribution of weasels (*Mustela nivalis*) in a phylogeographic suture zone. *Biological Journal of the Linnean Society* 106, 57–69.

McDonald, R.A. (2002). Resource partitioning among British and Irish mustelids. *Journal of Animal Ecology* 71, 185–200.

McDonald, R.A. & Harris, S. (2002). Population biology of stoats *Mustela erminea* and weasels *Mustela nivalis* on game estates in Great Britain. *Journal of Applied Ecology* 39, 793–805.

McDonald, R.A, Harris, S., Turnbull, G., Brown, P. & Fletcher, M. (1998). Anticoagulant rodenticides in stoats (*Mustela erminea*) and weasels (*Mustela nivalis*) in England. *Environmental Pollution* 103, 17–23.

McDonald, R.A., Webbon, C. & Harris, S. (2000). The diet of stoats (*Mustela erminea*) and weasels (*Mustela nivalis*) in Great Britain. *Journal of Zoology* 252, 363–371.

McGreal, F. (2010). Stoat obtaining fish, Clew Bay, Co. Mayo. *Irish Naturalists' Journal* 31, 67.

McHenry, E., O'Reilly, C., Sheerin, E., Kortland, K. & Lambin, X. (2016). Strong inference from transect sign surveys: combining spatial autocorrelation and misclassification occupancy models to quantify the detectability of a recovering carnivore. *Wildlife Biology* 22, 209–216.

McManus, A., Holland, C.V., Henttonen, H. & Stuart, P. (2021). The invasive bank vole (*Myodes glareolus*): a model system for studying parasites and ecoimmunology during a biological invasion. *Animals* 11, 2529.

McMurtrie, P., Edge, K., Crouchley, D. et al. (2011). Eradication of stoats (*Mustela erminea*) from Secretary Island, New Zealand. In C.R. Veitch, M.N. Clout & D.R. Towns (eds.), *Island invasives: eradication and management*. IUCN, Gland, Switzerland, pp. 455–460.

McNicol, C.M., Bavin, D., Bearhop, S. *et al.* (2020). Translocated native pine martens *Martes martes* alter short-term space use by invasive non-native grey squirrels *Sciurus carolinensis*. *Journal of Applied Ecology* 57, 903–913.

Mergey, M., Helder, R. & Roeder, J.-J. (2011). Effect of forest fragmentation on space-use patterns in the European pine marten (*Martes martes*). *Journal of Mammalogy* 92, 328–335.

Miller, C., Elliot, M. & Alterio, N. (2001). Home range of stoats (*Mustela erminea*) in podocarp forest, South Westland, New Zealand: implications for a control strategy. *Wildlife Research* 28, 165–172.

Mills, L.S., Bragina, E.V., Kumar, A.V. *et al.* (2018). Winter color polymorphisms identify global hot spots for evolutionary rescue from climate change. *Science* 359, 1033–1036.

Moeller, A.K., Lukacs, P.M. & Horne, J.S. (2018). Three novel methods to estimate abundance of unmarked animals using remote cameras. *Ecosphere* 9, e02331.

Moll, R.J., Kilshaw, K., Montgomery, R.A. et al. (2016). Clarifying habitat niche width using broad-scale, hierarchical occupancy models: a case study with a recovering mesocarnivore. *Journal of Zoology* 300, 177–185.

Moller, H. & Alterio, N. (1999). Home range and spatial organisation of stoats (*Mustela erminea*), ferrets (*Mustela furo*) and feral house cats (*Felis catus*) on coastal grasslands, Otago Peninsula, New Zealand: implications for yellow-eyed penguin (*Megadyptes antipodes*) conservation. *New Zealand Journal of Zoology* 26, 165–174.

Monterroso, P., Godinho, R., Oliveira, T. et al. (2019). Feeding ecological knowledge: the underutilised power of faecal DNA approaches for carnivore diet analysis. *Mammal Review* 49, 97–112.

Montgomery, W.I., Provan, J., McCabe, A.M. & Yalden, D.W. (2014). Origin of British and Irish mammals: disparate post-glacial colonisation and species introductions. *Quaternary Science Reviews* 98, 144–165.

Moors, P.J. (1975). The annual energy budget of a weasel (*Mustela nivalis* L.) population in farmland. PhD Thesis, University of Aberdeen.

Moriarty, K.M., Linnell, M.A., Thornton, J.E. & Watts III, G.W. (2018). Seeking efficiency with carnivore survey methods: a case study with elusive martens. *Wildlife Society Bulletin* 42, 403–413.

Morin, D.J., Higdon, S.D., Lonsinger, R.C. et al. (2019). Comparing methods of estimating carnivore diets with uncertainty and imperfect detection. *Wildlife Society Bulletin* 43, 651–660.

Morris, T.A. (2002). *Europe and England in the Sixteenth Century*. Routledge, New York.

Mos, J. & Hofmeester, T.R. (2020). The *Mostela*: an adjusted camera trapping device as a promising non-invasive tool to study and monitor small mustelids. *Mammal Research* 65, 843–853.

Mulder, J.L. (1990). The stoat *Mustela erminea* in the Dutch dune region, its local extinction and a possible cause: the arrival of the fox *Vulpes vulpes*. *Lutra* 33, 1–21.

Mullins, J., Statham, M.J., Roche, T., Turner, P.D. & O'Reilly, C. (2010). Remotely plucked hair genotyping: a reliable and non-invasive method for censusing pine marten (*Martes martes* L. 1758) populations. *European Journal of Wildlife Research* 56, 443–453.

Murphy, E.C. & Dowding, J.E. (1994). Range and diet of stoats (*Mustela erminea*) in a New Zealand beech forest. *New Zealand Journal of Ecology* 18, 11–18.

Murphy, E.C. & Dowding, J.E. (1995). Ecology of the stoat in *Nothofagus* forest: home range, habitat use and diet at different stages of the beech mast cycle. *New Zealand Journal of Ecology* 19, 97–109.

Musacchio, J.M. (2001). Weasels and pregnancy in Renaissance Italy. *Renaissance Studies* 15, 172–187.

Newton, I. (1998). *Population Limitation in Birds*. Academic Press, London.

Norbury, G. & Jones, C. (2015). Pests controlling pests: does predator control lead to greater European rabbit abundance in Australasia? *Mammal Review* 45, 79–87.

Nordstrom, M., Hogmander, J., Nummelin, J. et al. (2002). Variable responses of waterfowl breeding populations to long-term removal of introduced American mink. *Ecography* 25, 385–394.

NSRF (2014). The Scottish Code for Conservation Translocations: Best Practice Guidelines for Conservation Translocations in Scotland. Scottish Natural Heritage, Inverness.

Nyholm, E.S. (1959). Stoats and weasels in their winter habitat. In C.M. King (ed.), *Biology of Mustelids: Some Soviet Research*. British Library Lending Division, Boston Spa, Yorkshire, pp. 118–131.

O'Donnell, C.F., Dilks, P.J. & Elliott, G.P. (1996). Control of a stoat (*Mustela erminea*) population irruption to enhance mohua (yellowhead) (*Mohoua ochrocephala*) breeding success in New Zealand. *New Zealand Journal of Zoology* 23, 279–286.

Oli, M.K. (2003). Population cycles of small rodents are caused by specialist predators: or are they? *Trends in Ecology & Evolution* 18, 105–107.

O'Mahony, D.T. (2014). Socio-spatial ecology of pine marten (*Martes martes*) in conifer forests, Ireland. *Acta Theriologica* 59, 251–256.

O'Mahony, D.T., O'Reilly, C. & Turner, P. (2006). National Pine Marten Survey of Ireland 2005. *COFORD Connects Environment* No. 7.

O'Mahony, D.T, O'Reilly, C. & Turner, P. (2012). Pine marten (*Martes martes*) distribution and abundance in Ireland: a cross-jurisdictional analysis using non-invasive genetic survey techniques. *Mammalian Biology* 77, 351–357.

O'Mahony, D.T., Powell, C., Power, J. et al. (2017). Non-invasively determined multi-site variation in pine marten *Martes martes* density, a recovering carnivore in Europe. *European Journal of Wildlife Research* 63, 48.

O'Reilly, C., Statham, M., Mullins, J., Turner, P.D. & O'Mahony, D.T. (2008). Efficient species identification of pine marten (*Martes martes*) and red fox (*Vulpes vulpes*) scats using a 5' nuclease real-time PCR assay. *Conservation Genetics* 9, 735–738.

O'Reilly, C., Turner, P., O'Mahony, D.T. et al. (2021). Not out of the woods yet: genetic insights related to the recovery of the pine marten (*Martes martes*) in Ireland. *Biological Journal of the Linnean Society* 132, 774–788.

Osinga, T., Thurfjell, H. & Hofmeester, T.R. (2022). Snow limits polecat *Mustela putorius* distribution in Sweden. *Wildlife Biology*, e01051.

O'Sullivan, P.J. (1983). The distribution of the Pine marten (*Martes martes*) in the Republic of Ireland. *Mammal Review* 13, 39–44.

PACEC (2014). The Value of Shooting: The economic, environmental and social contribution of shooting sports to the UK. British Association for Shooting and Conservation, Rossett.

Packer, J.J. & Birks, J.D. (1999). An assessment of British farmers' and gamekeepers' experiences, attitudes and practices in relation to the European polecat *Mustela putorius*. *Mammal Review* 29, 75–92.

Paez, M., Sundaram, A., Willoughby, M. & Janna, R. (2021). Comparison of minimally invasive monitoring methods and live trapping in mammals. *Genes* 12, 1949.

Paterson, W.D. & Skipper, G. (2008). The diet of pine martens (*Martes martes*) with reference to squirrel predation in Loch Lomond and the Trossachs National Park, Scotland. *Glasgow Naturalist* 25, 75–82.

Peacock, S. (2011). Projected 21st century climate change for wolverine habitats within the contiguous United States. *Environmental Research Letters* 6, 014007.

Pertoldi, C., Rødjajn, S., Zalewski, A. et al. (2013). Population viability analysis of American mink (*Neovison vison*) escaped from Danish mink farms. *Journal of Animal Science* 91, 2530–2541.

Petty, S.J. (1999). Diet of tawny owls (*Strix aluco*) in relation to field vole (*Microtus agrestis*) abundance in a conifer forest in northern England. *Journal of Zoology* 248, 451–465.

Pierce, R. (1996). Ecology and management of the black stilt *Himantopus novaezelandiae*. *Bird Conservation International* 6, 81–88.

Poulton, S., Birks, J.D.S., Messenger, J.E. & Jefferies, D.J. (2006). A quality-scoring system for using sightings data to assess pine marten distribution at low densities. In M. Santos-Reis, J. Birks, E.C. O'Doherty & G. Proulx (eds.), *Martes in Carnivore Communities*. Alpha Wildlife Publications, Sherwood Park, Alberta, pp. 177–202.

Pounds, C.J. (1981). Niche overlap in sympatric populations of stoats (*Mustela erminea*) and weasels (*Mustela nivalis*) in north-east Scotland. DPhil Thesis, University of Aberdeen.

Powell, R.A. (1982). Evolution of black-tipped tails in weasels: predator confusion. *American Naturalist* 119, 126–131.

Powell, R.A. (1994). Structure and spacing of *Martes* populations. In S.W. Buskirk, A.S. Harestad, M.G. Raphael & R.A. Powell (eds.), *Martens, Sables, and Fishers: Biology and Conservation*. Cornell University Press, New York, pp. 101–121.

Powell, R.A. (2001). Theory and methods in carnivore conservation. *Endangered Species Update* 18, 98–102.

Powell, R.A., Lewis, J.C., Slough, B.G. et al. (2012). Evaluating translocations of martens, sables and fishers: testing model predictions with field data. In K.B. Aubrey, J. Zielinski, M.G. Raphael, G. Proulx & S.W. Buskirk (eds.), *Biology and Conservation of Martens, Sables, and Fishers: A New Synthesis*. Cornell University Press, New York, pp. 93–137.

Power, J. (2015). Non-invasive genetic monitoring of pine marten (*Martes Martes*) and stone marten (*Martes foina*) in and around the Nietoperek bat hibernation site, Poland. PhD Thesis, Waterford Institute of Technology.

Previtali, A., Cassini, M.H. & Macdonald, D.W. (1998). Habitat use and diet of the American mink (*Mustela vison*) in Argentinian Patagonia. *Journal of Zoology* 246, 482–486.

Pulliainen, E. (1981). Winter habitat selection, home range, and movements of the pine marten (*Martes martes*) in a Finnish Lapland forest. In J.A. Chapman & D. Pursley (eds.), *Proceedings of the Worldwide Furbearer Conference*, Frostburg, Maryland. Johns Hopkins Press, Baltimore, pp. 1068–1087.

Pulliainen, E. & Ollinmaki, P. (1996). A long-term study of the winter food niche of the pine marten *Martes martes* in northern boreal Finland. *Acta Theriologica* 41, 337–352.

Putman, R.J. (2000). Diet of pine martens (*Martes martes* L.) in west Scotland. *Journal of Natural History* 34, 793–797.

Rae, S. (1999). The effect of predation by mink on ground-nesting birds in the Outer Hebrides. *Outer Hebrides (The Western Isles) Bird Report 1999*, 105–113.

Ratz, H. (2000). Movements by stoats (*Mustela erminea*) and ferrets (*M. furo*) through rank grass of yellow-eyed penguin (*Megadyptes antipodes*) breeding areas. *New Zealand Journal of Zoology* 27, 57–69.

Rebergen, A., Keedwell, R., Moller, H. & Maloney, R. (1998). Breeding success and predation at nests of banded dotterel (*Charadrius bicinctus*) on braided riverbeds in the central South Island, New Zealand. *New Zealand Journal of Ecology* 22, 33–41.

Reddiex, B., Hickling, G.J., Norbury, G.L. & Frampton, C.M. (2002). Effects of predation and rabbit haemorrhagic disease on population dynamics of rabbits (*Oryctolagus cuniculus*) in North Canterbury, New Zealand. *Wildlife Research* 29, 627–633.

Remonti, L., Balestrieri, A., Raubenheimer, D. & Saino, N. (2016). Functional implications of omnivory for dietary nutrient balance. *Oikos* 125, 1233–1240.

Remonti, L., Balestrieri, A., Ruiz-González, A. et al. (2012). Intraguild dietary overlap and its possible relationship to the coexistence of mesocarnivores in intensive agricultural habitats. *Population Ecology* 54, 521–532.

Reynolds, J.C. & Aebishcer, N. (1991). Comparison and quantification of carnivore diet by faecal analysis: a critique, with recommendations, based on a study of the fox *Vulpes vulpes*. *Mammal Review* 21, 97–122.

Reynolds, J.C., Goddard, H.N. & Brockless, M.H. (1993). The impact of local fox (*Vulpes vulpes*) removal on fox populations at two sites in southern England. *Gibier Faune Sauvage* 10, 319–334.

Robert, A., Colas, B., Guigon, I. et al. (2015). Defining reintroduction success using IUCN criteria for threatened species: a demographic assessment. *Animal Conservation* 18, 397–406.

Roche, R. & Merne, O. (1977). *Saltees: Islands of Birds and Legends*. Macmillan, Toronto.

Romanowski, J. & Lesinski, G. (1991). A note on the diet of stone marten in southeastern Romania. *Acta Theriologica* 36, 201–204.

Roth, T.C., Lima, S.L. & Vetter, W.E. (2006). Determinants of predation risk in small wintering birds: the hawk's perspective. *Behavioral Ecology and Sociobiology* 60, 195–204.

Rowcliffe, J.M., Carbone, C., Kays, R. et al. (2014). Density estimation using camera trap surveys: the Random Encounter Model. In P. Meek & A.G. Ballard (eds.), *Camera trapping in Wildlife Management and Research*. CSIRO Publishing, Melbourne, pp. 317–324.

Roy, S., Milborrow, J., Allan, J. & Robertson, P. (2014). Pine martens on the Isle of Mull – assessing risks to native species. NatureScot Commissioned Report No. 560.

Roy, S., Reid, N. & Mcdonald, R.A. (2009a). A review of mink predation and control for Ireland. *Irish Wildlife Manuals* No. 40. National Parks and Wildlife Service, Department of the Environment, Heritage and the Gaeltacht, Ireland.

Roy, S., Reid, N. & Mcdonald, R.A. (2009b). A review of mink predation and control for Ireland. Dublin, Ireland: National Parks and Wildlife Service, Department of the Environment, Heritage and the Gaeltacht, Ireland.

Ruell, E.W. & Crooks, K.R. (2007). Evaluation of noninvasive genetic sampling methods for felid and canid populations. *Journal of Wildlife Management* 71, 1690–1694.

Ruette, S., Vandel, J.M., Albaret, M. & Devillard, S. (2015). Comparative survival pattern of the syntopic pine and stone martens in a trapped rural area in France. *Journal of Zoology* 295, 214–222.

Ruffino, L., Hartley, S.E., Degabriel, J.L. & Lambin, X. (2018). Population-level manipulations of field vole densities induce subsequent changes in plant quality but no impacts on vole demography. *Ecology and Evolution* 8, 7752–7762.

Ruiz-Suárez, N., Melero, Y., Giela, A. *et al.* (2016). Rate of exposure of a sentinel species, invasive American mink (*Neovison vison*) in Scotland, to anticoagulant rodenticides. *Science of the Total Environment* 569, 1013–1021.

Ruiz-Gutierrez, V., Hooten, M.B. & Campbell Grant, E.H. (2016). Uncertainty in biological monitoring: a framework for data collection and analysis to account for multiple sources of sampling bias. *Methods in Ecology and Evolution* 7, 900–909.

Sainsbury, K.A., Shore, R.F., Schofield, H. *et al.* (2018). Long-term increase in secondary exposure to anticoagulant rodenticides in European polecats *Mustela putorius* in Great Britain. *Environmental Pollution* 236, 689–698.

Sainsbury, K.A., Shore, R.F., Schofield, H. *et al.* (2020). Diets of European polecat *Mustela putorius* in Great Britain during fifty years of population recovery. *Mammal Research* 65, 181–190.

Sales, N.G., Mckenzie, M.B., Drake, J. *et al.* (2020). Fishing for mammals: Landscape-level monitoring of terrestrial and semi-aquatic communities using eDNA from riverine systems. *Journal of Applied Ecology* 57, 707–716.

Salo, P., Korpimäki, E., Banks, P.B., Nordström, M. & Dickman, C.R. (2007). Alien predators are more dangerous than native predators to prey populations. *Proceedings of the Royal Society of London B* 274, 1237–1243.

Salomonsen, F. (1939). *Moults and Sequence of Plumages in the Rock Ptarmigan* (Lagopus mutus [Montin]). P. Haase & Son, Copenhagen.

Sandell, M. (1986). Movement patterns of male stoats *Mustela erminea* during the mating season: differences in relation to social status. *Oikos* 47, 63–70.

Sandell, M. & Liberg, O. (1992). Roamers and stayers: a model on male mating tactics and mating systems. *American Naturalist* 139, 177–189.

Santos, M.J., Matos, H.M., Baltazar, C., Grilo, C. & Santos-Reis, M. (2009). Is polecat (*Mustela putorius*) diet affected by "mediterraneity"? *Mammalian Biology* 74, 448–455.

Scott, R. (2011). *Atlas of Highland Land Mammals*. Highland Biological Recording Group, Inverness.

Seddon, P.J. (2010). From reintroduction to assisted colonization: moving along the conservation translocation spectrum. *Restoration Ecology* 18, 796–802.

Sheehy, E. (2013). The role of the pine marten in Irish squirrel population dynamics. PhD Thesis, NUI Galway.

Sheehy, E. & Lawton, C. (2014). Population crash in an invasive species following the recovery of a native predator: the case of the American grey squirrel and the European pine marten in Ireland. *Biodiversity and Conservation* 23, 753–774.

Sheehy, E. & Lawton, C. (2015). Predators of red and grey squirrels in their natural and introduced ranges. In C. Shuttleworth, P.W.W. Lurz & M. Hayward (eds.), *Red Squirrels: Ecology, Conservation & Management in Europe*. European Squirrel Initiative, pp. 211–232.

Sheehy, E., Sutherland, C., O'Reilly, C. & Lambin, X. (2018). The enemy of my enemy is my friend: native pine marten recovery reverses the decline of the red squirrel by suppressing grey squirrel populations. *Proceedings of the Royal Society of London B* 285, 20172603.

Shore, R., Birks, J., Afsar, A., Wienburg, C. & Kitchener, A. (2003). Spatial and temporal analysis of second-generation anticoagulant rodenticide residues in polecats (*Mustela putorius*) from throughout their range in Britain, 1992–1999. *Environmental Pollution* 122, 183–193.

Sidorovich, V.E. (2000). Seasonal variation in the feeding habits of riparian mustelids in river valleys of NE Belarus. *Acta Theriologica* 45, 233–242.

Sidorovich, V.E., Krasko, D., Sidorovich, A., Solovej, I. & Dyman, A. (2006). The pine marten's *Martes martes* ecological niche and its relationships with other vertebrate predators in the transitional mixed forest ecosystems of northern Belarus. In M. Santos-Reis, J. Birks, F.C. O'Doherty & G. Proulx (eds.), *Martes in Carnivore Communities*. Alpha Wildlife Publications, Sherwood Park, Alberta, pp. 109–126.

Sidorovich, V.E. & Macdonald, D.W. (2001). Density dynamics and changes in habitat use by the European mink and other native mustelids in connection with the American mink expansion in Belarus. *Netherlands Journal of Zoology* 51, 107–126.

Simms, D. (1979). North American weasels: resource utilization and distribution. *Canadian Journal of Zoology* 57, 504–520.

Simpson, V. (2006). Patterns and significance of bite wounds in Eurasian otters (*Lutra lutra*) in southern and south-west England. *Veterinary Record* 158, 113–119.

Siriwardena, G.M. (2004). Possible roles of habitat, competition and avian nest predation in the decline of the Willow Tit *Parus montanus* in Britain. *Bird Study* 51, 193–202.

Siriwardena, G.M. (2006). Avian nest predation, competition and the decline of British marsh tits *Parus palustris*. *Ibis* 148, 255–265.

Sleeman, D. (1992). Diet of Irish stoats. *Irish Naturalists' Journal* 24, 151–153.

Slough, B.G. (1989). Movements and habitat use by transplanted marten in the Yukon Territory. *Journal of Wildlife Management* 53, 991–997.

Smal, C.M. (1988). The American mink *Mustela vison* in Ireland. *Mammal Review* 18, 201–208.

Smal, C.M. (1991). Population studies on feral American mink *Mustela vison* in Ireland. *Journal of Zoology* 224, 233–249.

Smith, S.A. & Gilbert, J. (2001). *National Inventory of Woodland and Trees: England*. Forestry Commission, Edinburgh.

Solow, A., Roy, S., Bell, C., Milborrow, J. & Roberts, D. (2013). On inference about the introduction time of an introduced species with an application to the pine marten on Mull. *Biological Conservation* 159, 4–6.

Sorace, A., Petrassi, F. & Consiglio, C. (2004). Long-distance relocation of nestboxes reduces nest predation by pine marten *Martes martes*. *Bird Study* 51, 119–124.

Stetz, J.B., Seitz, T. & Sawaya, M.A. (2015). Effects of exposure on genotyping success rates of hair samples from brown and American black bears. *Journal of Fish and Wildlife Management* 6, 191–198.

Storch, I., Lindstrom, E. & De Jounge, J. (1990). Habitat selection and food habits of the pine marten in relation to competition with the red fox. *Acta Theriologica* 35, 311–320.

Strann, K.B., Yoccoz, N.G. & Ims, R.A. (2002). Is the heart of Fennoscandian rodent cycle still beating? A 14-year study of small mammals and Tengmalm's owls in northern Norway. *Ecography* 25, 81–87.

Stringer, A.P., MacPherson, J., Carter, S. et al. (2018). *The feasibility of reintroducing pine martens (Martes martes) to the Forest of Dean and lower Wye Valley*. Gloucestershire Wildlife Trust, Vincent Wildlife Trust & the Forestry Commission.

Suffice, P., Asselin, H., Imbeau, L., Cheveau, M. & Drapeau, P. (2017). More fishers and fewer martens due to cumulative effects of forest management and climate change as evidenced from local knowledge. *Journal of Ethnobiology and Ethnomedicine* 13, 1–14.

Summers, R.W., Green, R.E., Proctor, R. et al. (2004). An experimental study of the effects of predation on the breeding productivity of capercaillie and black grouse. *Journal of Applied Ecology* 41, 513–525.

Sundell, J. & Norrdahl, K. (2002). Body size-dependent refuges in voles: an alternative explanation of the Chitty effect. *Annales Zoologici Fennici* 39, 325–333.

Sundell, J. & Ylönen, H. (2008). Specialist predator in a multi-species prey community: boreal voles and weasels. *Integrative Zoology* 3, 51–63.

Tabak, M.A., Falbel, D., Hamzeh, T. et al. (2022). CameraTrapDetectoR: Automatically detect, classify, and count animals in camera trap images using artificial intelligence. *bioRxiv*, 479461.

Tapper, S. (1979). The effect of fluctuating vole numbers (*Microtus agrestis*) on a population of weasels (*Mustela nivalis*) on farmland. *Journal of Animal Ecology* 48, 603–617.

Tapper, S. (1992). *Game Heritage: An Ecological Review from Shooting and Gamekeeping Records*. Game Conservancy, Fordingbridge.

Teerink, B. (2003). *Hair of West-European Mammals: Atlas and Identification Key*. Cambridge University Press.

Thom, M.D., Johnson, D.D. & Macdonald, D.W. (2004). The evolution and maintenance of delayed implantation in the Mustelidae (Mammalia: Carnivora). *Evolution* 58, 175–183.

Thompson, C.M., Royle, J.A. & Garner, J.D. (2012). A framework for inference about carnivore density from unstructured spatial sampling of scat using detector dogs. *Journal of Wildlife Management* 76, 863–871.

Thompson, W. (1856). *The Natural History of Ireland*. Reeve, Benham and Reeve, London.

Thomson, A.P. (1951). A history of the ferret. *Journal of the History of Medicine and Allied Sciences* 6, 471–480.

Thomson, D.L., Green, R.E., Gregory, R.D. & Baillie, S.R. (1998). The widespread declines of songbirds in rural Britain do not correlate with the spread of their avian predators. *Proceedings of the Royal Society of London B* 265, 2057–2062.

Tosh, D.G., Lusby, J., Montgomery, W.I. & O'Halloran, J. (2008). First record of greater white-toothed shrew *Crocidura russula* in Ireland. *Mammal Review* 38, 321–326.

Trout, R., Chasey, D. & Sharp, G. (1997). Seroepidemiology of rabbit haemorrhagic disease (RHD) in wild rabbits (*Oryctolagus cuniculus*) in the United Kingdom. *Journal of Zoology* 243, 846–853.

Trout, R., Langton, S., Smith, G. & Haines-Young, R. (2000). Factors affecting the abundance of rabbits (*Oryctolagus cuniculus*) in England and Wales. *Journal of Zoology* 252, 227–238.

Twining, J.P., Montgomery, W.I., Price, L., Kunc, H.P. & Tosh, D.G. (2020a). Native and invasive squirrels show different behavioural responses to scent of a shared native predator. *Royal Society Open Science* 7, 191841.

Twining, J.P., Montgomery, W.I. & Tosh, D.G. (2020b). The dynamics of pine marten predation on red and grey squirrels. *Mammalian Biology* 100, 285–293.

Twining, J.P., Montgomery, W.I. & Tosh, D.G. (2021). Declining invasive grey squirrel populations may persist in refugia as native predator recovery reverses squirrel species replacement. *Journal of Applied Ecology* 58, 248–260.

Twining, J.P., Sutherland, C., Reid, N. & Tosh, D.G. (2022). Habitat mediates coevolved but not novel species interactions. *Proceedings of the Royal Society of London B* 289, 20212338.

Vaisfeld, M. (1972). Ecology of the ermine in the cold season in the European North. In C.M. King (ed.), *Biology of Mustelids: Some Soviet Research*. New Zealand Department of Scientific and Industrial Research, Wellington, pp. 1–10.

Valkama, J., Korpimäki, E., Arroyo, B. et al. (2005). Birds of prey as limiting factors of gamebird populations in Europe: a review. *Biological Reviews* 80, 171–203.

Van Den Berge, K. & Gouwy, J. (2006). Hot spot for pine marten (*Martes martes*) and first record of a natal den in Flanders (Belgium). *Lutra* 54, 99–109.

van der Merwe, M. & Brown, J.S. (2008). Mapping the landscape of fear of the Cape ground squirrel (*Xerus inauris*). *Journal of Mammalogy* 89, 1162–1169.

van Wijngaarden, A. & Morzer Bruijns, M.F. (1961). The ermine *Mustela erminea* L., on the island of Terschelling. *Lutra* 3, 35–42.

Veale, A.J. (2013). The invasion ecology and molecular ecology of stoats (*Mustela erminea*) on New Zealand's islands. PhD Dissertation, University of Auckland.

Veale, A.J., Edge, K.A., McMurtrie, P. *et al.* (2013). Using genetic techniques to quantify reinvasion, survival and *in situ* breeding rates during control operations. *Molecular Ecology* 22, 5071–5083.

Veale, A.J., Hannaford, O.D., Russell, J.C. & Clout, M.N. (2012). Modelling the distribution of stoats on New Zealand offshore islands. *New Zealand Journal of Ecology* 36, 38–47.

Veale, A.J., Holland, O., McDonald, R.A., Clout, M. & Gleeson, D. (2015). An invasive non-native mammal population conserves genetic diversity lost from its native range. *Molecular Ecology* 24, 2156–2163.

Veale, E.M. (1966). *The English Fur Trade in the Later Middle Ages*. Oxford University Press.

Velander, K.A. (1983). *Pine Marten Survey of Scotland, England and Wales: 1980–1982*. Vincent Wildlife Trust, London.

Velander, K.A. (1986). *A Study of Pine Marten Ecology in Inverness-shire*. Nature Conservancy Council, Peterborough.

Virgós, E., Zalewski, A., Rosalino, L.M. et al. (2012). Habitat ecology of *Martes* species in Europe. In K.B. Aubry, W.J. Zielinski, M.G. Raphael, G. Proulx & S.W. Buskirk (eds.), *Biology and Conservation of Martens, Sables, and Fishers: A New Synthesis*, pp. 255–266.

Walsingham, L. & Payne-Gallwey, R. (1887). *The Badminton Library of Sport and Pastimes: Shooting*. Longmans Green & Co., London.

Walton, K.C. (1968). Studies on the biology of the polecat *Mustela putorius* (L). MSc Thesis, University of Durham.

Warren, P.K. & Baines, D. (2002). Dispersal, survival and causes of mortality in black grouse *Tetrao tetrix* in northern England. *Wildlife Biology*, 8, 91–97.

Wauters, L.A., Gurnell, J., Martinoli, A. & Tosi, G. (2002a). Interspecific competition between native Eurasian red squirrels and alien grey squirrels: does resource partitioning occur? *Behavioral Ecology and Sociobiology* 52, 332–341.

Wauters, L.A., Tosi, G. & Gurnell, J. (2002b). Interspecific competition in tree squirrels: do introduced grey squirrels (*Sciurus carolinensis*) deplete tree seeds hoarded by red squirrels (*S. vulgaris*)? *Behavioral Ecology and Sociobiology* 51, 360–367.

Wearn, O.R. & Glover-Kapfer, P. (2019). Snap happy: camera traps are an effective sampling tool when compared with alternative methods. *Royal Society Open Science* 6, 181748.

Weber, D. (1989). The ecological significance of resting sites and the seasonal habitat change in polecats (*Mustela putorius*). *Journal of Zoology* 217, 629–638.

Westman, K. (1968). On the occurrence of American and European mink in Finland. *Suomen Riista* 20, 50–61.

Westra, S.A. (2019). A nine month small mustelid survey across four research sites in the Netherlands. *Lutra* 62, 89–107.

Wildhagen, A. (1965). Present distribution of North American mink in Norway. *Journal of Mammalogy* 37, 116–118.

Williams, W.P. & Jones, W.A. (1873). *A Glossary of Provincial Words and Phrases in use in Somersetshire*. Longmans, Green, Reader and Dyer, London.

Wilson, R.P., Rose, K.A., Gunner, R. et al. (2021). Animal lifestyle affects acceptable mass limits for attached tags. *Proceedings of the Royal Society of London B* 288, 20212005.

Wolf, C.M., Griffith, B., Reed, C. & Temple, S.A. (1996). Avian and mammalian translocations: update and reanalysis of 1987 survey data. *Conservation Biology* 10, 1142–1154.

Wright, J. (1898–1905). *The English Dialect Dictionary*. Henry Frowde, Oxford.

Wright, P.G., Croose, E. & MacPherson, J.L. (2022). A global review of the conservation threats and status of mustelids. *Mammal Review* 52, 410–424.

Yalden, D. (1999). *The History of British Mammals*. A & C Black, London.

Yamaguchi, N. & Macdonald, D.W. (2003). The burden of co-occupancy: intraspecific resource competition and spacing patterns in American mink, *Mustela vison*. *Journal of Mammalogy* 84, 1341–1355.

Yazan, Y.P. (1970). Relations between the marten (*Martes martes*), sable (*Martes zibellina*), and kidas (*M. martes* × *M. zibellina*) as predators, and the squirrel

(*Sciurus vulgaris*) as a prey. *Transaction International Congress of Game Biologists* 9, 530–538.

Zalewski, A. (2000). Factors affecting the duration of activity by pine martens (*Martes martes*) in the Białowieża National Park, Poland. *Journal of Zoology* 251, 439–447.

Zalewski, A. (2005). Geographical and seasonal variation in food habits and prey size of European pine martens. In D.J. Harrison, A.K. Fuller & G. Proulx (eds.), *Martens and Fishers* (Martes) *in Human-Altered Environments*. Springer, Boston, pp. 77–98.

Zalewski, A. & Jędrzejewski, W. (2006). Spatial organisation and dynamics of the pine marten Martes martes population in Białowieza Forest (E Poland) compared with other European woodlands. *Ecography* 29, 31–43.

Zalewski, A., Jędrzejewski, W. & Jędrzejewska, B. (1995). Pine marten home ranges, numbers and predation on vertebrates in a deciduous forest (Białowieza National Park, Poland). *Annales Zoologici Fennici* 32, 131–144.

Zalewski, A., Szymura, M., Kowalczyk, R. & Brzeziński, M. (2021). Low individual diet variation and high trophic niche overlap between the native polecat and invasive American mink. *Journal of Zoology* 314, 151–161.

Zielinski, W.J., Linnell, M.A., Schwartz, M.K. & Pilgrim, K. (2020). Exploiting the winter trophic relationship between weasels (*Mustela* spp.) and their microtine prey as a survey method for weasels in meadow ecosystems. *Northwest Science* 93, 185–192.

Zielinski, W.J., Tucker, J.M. & Rennie, K.M. (2017). Niche overlap of competing carnivores across climatic gradients and the conservation implications of climate change at geographic range margins. *Biological Conservation* 209, 533–545.

Index

SPECIES INDEX

Note: all index entries refer to the British Isles unless otherwise described. Cross-reference targets in *italics* refer to subentries within the same main entry topic. Page numbers in **bold** refer to figures and figure captions. Page numbers in *italics* refer to tables.

Aegolius funereus (Tengmalm's Owl) 255
Alcedo atthis (Kingfisher) 57
American Marten (*Martes americana*) 17, 110, 136, **266**
  climate change impact 311
  distribution *265*, **267**
American Mink (*Neovison vison*) 1, 4, **4**, 10, 14, 161–177
  bird species, impact on *see* birds, American Mink impact (*General Index*)
  body shape **13**, **162**
  capture-mark-release-recapture 169, *169*
  competition with Eurasian Otter 163, 164
  competition with European Polecats 42, 163, 309
  competition with Stoats 42
  control/eradication attempts 167–174
    costs 161, 176
    effectiveness 183
    Europe 164
    habitat favourability model 176–177
    Ireland 174–177
    mainland Britain 167, **168**, 168–169, *169*, 170–173, 174
    Outer Hebrides 165–167, 175–176
    for protection of vulnerable species 175–176, 177
    replacement animals, source area 169–170, 175
  return of mink after 167–168, 169, 170, 175, 177
  Scotland 173–174
  'vacuum effect' 169, 170, 175
  see also trapping (below)
  COVID-19 susceptibility 322
  daily food requirements 239
  decline, with otter recovery 164
  delayed implantation 44, 45–46
  dens and sites of 163
  diet 163, 164, 212
  dietary niche 163, 164, **230**
  distribution 161–162, 163, 167
    Ireland 176, **176**
    Outer Hebrides 165
  escapes/releases from fur farms 14, 161, 163, 165, 174, 177
  establishment in UK 14, 161–165, 167–174
  fur colour 11, 13, **13**, **162**, 168
  fur trade and 161–163, **162**, 164, 290
    British mink farms **162**, 163, 164–165, 177, 251
    COVID-19 at fur farms 322
    in Europe 161–162
    Irish mink farms 174, **175**, 177
    see also fur farms (*General Index*)
  as generalist predators 163
  habitat 14, 163, 176
  Hebridean Mink Project **165**, 165–167, 175–176
  home range size, males *vs* females 303
  humane killing 172

SPECIES INDEX

interspecific aggression with Otters 164
as invasive species 161–165, 251
  due to loss of small carnivores 183
  Europe 164
  impact on native fauna 163, 164, 165–167, 174
  Ireland 174–177, **176**
  Isle of Mull 181
  mainland Britain 167–173
  Outer Hebrides 165–167, 175–176
  Scotland 164, 173–174
kits and litters 44
mating and breeding 44
prey, hunting 41–42, **42**
  impact on home range size 303
recolonisation prevention 173–174, 176
scats 186
semiaquatic nature 14, 163
size, and weight 14, 28, 163
tail **13**, 28
timing of births 45–46
trapping 166, 167, 168, 170
  cage traps 168, 170
  Ireland 174–175, 176
  live-capture traps 172
  mainland Britain 170–173
  male to female ratio 170, **171**
  'mink rafts' **165**, 170–173, **172**, **173**
  numbers (2003–2005), UK **168**, 168–169, 169
  Outer Hebrides 166, 167, 168, 175
  Scotland 173–174
Anarhynchus frontalis (Wrybill) 249
Aonyx reevei (Clawless Otter) 18
Apodemus flavicollis (Yellow-necked Mouse) 199, 199, 246
Apodemus sylvaticus see Wood Mouse (*Apodemus sylvaticus*)
Ardea cinerea (Grey Heron) 6, **7**, **8**
Arvicola amphibius see Water Vole (*Arvicola amphibius*)

Badgers 2, 3, 15
  European *see* European Badger (*Meles meles*)
Baltic Squirrel 50, 52
Banded Dotterel (*Charadrius bicinctus*) 249
Bank Vole (*Myodes glareolus*) 41
  increase with clear-felling 130
  in Ireland 220
  Kielder Forest 243–244
  mortality in winter 246
  in Pine Marten diet 44, 130, 226, 252
  in Wales 199, 199, 200
  in Weasel diet 215, 216

Beech Marten *see* Stone Marten (*Martes foina*)
Black Grouse (*Tetrao tetrix*) 257, 258
Black Stilt (*Himantopus novaezelandiae*) 249
Black Woodpecker (*Dryocopus martius*) 255
Blackbird (*Turdus merula*) 255
Blue Tit (*Cyanistes caeruleus*) 255, **255**
Brown Hare (*Lepus europaeus*) 62–63
Brown Rat (*Rattus norvegicus*) **216**, 219–220, **231**
  costs and rodenticide use 231–232
Bullfinch (*Pyrrhula pyrrhula*) 62–63
Buteo buteo (Buzzard) **241**
Buzzard (*Buteo buteo*) **241**

Canis lupus lupus (Eurasian Wolf) 48
Capercaillie (*Tetrao urogallus*) 258
Castor fiber (Eurasian Beaver) 72
Charadrius bicinctus (Banded Dotterel) 249
Circus approximans (Harrier Hawk) 149
Clawless Otter (*Aonyx reevei*) 18
Columba palumbus (Woodpigeon) 226
Common Kestrel (*Falco tinnunculus*) 246
Common Shrew (*Sorex araneus*) 199, 199
Common Weasel (*Mustela nivalis vulgaris*) 5, 314, 315
  fur colour, winter 314, 315
  see also Weasel (*Mustela nivalis*)
Corncrake (*Crex crex*) **160**, 160–161, 179
Cottony Cushion Scale (*Icerya purchasi*) **270**, 271
Crex crex (Corncrake) **160**, 160–161, 179
Crocidura russula (Greater White-toothed Shrew) 220
Curlew (*Numenius arquata*) **166**, 248

Dryocopus martius (Black Woodpecker) 255

Enhydra lutris (Sea Otter) 1, **49**
Eurasian Beaver (*Castor fiber*) 72
Eurasian Lynx (*Lynx lynx*) 48
Eurasian Otter (*Lutra lutra*) 1, 2, **4**, **13**
  AIHTS implementation 72
  competition with mink 163, 164
  decline/extinctions 164
    mink invasion and 163
  evolution 18
  fur colour **13**
  habitat **13**
  hunting 1, **164**
  in interglacial periods in Britain 18
  number in Mesolithic 19
  size 3, **4**, **13**
  surveying/monitoring, eDNA analysis 206
  as vermin, control (Tudor Period) 57

Eurasian Wolf (*Canis lupus lupus*) 48
European Badger (*Meles meles*) 2
AIHTS implementation 72
in interglacial periods in Britain 18
size 3, **4**
as vermin, control (Tudor Period) 57
European Mink (*Mustela lutreola*) 164
European Pine Marten (*Martes martes*) see Pine Marten (*Martes martes*)
European Polecat (*Mustela putorius*) 10–14, **11**, 43, 57, 69, 82, 88, **221**, **309**
  absence from Ireland 11, 208
  anticoagulant rodenticide residues 232, 234, 309
    habitat, geographic region 232–233
  Bern Convention, Appendix III 308
  climate change impact 311
  competitors 321
    American Mink 42, 163, 309
    Pine Marten, competition limited 321
  daily food requirements 239
  decline/extinctions 59, 70–71, 78, 223, 308
    current/future rate, concerns 308
    drivers of 309
    mink invasion and 163
    recovery (last 50 years) *see below*
  dens and resting sites 43, 163
  diet 43, 163, 212, 220–224, **224**, 231, 321
    from 1960s–2010s, changes 223, **224**
    amphibians 163, 221, **224**
    birds 223–224, **224**
    garden visits and 88
    rabbits 163, 220–221, **221**, 223, 224, **224**
    rodenticide exposure 232–233
    rodents 163, 221, **224**, 231, 232
    seasonal variation 223
    sexual dimorphism 221, 222
    see also prey, hunting (below)
  dietary niche **230**
    changes (1960s–2010s) 223, 224, **224**
  distinguishing features 11, **11**, 86, 88
  distribution (1800–1915) **59**, 70–71
  distribution (1960s–2010s) 83, 83–85
  distribution (2014–2015) 83, **83**, 89, **89**
  distribution (Wales) **69**, 71, 83, **83**, 89, 164, 223
  distribution (future), climate change impact 311
  domestic gene (Ferret) introgression 87, 88, 89
  domestication, and role 276–277
  evolution 19
  Ferret hybridisation with 84, 85, 86, 87, 88, 90, 310

  minimising 90
  see also Polecat-Ferret
  Ferrets, distinguishing from 86, 88
  folklore about 294
  foraging strategy 221, 223
  fur colour 11, **11**, 58, 82, 86, **88**, **309**
  fur trade 50, **65**, 308, 310
  fur uses, Middle Ages 50, 55, 70
  gait 43
  gamekeepers killing 71, 79
  as generalist predator 220
  genetic diversity 87
  genetic 'purity', distribution 87, 89–90
  guard hairs 11, **11**
  habitats 43, 221, 223, 310, 321
    farmland and grassland 43, **82**, 89, 163, 179, **222**, 223
    loss/change, impact of 309
    population density estimates 307
    wetlands 43, 163, 179
    woodland 43
  identification, guidelines 86
  IUCN Red List assessment 308–309
  kits, litters, young 44, **58**, 91
  legal protection 89, 308
  lifespan 43
  mating and breeding 43, 44, 87, 91
  nocturnal 10, 43, **230**
  number in Mesolithic 19
  population density/number
    area-specific 307–308
    lack of accurate information 308–309
    prey abundance affecting 309
    Scotland 307–308
  population estimate (2018), Britain 83, 307–308
  population trends 89, 307–310
  as predator, control of 70–71, 79
    see also as vermin, control (below)
  predator–prey interactions, Water Voles 248
  prey, hunting 42, 43, **221**, **222**, 321
    decline due to habitat change 309
    mice and voles 81
    rabbit decline impact 81–82
    rabbits 42, 43, 58–59, 80, 81, 221, **221**
    rodents, in farmed areas 310
    see also diet (above)
  'pure' *vs* 'impure' population 89–90
  recovery from decline, recolonisation 78, 79, 83–85, 89, 223
    diffusion model pattern 84
  road traffic collisions 310
  scats 186

sense of smell and sight 43
size and weight 3, **4**, 10, 28, 43
surveying/monitoring *see* surveying and monitoring (*General Index*)
tail length 28
territorial nature 297–298
as vermin, control 56, 57, **57**, 58–59, 70, 310
  20th century, impact 69
  persistence despite, large litters **58**
  Scotland 68
  ways of killing (Middle Ages) 59–60, **60**
European Rabbit (*Oryctolagus cuniculus*) **36**, 37
  American Mink predation of 42
  biological control by Weasels/Stoats 247
  as European Polecat prey **42**, 43, 80, 221, **221**
  National Gamebag Census 81–82, **82**
  in New Zealand 148
  Stoat predation of **154**, 248–249
  see also rabbits (*General Index*)
European Stoat 220
  see also Stoat (*Mustela erminea*)

Falco tinnunculus (Common Kestrel) 246
Ferret (*Mustela putorius furo*) 11, 59, 84–85
  albino 85
  breeding in New Zealand 151, 155
  COVID-19 susceptibility 322
  diet **231**
  distinguishing from European Polecat 86, 88
  domestication 11, 84, **85**, 87–88, 178
  eating after killing rabbit 271
  escape/releases 85, 90, 178
  European Polecat hybridisation 84, 85, 86, 87, 88, 90, 310
    see also Polecat-Ferret
  feral 178–180
    impact on Irish Stoat 180
    in Ireland 11, 178–179
    New Zealand 178, 180, 250
    Rathlin Island, Northern Ireland 178–179, 180
  fur colour 85, **85**, 87
  habitats 179
  identification, guidelines 86
  importing by New Zealand 150–151, 271
  introduction into Britain 84–85
  origin/historical aspects 84
  prey, hunting 271
  for rabbit hunting, use **84**, 84–85, **178**, 277
  removal trial, impact 180
  swimming 157
Ficedula hypoleuca (Pied Flycatcher) 256

Field Vole (*Microtus agrestis*) 6, 39, 77, **245**
  annual production (Britain) 246, 272
  habitat 216
  increase after clear-felling of trees 130
  Kielder Forest, dynamics in 243–245, 245, 246
  population cycles 225, 241, 242–244, **245**
    food supply effect 246
    predator effect 242–243, 244, 245, 246
    Stoats effect 248
    Weasel coexistence 244, 245–246
    see also prey population cycles (*General Index*)
  predation/predators of
    generalist and specialist 246
    Pine Martens 44, 225, 252
    Stoats 217, 248, 272
    Weasels 215–216, 240, **241**, 242–243, 246
  in Wales 199, *199*
Fisher (*Pekania pennanti*) 2, **266**
  climate change impact 311
  distribution 265, **267**
  microsatellites (DNA) 119
  translocation and breeding 119
Fossil Stoat (*Mustela palermínea*) 31

Gallirallus australis (Weka) 149
Greater White-toothed Shrew (*Crocidura russula*) 220
Grey Heron (*Ardea cinerea*) 6, **7**, **8**
Grey Partridge (*Perdix perdix*) 72, 179
Grey Squirrel (*Sciurus carolinensis*) 2, **259**
  American Marten distribution and 265, **267**
  breeding and litter number 272
  decline, Pine Marten recovery link 259, 261, 262, 263, 269, 272
  distribution (Britain, 1945, 2010) 260, **260**
  distribution (Ireland) *see* Ireland (*General Index*)
  distribution (USA) 265, **267**
  Fisher distribution and 265, **267**
  habitats 264
  as invasive non-native species 147, 259, 260
  niche overlap with Red Squirrels 264, 265
  as Pine Marten prey **263**, 264, 265, 272, 273
    direct predation 264, 265, 273
    Ireland 272
    susceptibility due to naivety 265–269, 272
  Pine Martens and 259–274
    as biological control method 270–272, 273
    citizen science survey 269
    density importance, for squirrel decline 268
    distribution relationship, Ireland 261–262, 268

interaction types and complexity 272–273
'landscape of fear' 265, 268
as management tool 272–274
recovery, Grey Squirrel decline 259, 261, 262, 263, 269, 272, 273
reintroductions, squirrel decline with 272–273
response to Pine Marten scent 267
single/lone marten release problem 273–274
translocations to Wales effect 268, 272–273
Red Squirrel body mass/energy use *vs* 265
response to Red Fox scent 267
spatial ecology 127
trapping to control 274
vigilance, impact on fitness 265
young, as Pine Marten prey 265
Gulo gulo see Wolverine (*Gulo gulo*)

Harrier Hawk (*Circus approximans*) 149
Hazel Dormouse (*Muscardinus avellanarius*) 199, 199, 256
Himantopus novaezelandiae (Black Stilt) 249
House Mice (*Mus musculus*) 219

Iberian Lynx (*Lynx pardinus*) 221
Icerya purchasi (Cottony Cushion Scale) **270**, 271
Irish Hare (*Lepus timidus hibernicus*) 26
Irish Stoat (*Mustela erminea hibernica*) 4, **25**, 26, **27**, **219**
  diet 219–220
  distribution 219
  feral Ferret impact 180
  few carnivore competitors **27**
  folklore 291
  fossil specimens 26
  fur colour 26
  island populations 39
  killing, folklore 291
  legal protection 73, 220
  mating and breeding 44
  saliva, poisonous 291
  sexual dimorphism 26, 220
  size, comparison with British Stoat 26, 220
  surveying/monitoring, Mostela camera trap 202–203, **203**

Kakapo (*Strigops habroptilus*) **157**
Kingfisher (*Alcedo atthis*) 57

Lagopus muta (Ptarmigan) **319**, 320
Lapwing (*Vanellus vanellus*) 180

Least Weasel (*Mustela nivalis nivalis*) 5, **31**, 314, **314**, 315
  fur colour in winter **314**, 314–315
  see also Weasel (*Mustela nivalis*)
Lemming (*Myopus chisticolor*) 241
  decline, Greenland 247
  in Pine Marten diet 226
  population cycles *see* prey population cycles (*General Index*)
  Stoats hunting of **247**
Lepus europaeus (Brown Hare) 64
Lepus timidus (Mountain Hare) **319**, 320
Lepus timidus hibernicus (Irish Hare) 26
Lutra lutra see Eurasian Otter (*Lutra lutra*)
Lutrinae (otters) 15
Lynx lynx (Eurasian Lynx) 48
Lynx pardinus (Iberian Lynx) 221

Mammuthus primigenius (Woolly Mammoth) 18, 19
Martes 4
  evolution, prey types and 17–18
  evolutionary diversification, Pleistocene 17
Martes americana see American Marten (*Martes americana*)
Martes foina see Stone Marten (*Martes foina*)
Martes martes see Pine Marten (*Martes martes*)
Martes melampus 17
Martes zibellina 17
Meerkat (*Suricata suricatta*) 13
Meles meles see European Badger (*Meles meles*)
Melinae (badgers) 15
Mellivorinae (honey badgers) 15
Mephitidae 15
Mephitinae (skunks) 15
Microtus agrestis see Field Vole (*Microtus agrestis*)
Milvus milvus (Red Kite) **70**, 71
Mink 10–14
  American *see* American Mink (*Neovison vison*)
  European (*Mustela lutreola*) 164
  prey, hunting 41–42
Mohua (*Mohoua ochrocephala*) 218
Mountain Hare (*Lepus timidus*) **319**, 320
Mus musculus (House Mice) 219
Muscardinus avellanarius (Hazel Dormouse) 199, 199, 256
Muskrat (*Ondatra zibethicus*) 163
Mustela 4
  evolution, prey types and 17–18
  meaning of genus name 29
Mustela erminea see Stoat
Mustela erminea hibernica see Irish Stoat (*Mustela erminea hibernica*)

SPECIES INDEX · 349

Mustela furo see Ferret (*Mustela putorius furo*)
Mustela nivalis see Weasel (*Mustela nivalis*)
Mustela nivalis nivalis see Least Weasel (*Mustela nivalis nivalis*)
Mustela nivalis vulgaris see Common Weasel (*Mustela nivalis vulgaris*)
Mustela palerminea (Fossil Stoat) 31
Mustela putorius furo see Ferret (*Mustela putorius furo*)
Mustela robusta 19
Mustelidae 15
   chronogram **16**
   classification 15
   members, species and genera 2
   size 2
Mustelinae (martens and weasels) 15
Mustrela lutreola (European Mink) 164
Myodes glareolus (Bank Vole) 39, 199, **199**, 200
Myopus chisticolor see Lemming (*Myopus chisticolor*)

Neovison vison see American Mink (*Neovison vison*)
North Pacific Sea Otter 3
Novius cardinalis (Vedalia Ladybird Beetle) 270, 271
Numenius arquata (Curlew) 166, 248

Ondatra zibethicus (Muskrat) 163
Oryctolagus cuniculus see European Rabbit (*Oryctolagus cuniculus*)
Otter
   Eurasian *see* Eurasian Otter (*Lutra lutra*)
   North Pacific Sea 3
   Sea Otter (*Enhydra lutris*) 1, **49**

Pacific Rat (*Rattus exulans*) 157
Pekania pennanti see Fisher (*Pekania pennanti*)
Perdix perdix (Grey Partridge) 72, 179
Phasianus colchicus (Pheasant) 256
Pheasant (*Phasianus colchicus*) 256
Pied Flycatcher (*Ficedula hypoleuca*) 256
Pine Marten (*Martes martes*) 1, **2**, 4, 10–14, **23**, 48, 93, **182**, **262**, **299**, **322**
   acute/chronic stress, behaviour 112–113, 118–119, 127
   adaptation to cold climate 20, **21**, **22**, 300
   age assessment **114**, 114–115
   agility 13, 14, 43–44, **44**, 48
   AIHTS implementation 72
   anticoagulant rodenticide residues 232, 234
   behaviour, research 127
   bib *see fur colour (below)*
   bird nest box protection from 255–256
   body shape 10, 43, 44, **44**
   breeding *see mating and breeding (below)*
   carrying capacity, natural 273
   climate/weather
      behavioural response to 300
      cold, adaptation to 20, **21**, **22**, 300
      impact on home range size 300
      impact on population density 301
   coexistence with prey and bird species 107–108
   collective noun ('a richness') 47
   competitors 321
   conservation in Britain 145–146
   COVID-19 infection 322
   dens 105, **119**, 120, 130, 302
   detection techniques 94
   diet (omnivorous) 44, 105, **182**, 212, 224–231, 252, 272
      Bank Vole 44, 130, 226, 252
      bats **256**, 256–257
      bees, wasps, beetles 226, 227
      birds 44, 182, 226, 227, 252, 254, 255, 256
      carrion 226, 227
      eating at hides, or homes **182**, **299**, **322**
      Field Vole 225, 252, 272
      food selection **210**, 252
      fruit 44, 225–226, 227, 227, **228**, 229, 230–231, **299**
      game birds 256
      geographical variation 226–227
      Grey Squirrel 264, 265, 272
      in Ireland 229, 299–300
      larger prey 227, 228
      macronutrient energy ratio 229–230
      mushrooms 227
      Red Squirrel 253–254, 264, 265
      scat analysis 210–211, 227
      in Scotland 225–226, 227
      seasonal variation 227–228, 229
      small mammals 225, 226, 227, 227
      Stone Marten presence effect 228–229
      various (other) species 226, 229
      Wood Mouse **225**
   dietary niche 230
   distribution (Britain) 69, 91–94, 97, 260
      1850 **61**
      1900s 69, 91, 92, **92**
      1994 survey 92
      2022 **146**
      Forest of Dean 145
      pre-/post-legal protection (1988) **92**

pre-1800 22, 69
Scotland *see* Scotland (*General Index*)
southern Britain 94–95
Wales (after reintroductions) 139–141, **140**
Wales (before decline) 69, 94, 103
distribution (Europe) **21**, 22
Ireland *see* Ireland (*General Index*)
distribution (global, 2008) **253**
ecology, information for reintroductions 105
energy requirements 228, 229
European Polecat competition limited 321
evolution 17, 18, 43, 47, **262**
exploitation 52–53, 55
  see also fur trade (below)
extinctions/decline in numbers 60, 69–70, 78
  by 1850 60, 61, **61**
  Ireland 70, 98–99, 100, 101
  prevention in Wales 106
  south of Scotland, prevention 103
  southern Britain 69, 91, 94, 97
folklore about 293
  in Ireland 292–293
footprint **3**
foraging strategy 228, 229
fossils 47
foxes controlling 182
fur colour **2**, **12**, 13, 20, **21**, **23**, 194
  bib, apricot colouring **2**, **12**, 13, **23**, **93**, **299**
  bib, reintroductions to Wales and 125, 137, **137**, 194, **195**, 196
  bib patterns 13, 194, **195**
  bib photography ('jiggler') 194–196, **195**
fur trade and 47, 61, 69, **285**
  in Ireland 98
  Middle Ages 49–50, 51, 52–53, 55
gait 13
as generalist predator 14, 105
genetic diversity, low 295
Grey Squirrel and *see under* Grey Squirrel (*Sciurus carolinensis*)
habitat 22, **110**, 129, **129**, 225
  effect on home range size 298, 299
  forest fragmentation impact 311
  grassland in woodland, for hunting **230**, 301
  Ireland 99, **101**, 102, 272, 297, 302
  loss/change, impact of 311
  range of, use 301–302
  for reintroductions 129, **129**
  rocky upland areas 301
  scrub (Hazel) **101**
  trees/woodland 13, 22, **48**, **110**, **129**, 225, 301, 321

haplotypes 97, 109
home range size 128–129, 298, **299**
  habitat effect 298, 299
  Ireland 298–300, 299
  males *vs* females 298
  Scotland 298, 299
  seasonal variation 300
hunting of 52–53, 55, 60–61, 91–92
  19th century 60–61
  in Ireland (1500s) 98
  see also as vermin, control (below)
identification by markings 194
information centre (Wales) **141**, 142
in interglacial periods in Britain 18
as invasive species on Skye, Mull 93, 180–183, 184
  entry route/method 181–182
in Ireland *see* Ireland (*General Index*)
Irish folklore 292–293
kits, and litters
  birth timing 45, 118, **121**, **122**, 271
  number/size 43, 44, 271
legal protection for 64, 92, 95, 182, 310
lifespan 44–45, 91
macronutrient intake 229–230
males, reproductive capacity 112
mating and breeding 43, 44, 44–45, 46, 91, **112**, 118, 271
  breeding strategy 45, 46, 271
  delayed implantation 43, 45, 46, 118, 271
  slowness, impact 91, 271
  timing 45, 118
  translocated animals 118–124
in Mesolithic 19, 47, **48**, 70, 181
origins in England and Wales (DNA) 97
overharvesting, for reintroductions 110
paws/feet/toes 20, **22**, 43–44, **44**
persecution
  hunting/fur trade *see above*
  in Ireland 98–99
  as vermin, control *see below*
personality 130–132
as pets 293
population densities 300–302
  continental Europe 301
  Ireland 301
  Scotland 191–193, 297, 300–301
population numbers 91, 93, 182, 262, 297
  decline by 1900s 70
  in Mesolithic 19, 70
  recovery *see recovery from decline (below)*
  poster to encourage reports of **95**

as predator, control 69–70, 91
predator–prey interactions *see* predator–prey
  interactions (*General Index*)
prey, abundance, little impact on number
  271–272
prey, hunting
  birds 44, 182, 226, 227, 252, 254, 255, 256
  climate change impact on prey 321
  Grey Squirrel **263**, 264
  prey types 44, 105, 182, 225
  see also diet (*above*)
radio-collared *see* radio collar (*General Index*)
recovery from decline, recolonisation 78,
    91–94, **92**
  Ireland 101–102, 272
  Ireland, Grey Squirrel decline 259, 261, 262,
    264, 269
  legal protection impact 92–93
  northern Scotland 91–92
  southern Scotland and Borders 92–93
Red Squirrel coevolution **262**
Red Squirrel number, positive effect on
  261–262, 264, 265, 269
reintroductions/translocations 274
  to England 2, 97, 105, 274
  Grey Squirrel management 272–273
  single/lone, release 273–274
  unlicensed, ill-informed 272–273
  to Wales 103–135, **262**
  see also reintroduction (*General Index*)
releases
  on Isle of Mull 93, 181
  in Scottish Borders 93
'relict' populations in England/Wales 97
reported sightings
  in England and Wales 95–97, **96**
  limitations 94, 96
  questionnaire on 96
roadkill, mortality 105, 112, 125, 145, 207, 274
scats **187**, 187–188
  analysis of diet 210–211
  conditions affecting finding of 189
  DNA analysis 188, 190, 192–193
  examination, after reintroductions 131, **132**,
    139
  Grey Squirrel in 264
  not found, false negative results 188–189
  Red Fox scat similarity 188, 210–211
  Red Squirrel in 254
  sample collection, for DNA analysis 190, 194
  scent 187–188
  surveys 94, 97, 139, 187–189, 190, 193, 193, 194

scent, Grey Squirrel and Red Squirrel
  response 267–268
in Scotland *see* Scotland (*General Index*)
second-rarest carnivore in Britain 19
sitting upright on hind legs 13
size 3, **4**, 10, 13, 28
  comparisons with other mustelids **4**
  latitude effect 228
spatial ecology **126**, 127
Stone Marten coexistence 228–229
Stone Marten differences 20
strategic recovery plan 145
strychnine poisoning 70, 101
surveying/monitoring *see* surveying and
  monitoring (*General Index*)
as 'sweet mart' 14, 61
swimming ability 181–182
tail, and length of 13, 28
teeth 114, **114**
territorial habit 108, 297–298
as tourist attraction 183
trapping for reintroductions 113, **114**
trapping/snaring 60–61, **62**, 91, 99, 100
tree climbing 13, 14, 43, **44**, **48**, **98**, 264, **301**
as vermin, control 56, 60–61, 91
  19th century 61
  20th century, impact 69
  Ireland 99
  price per head 57, 61
  Scotland 68
  trapping methods 60–61, **62**
  Tudor Period 56, 60
  value of skins 61
weight 28
young, rearing, and food supply 118, 271
Polecat-Ferret (*Mustela putorius* x *Mustela putorius*
  furo) 10, 11, **12**, 86–90, **90**
  distinguishing features **12**
  distribution **89**
  identification, guidelines 86
  see also European Polecat (*Mustela putorius*);
    Ferret (*Mustela putorius furo*)
Ptarmigan (*Lagopus muta*) **319**, 320
Pygmy Shrew (*Sorex minutus*) 199, *199*, 219, **219**
Pyrrhula pyrrhula (Bullfinch) 62–63

Rattus exulans (Pacific Rat) 157
Rattus norvegicus see Brown Rat (*Rattus norvegicus*)
Red Fox (*Vulpes vulpes*) 33, 169, 181
  diet
    Field Vole 246
    seasonal variation 230

Pine Marten escape from 301
scats 188, 209–210
scent, Grey Squirrel response 267–268
Red Kite (*Milvus milvus*) 70, 71
Red Squirrel (*Sciurus vulgaris*) 261
  distribution (Britain, 1945, 2010) 260
  distribution (global, 2008) 253
  distribution (Ireland, 1945, 2010) 260, 260–261, 264
  Grey Squirrel body mass/energy use *vs* 265
  habitats 264
  niche overlap with Grey Squirrels 264, 265
  Pine Marten coevolution 262
  Pine Marten predation of 253–254, 264, 265
  recolonisation, Ireland 262
  recovery, with decline in Grey Squirrel 259, 265
    Pine Marten relationship 261–262, **262**, 264, 265, 269
  response to Pine Marten scent 268
Rodolia cardinalis (Vedalia Ladybird Beetle) 270, 271

Sable (*Martes zibellina*) 49
Sciurus carolinensis see Grey Squirrel (*Sciurus carolinensis*)
Sea Otter (*Enhydra lutris*) 1, **49**
Sorex araneus (Common Shrew) 199, *199*
Sorex minutus (Pygmy Shrew) 199, *199*, 219, **219**
Squirrel
  Baltic, fur 50, 52
  Grey *see* Grey Squirrel (*Sciurus carolinensis*)
  Red *see* Red Squirrel (*Sciurus vulgaris*)
Stoat (*Mustela erminea*) 10, 44, **75**, **283**
  American Mink competition 42
  anticoagulant rodenticide residues 232, 233, 234
  autumn moult 316
  bird species, impact on 107
  body shape 9, **9**
  breeding strategy 34–35, 74, 271, 284
  camouflage, and lack of 312, **313**, 317, 320, **320**, 321
  cats, competition/avoidance 155
  colloquial name (Cain) 62
  competition with large predators 155
  daily food requirements 239
  daytime activity 5
  decline in numbers 37, 76, 78, 208, 296, 297
    myxomatosis and 37, 71, 78, 248–249, 296
  diet 34, 35, 37, 158, 212, 214–219
    birds 107, **160**, 218

Day's study 214–215
eggs 218
rabbit 214–215, 216, 217–218
rodents 29, 34, 214–215, 218, 233
seasonal variation 213, 214, 217–218
sexual dimorphism and 217–218, 222
see also prey, hunting (below)
dietary niche 230
distribution 76, 306, 307
  global **31**, 306
diurnal **230**, 305
evolution 31, 154
exploitation/persecution 50, 55
  for ermine *see* ermine (*General Index*)
  hunting *see below*
  survival from 74–78
extinction, resilience against 74–78
failure to control rabbits 152–154, 249, 250, 271
Ferret avoidance 155
foraging strategy/bouts 304
fossils 31
fur colour 5, 9, **9**, 75, **283**, 291
  black tip to tail 5, 9, **9**, 10, **283**, 316
  brown in winter (England) 9, 312, **312**, 316, 317, 318
  Pied colouring 317, 318, **318**
  summer, thermal properties 320
fur colouring, winter whitening 9, **312**, 312–321, **313**, 320
  adaptation for change in climate 317, 318, 320
  camouflage 312, **313**, 317, 321
  genetic control 316
  geographic distribution 317
  melanocyte-stimulating hormone 316
  minimum temperature for 316
  New Zealand and 318, 320
  Pied Stoat 317, 318, **318**
  predation avoidance 321
  snow cover decline, impact 313–314
  thermal properties 320
  transition zone 312, **312**
  see also ermine
fur trade, Middle Ages 50, 55
  see also ermine
gait 5, 9
genetic bottleneck 296
genetic diversity, low 295, 297
guard hairs 320
habitat **36**, 37, 306, 307
home range size 304–307, 306
  different countries 305–306, 306

SPECIES INDEX · 353

of juveniles 304–305
of large assertive males 305
males *vs* females 304–306
radio tracking 304, 305
'roamers' and 'stayers' 305
seasonal variation 305
hunting/persecution 62
 on game estates 76–77, 78
 see also trapping/killing (below)
'in ermine' (white winter coat) 9
 see also ermine (*General Index*)
in interglacial periods in Britain 18
as invasive species 183–184
 New Zealand *see* New Zealand (*General Index*)
 UK offshore islands (Orkney) 160–161
Irish folklore 291–293
island populations 35, 39–41, 40, 183–184
 New Zealand 156, **156**, 157, 158, 183
killing, bad luck 291, 292
kits, and litters 44, 46, 284
legal protection and 72
lifespan 34
local extinction, recolonisation 38
mating, and breeding 34, 44
 breeding strategy 34–35, 74, 271, 284
 delayed implantation 34–35, 37, 46, 284, 305
 males, number of females per 305
movements 5, 9, 304, 321
moving in packs **291**, 292
nests under snow 197
in New Zealand *see* New Zealand (*General Index*)
nocturnal in winter 305
Pied Stoat 317, 318, **318**
population bottleneck 295–296, 297
population growth rate, factors affecting 74–75, 76, 307
population models 74–75
population size 34, 38, 39, 76, 78, 307
 Britain (1961–2001) 78
 Britain (1970–2016) 208
 Britain (1995) 78, 307
 Britain (2018) 78
 fluctuations 76, 78
 in Mesolithic 19
 numerical response 218–219
 survival despite high mortality 76
precociousness 284
predation avoidance, white coat role 321
predator-deflection marks 9–10
predator–prey interactions *see* predator–prey interactions (*General Index*)

as predators, control of 67–69, **68**, 72, 73–74
 increase, 1966–2016 76
 National Gamebag Census data 76–77, 78
 proportion to be removed 76
prey, hunting 34, 35–39, 154, **215**, 304, 305
 in bursts, rest periods 271, 305
 comparison with Weasel 28
 Field Voles 29, 248
 game birds 248
 large prey 35–37, **36**, 154, 155, 216–217
 Lemming **247**
 method (constant movement) 321
 in New Zealand 154–156
 preference, nutritional value 212–214
 rabbits **36**, 37, 77, 78, 81, 152, **153**, 155, 214–215, 248–249, 250
 rodents 29, 34, 214–215, 218, 233
 switching between different sizes 38
 timing 271
 Water Voles 35, 36–37, 39, 248
 Weasels 38
 see also diet (above)
as prey for larger predators **241**, 321
 birds of prey 9, 10
 prey reduction, effect on reproductive success 234
procession of **291**
productivity (annual) 34
pulse rate (resting) 33
scats 186, **186**
sexual dimorphism 26, 28, 28–29, **217**, 217–218
size **4**, 9, 28, 28, 29
 comparison with Irish Stoat 26
 by sex **217**
skin, in Irish folklore 292
space use 302–307
as specialist predator 28–31, 218
spinal flexibility **34**
status associated 288–291
stomach capacity 212–213
surveying/monitoring *see* surveying and monitoring (*General Index*)
swimming ability 35, 156, **156**
tail, black tip to 5, 9, **9**, 10, **283**, 316
tail length 26, 28
 male *vs* female 26
temperature range for 26
territorial habit 297–298, 304
on Terschelling 38–39
trapping/killing 67, 72, 76, 151–152, 296
 in New Zealand 158–159
 timing, effect on population model 75–76

tunnel traps 72, 73–74
upright on hind legs **283**
as vermin, control 56, 61–62, 67
  Scotland 68
Weasel coexistence 38–39, 154
Weasel comparison 28–29
Weasel competition 28, 37–39, 43, 154
weight 28, 37
young, appearance **75**
young females
  mating age 34–35
  survival, impact on population growth rate 75–76
Stone Marten (*Martes foina*) 20–24, **23**
  in Britain, and exterminated? 20, 22, 24
  damage caused by 20
  diet
    bats 256–257
    seasonal variation 230
  distribution (Europe) **21**, 22, 24
  feet/paws 20
  fur, and fur trade (*foynes*) 49–50
  fur colour 20, 22, **23**, 24
  habitat 20, 22
  imported from Italy 24
  Pine Marten coexistence 228–229
  Pine Marten differences 20, 22
  sub-fossil specimens 22, 24
  use for rodent control 24
Strigops habroptilus (Kakapo) **157**
Suricata suricatta (Meerkat) 13

Tengmalm's Owl (*Aegolius funereus*) 255
Tetrao tetrix (Black Grouse) 257, 258
Tetrao urogallus (Capercaillie) 258
Turdus merula (Blackbird) 255

Vanellus vanellus (Lapwing) 180
Vedalia Ladybird Beetle (*Rodolia cardinalis; Novius cardinalis*) **270**, 271
Vole
  Bank *see* Bank Vole (*Myodes glareolus*)
  Field *see* Field Vole (*Microtus agrestis*)
  Water *see* Water Vole (*Arvicola amphibius*)
Vulpes vulpes see Red Fox (*Vulpes vulpes*)

Water Vole (*Arvicola amphibius*) 35, 38–39, 247, **248**
  conservation projects 173
  mink control at UK important site 168
  mink hunting 41–42
  mortality, stoats hunting 35, 36–37, 39, 248
  population cycles 248

Weasel (*Mustela nivalis*) 3, 5–10, **32**, **77**
  absence from Ireland 39
  ancient beliefs 276
  anticoagulant rodenticide residues 232, 233
  beliefs about 2, 276, 293–294
  benefits to humans 62–63
  body shape 5, 29–30
    cold climate, susceptibility 32
  breeding rate, *vs* prey breeding rate 247
  breeding strategy 28, 29, 33–34, 46, 74, 283
    rodent density and 29, 33, 34, 37, 154, 238, 247
  burrows/nests 9, 32, **32**
  cave painting 275, **275**
  childbirth, as talisman for 277–281
  Common *see* Common Weasel
  conception, beliefs on 2, 279–280, **280**, 281
  daily food requirements 32–33, 239
  decline/extinction
    1970–2016 trends 208
    local extinction, recolonisation 38
  defecation rate 33
  diet 37, 212, 214–219
    birds 215, 216, 218
    Day's study 214–215
    eggs 218
    Field Voles 215, 216, 240, **241**, 246
    King's study 215–216
    nutritional value (prey), weights 212–213
    in open habitats 216
    rabbits 214–215, 218
    rodents 215, 216, **216**, 218, 233, 240
    seasonal variation **213**, 214, 218, 238
    sexual dimorphism not detected 218
    in woodland 215–216
    see also prey, hunting (below)
  dietary niche **230**
  distribution 14, 76, 77, 307
    Wales 200, **200**, 307
  diurnal nature 5–6
  energy balance 32–33
  evolution 18, 18–19, 30–31, 154, 317
  exploitation/persecution 50, 55
    fur uses/trade, Middle Ages 50, 55
    survival from 74–78
  exported to New Zealand *see* New Zealand (*General Index*)
  extinction, resilience against 74–78
  female, age at maturity 33
  fertility beliefs and 278
  fierceness/bravery 5, 6, **7**, **8**, 9
  flexibility **29**, 29–30

SPECIES INDEX · 355

folklore about *see* mythology and folklore (*General Index*)
food requirements 32–33, 239
fossil 31, 41
fur colour 5, **6**, **7**, **8**, **77**
   'gular spots' 5, **6**
   summer, thermal properties 320
fur colouring, winter 312–321, **314**
   camouflage **314**, 321
   Europe *vs* Britain 314
   evolution and 317
   genes associated 315, 316–317
   geographic distribution **315**, 315–316
   melanocyte-stimulating hormone 316
   minimum temperature for 316
   nivalis and *vulgaris* morphs 314–315, **315**, 317
   predation avoidance 321
   thermal properties 320
   transition zone 317
fur layer, and body fat 32, 320
fur uses/trade, Middle Ages 50, 55
gait 5, 33
global distribution **31**
Grey Heron, battle with 6, **7**, **8**
habitat 5, 6, 216, **230**, 303
   farmland and grassland 6, 216, **230**, 303, 303
   impact on home range size 303, 303–304
   woodland 216, 303
in history
   Ancient Greek beliefs 276, 277, 279, 281, 282
   Renaissance 284–288
Holarctic distribution 314
home range size 302
   England 303
   habitat impact 303–304
   males *vs* females 302–303, 303
   methods of estimating 303–304
   radio-tracking 304
   Scotland 303
hunting of *see as* predators, control (below)
in interglacial periods in Britain 18
as invasive species, New Zealand *see* New Zealand (*General Index*)
island populations 39–41, *40*
IUCN Red List classification 207
jaws, teeth and bite force 30
kits, and litters 33, *44*, 283
   birth process 2, 279–280, **280**, 281
   care of kits 33, **278**, 278–279
   metabolic rate, heat loss 278–279
   moving 279
   Least *see* Least Weasel (*Mustela nivalis nivalis*)

legs, length 5, 29, **35**
lifespan 34
'live fast, die young' strategy 33, 46, 76
as maligned animal 282
mating and breeding 34, *44*
   breeding rate, *vs* prey breeding rate 238, 247
   gestation period 33, 46
   mating behaviour 282–283
meaning of species name 29
metabolic rate 212, 279
moulting (spring/autumn) 32, 316
movements 5, 6, 33, 321
nests and lining for 32
in New Zealand *see* New Zealand (*General Index*)
nivalis and *vulgaris* morphs 314–315
   crossbreeding 315
   evolution 317
   geographic distribution 317
   winter whitening and 314–315, **315**, 317
numerical response 238, 242, 244
as pets 276, **276**
population fluctuations/density 33, 34, 38, 76–78, 246–247, 302, 307
   decline in numbers (1970–2016) 207
   increase during/after myxomatosis 37, 77
   maximum density 34
   NGC data 76–77, **77**, 77–78
   number in Mesolithic 19
   numerical response 238, 242, 244
   prey density controlling 33, 76, 238, 242, 244, 246–247
   replacement after removals 76
population growth rate, factors affecting 74–75
population models 74–75
population size (numbers), Britain 307
predation avoidance, white coat role 321
predator-deflection mark absence 10
predator–prey interactions *see* predator–prey interactions (*General Index*)
as predators, control of 67–69, **68**, 72, 73–74
   hunting/killing on game estates 76–77, **77**
   National Gamebag Census data 76–77, **77**, 77–78
   proportion to be removed 76
   see also trapping/killing (below)
prey, carrying 30, **30**, **35**, **38**
prey, hunting 2, 6, 9, 241–242
   attributes for 29–30
   comparison with Stoat 28
   killing with bite to neck 282
   low prey density 38

method (constant movement) 321
nutritional value, weights 212–213
poultry 282
preference, nutritional value 212–214
rodent population cycles and 241–242, 243
rodents 14, 17, 28, 29, 33, 35, 37, **38**, 62–63, 154, 240, 247
rodents, functional response 238, 242
as solitary hunter 33
voles 238, 240, **241**, 242, 243, 246
in winter 242, **242**
see also diet (above)
prey, reduction, effect on reproductive success 234
as prey for larger predators **241**, 243, 321
productivity (rate of population increase) 34
pulse rate (resting) 33
reassessment needed 302
scats 186, **186**
dietary information 212
sexual dimorphism 28, 28–29, **217**, 217–218
shelter for 9, 32, **32**
size 3, **4**, 5, 28, 28
advantages of 5, 29, 30–31
cold climate, implications 31–32
evolution 30
by sex **217**
skull 5, 30
space use 302–307
as specialist predator 28–31, 212, 218, 246
spinal flexibility 29, **29**
Stoat coexistence 38–39, 154
Stoat comparison 28–29
Stoat competition 28, 37–39, 43, 154
stomach capacity 212–213

surveying/monitoring see surveying and monitoring (*General Index*)
tail size and colour 5, **6**, 10, 28
temperature and heat loss 32, 278–279
territorial habit 297–298
trapping/killing 67, 76, 78, 151–152
home range size estimates 303, 304
tunnel traps 72, 74
as unclean, not to be eaten 280
as vermin, control 56, 62–63, 67
Scotland 68
way of life 31–33
weight 3, 28
as witch, malevolent creature 281–284
young
carrying 279, **279**
survival, impact on population growth rate 75
Weka (*Gallirallus australis*) 149
Wildcat (*Felis silvestris*) 19, 321
in Ireland 292–293
Wolverine (*Gulo gulo*) 2, 3
climate change threat 311, **311**
evolution 18
Wood Mouse (*Apodemus sylvaticus*) 199, 199, 200, 215, 216
Irish Stoat diet 219
Kielder Forest 244
Pine Marten diet 44, **225**
Woodpigeon (*Columba palumbus*) 226
Woolly Mammoth (*Mammuthus primigenius*) 18, 19
Wrybills (*Anarhynchus frontalis*) 249

Yellow-necked Mouse (*Apodemus flavicollis*) 199, 199, 246

GENERAL INDEX

Note: page numbers in **bold** refer to figures or figure captions. Page numbers in *italics* refer to tables. Scientists mentioned in the main text are indexed, but not those only cited in parentheses as a reference.

Ad-Damiri (1341–1405) 277
adaptive management, invasive non-native species 184
adaptive radiation 13
Aelian 278, 294
Aesop's Fables 282
AIHTS (International Humane Trapping Standards) 72
Alcmene, story 281–282
Allbones, Henry 152
Allbones, Walter 152
anal scent glands 1, 2–3, 10
animal welfare
    live trapping *vs* camera traps 205
    predator (mustelids) control issues 180
Anthropocene 148
anticoagulant rodenticides 231–232, 233
Antoninus 279–280
aposematic markings, polecats 11
Aristophanes 277
Aristotle 276, 278
art, involving mustelids/or furs 50, 52, 51, 54, **65**, 284, **285, 286**

Bacon, Alice 114
bait, poisoned 67, 72
baiting sports 61
bats, Pine Marten predation of **256**, 256–257
Bavin, David 106, 107, 117, 119, 127, 130
    behavioural assessment, released animals 131–132, **132**
beliefs about Weasels 277–281, **280**, 293–294
    ability to revive dead kits **278**, 278–279
    childbirth 277–278
    conception and 2, 279–280, **280**, 281
    fertility 278
Bergmann's Rule 220, 228
Bettini, Maurizio 281
biological control 270–271
    of European Rabbit by Weasels/Stoats 247
    of Grey Squirrels by Pine Martens 270–272, 273
biosecurity 322
Biotrack radio collar 115, 116, 132
birds
    American Mink impact 106, 164
        Ireland 174

    Outer Hebrides 165–167, **166**
    breeding populations 254
    consideration before reintroductions 108
    decline due to habitat changes 255
    European Polecat diet 223–224, **224**
    feral Ferret impact concerns 179
    Pine Marten impact on other nest predators 254
    as Pine Marten prey 44, 108, 182, 226, 252, 254
        common birds (Blackbirds) 255
        model, no significant impact 254–255
        on Mull, concerns 182, 184
        nesting birds/eggs 252, 255–256
        Pied Flycatcher 256
        Scotland 227, 227
    Stoat diet, and impact on 107, **160**, 214, 248
        New Zealand Stoats 150, 155, 158, 183–184, 249
    Weasel diet 215, 216, 218
birds of prey 70, 71, 149, **241**
    Stoats at risk from 9, 10
Birks, Johnny 84, 95, 96, 294
birth, timing 45–46
    American Mink 45–46
    Pine Martens 45, 118, **121, 122**, 271
    see also specific mustelids (Species Index)
bite force, teeth 30
body shape 3, 5, 10
bones, dietary analysis 214
breeding strategies
    Pine Marten 45, 46, 271
    Stoat 34–35, 74, 271, 284
    Weasel 28, 29, 33–34, 46, 74, 283
    see also specific mustelids (Species Index)
Bridges, Josie 119, 120, 196
bromadiolone 233
Buckland, Frank 150

cage traps 113, 168, 170
Cairngorms Water Vole Conservation Project 173
camera traps
    abundance estimation 196
    advantages 204–205
    disadvantages 196, 205
    failure with Stoats and Weasels 196–197
    live trapping *vs* 204

Mostela *see* Mostela camera traps
Pine Martens survey 194, 196, 205
camouflage, winter whitening 312, **313**, **314**, 317, **319**, 321
  lack of **311**, **320**, 321
  see also Stoat (*Species Index*); Weasel (*Species Index*)
Canada, Stoat home range size 305–306, 306
canine distemper virus (CDV) 116
capture-mark-recapture method
  American Mink 169, 169
  surveying/monitoring 185
carnivores (small mustelids as) 14
  evolution 17
  on large islands 39
  smallest, Weasel as 3, 5
  see also specific mustelids (*Species Index*)
cats, feral
  in Ireland, folklore 291–292
  on Isle of Mull 181
  in New Zealand 155
  Stoat competition/avoidance 155
cave painting, Weasel 275, **275**
Charles III, King 289
childbirth
  ermine fur/mustelids to aid 284, **286**
  weasel as talisman 277–281
Chitty, Dennis 243
Chitty effect 243
citizen science 57, 196, 207, 208
  European Polecat surveys 309
  squirrel and Pine Marten interactions 269
classification, of small mustelids 15, **16**
  clades and subfamilies 15, **16**
climate
  global, evolutionary divergence and 17
  Pine Martens and *see* Pine Martens (*Species Index*)
  small size advantage 30–31
  Stoat winter whitening and 317, 318, 320
  warming in Mesolithic 19–20
  Weasel body shape/size and 31–32
climate change 184, 311, 321
  impact on American Marten/Fisher 311
  impact on Wolverines 311, **311**
  increase in polecat habitat 311
  reduced snow cover 311, 313–314, 321
  Stoat fur colour and 313–314, 317, 318, 320, 321
  Weasel fur colour and 321
clothing, fur used for 1, **65**
  ermine *see* ermine
  Middle Ages **50**, 50–52, **51**, **52**
  peerage and 288

rabbit fur 52
Renaissance 284, **285**, **286**
royalty 51, 52, 288–290, **289**
status associated 50, 51, 288–291
Tudor Period **55**, 289–290, **290**
see also fur trade/use; *specific mustelids*
coadaptation, predator–prey interactions 249–250
coevolution, predators and prey 249–250, 252, 253, **262**
coexistence, small mustelids 41–43, 209
  diet preferences/partitioning 209
  dietary niches and **229**, 230–231
  of Pine Martens and bird species 107–108
  of Stoats and Weasels 38–39, 154
  of Stone and Pine Martens 228–229
  of Weasels and Field Voles 244, 245–246
cold climates
  Pine Marten adaptation 20, **21**, **22**, 300
  small animal susceptibility 31–32
  see also fur coat, winter whitening
colour, of fur *see* fur coat, colouring of
community
  citizen science participation *see* citizen science participation, vermin control 57
  Pine Marten reintroductions to Wales 141–142, 144
competition between small mustelids 27, 28
  American Mink and Eurasian Otter 163, 164
  American Mink and European Polecats 42, 163, 309
  breeding strategies to reduce 46
  exploitation competition 38
  interference competition 38
  intra-sexual 297–298
  intraspecific, sexual dimorphism reducing 217
  Pine Martens 321
  Stoats and American Mink 42
  Stoats and feral cats 155
  Stoats and Weasels 28, 37–39, 43, 154
conservation translocation 103–104, 310
  aim, species restoration 104
  complex high-risk activity 104
  considerations before 104
  disease as risk of 108
  habitat suitability 104
  impact on prey/other species 108
  IUCN guidelines 104
  risks associated 107–108
  unlicensed, ill-informed 272–273
  see also reintroduction(s)
coronation robes 288, 289, **289**

GENERAL INDEX · 359

Costa, Mafalda 86–87
COVID-19 pandemic 322
crypsis *see* camouflage

da Vinci, Leonardo **286**, 287–288
Davison, Andrew 86
de Medici, Christina 284
defecation rate, Weasel 33
delayed implantation (DI) 45–46, 118
  advantages 45
  American Mink *44*, 45–46
  disadvantages 46
  European Polecats (research) 46
  evolution 45, 46
  Pine Martens 43, 45, 46, 118, 271
  Stoats 34–35, 37, 46, 284, 305
Delille, Paul Armand 81
demographic stochasticity 124
Denman, Huw **110**, 188
Denmark, rodenticide residues in mustelids 234
density of mustelids
  difficulty in obtaining data 239
  prey density and 33, 76, 236–238, **237**, 242, 244, 246–247
  see also specific mustelids (Species Index)
Devensian glaciation 18–19
diet(s) 14, 209–234
  interpretation of data 214
  nutritional value of prey and size 212–214
  seasonal variation **213**, 214, 217–218
  study methods 209
    carbon/nitrogen isotopes 211–212
    scat analysis *see* scats
    stomach contents (carcasses) 209, 212, 214
  see also specific mustelids (Species Index)
dietary niches 224, **224**, **230**, 230–231
dietary studies 239
Difenacoum 233
disease(s)
  myxomatosis *see* myxomatosis
  rabbit haemorrhagic disease (RHD) 221–222, 249, 250, 322
  risk with translocations 108
  surveillance 322
  threat to mustelids 321
disease risk analysis (DRA) 108
distribution of small mustelids 14
  global (Stoat and Weasel) **31**
  on islands *see* islands
  see also specific mustelids (Species Index)
diurnal mustelids 5–6, **230**, 305
DNA degradation 190, 197, 206, 211

DNA sequencing/analysis 15, 86
  environmental (eDNA) *see* environmental DNA (eDNA) analysis
  Pine Marten hair samples 190, 191–192
  Pine Marten origins, England/Wales 97
  Pine Marten scats 188, 190, 192
  scats in dietary analysis 211
  Stoat and Weasel hair samples 197, 199, 201
Dutch Small Mustelid Foundation 201–202

ecosystems 184
eDNA analysis *see* environmental DNA (eDNA) analysis
Edward IV, King 52
Edward VII, King **66**
Elizabeth I, Queen 56–57, 290, **290**
Elizabeth II, Queen 288–289
embryonic diapause 34, 45
enclosure(s), impact 63–64
Enclosure Acts 63–64
energy balance/requirements 31–33
  Pine Marten requirements 228, 229
  Red Squirrel *vs* Grey Squirrel 265
England
  European Polecat, population number 307
  European Polecat distribution 83, **83**
  Pine Marten origins 97
  Pine Marten reintroductions *see* reintroduction, Pine Martens to England
  Pine Marten sightings 95–97, **96**
  Weasel home range size 303
  see also specific topics and mustelids
environmental DNA (eDNA) analysis 205–206, 208
  'capture' 207
  false negatives 206
  false positives 206–207
  limitations 206
  occupancy models to analyse 207
  recommendations for use 208
  solitary animals, low detection rate 206
  traditional methods combined with 207–208
  water-based 205–206
environmental stochasticity 124
ermine 9, 50, 51, 52, 284, 290
  art **286**, 287–288
  costs 290
  royalty wearing 51, 52, **52**, 288–290, **289**, **290**
  status associated 51, 52, 288–290
  see also Stoat, fur colouring, winter whitening (Species Index)
Ermine Portrait, of Elizabeth I 290, **290**
ermine tails 289–290

ethics, predator control 180, 184
EU Invasive Alien Species (IAS) Regulation 162
evolution, small mustelids 15–26
　adaptive radiation 13
　Britain separation from Europe 18, 19–20
　divergence (Miocene and Pliocene) 17–18
　fur coat 48–49
　Mesolithic, warming climate 19–20
　origin in Eurasia 15, 17
　Pine Martens 17, 18, 43, 47
　primeval forests 18, 19
　small mustelids in British Isles 18–19
　Stoats 131
　Weasels 18, 18–19, 30–31, 317
　Wolverines 18
exploitation 1
　fur/pelts for clothing 49–55
　see also clothing, fur used for; fur trade/use; specific mustelids (Species Index)
extinction
　concerns 297
　see also specific mustelids (Species Index)

faecal glucocorticoid metabolites (fGCMs) 127, 131, 132, **132**
Fairley, James 99
fashion, and fur coats 50–51
feet 3
　Pine Marten 20, **22**, 43–44, **44**
　polecats 11
　Stone Marten 20
Fenn 'humane trap' 72
ferrets see Ferret (Species Index)
Finland
　American Mink, impact on birds 164
　removal of Stoats, impact on grouse 240
　rodenticide residues in mustelids 234
　Stoat home range size 306
　vole population, predator removal 244–245, **245**, 245–246
fish, farmed, mink damage to 167
fitch (fitchew) (Polecat fur) 50
flea furs 284
Fleming, John 22
flexibility, spinal 29, **29**, **34**
folklore see mythology and folklore
food requirements 239
　Weasel 32–33, 239
food supply, rodent population cycles 246, 250–251
footprints 3, **3**, 197
Forest of Dean, Pine Martens 141, 144–146, **146**

fossils 15, 17, 18, 19
　Irish Stoat and Irish Hare 26
　Pine Marten 47
　Stoat 31
　Stone Marten in Britain 22, 24
　Weasel 31, 41
'foul mart' 14
foynes (Stone Marten fur) 49
France, rodenticide residues in mustelids 234
fruit
　Pine Marten diet 44, 225–226, 227, 227, **228**, 229, 230–231, **299**
　switch in diet with season 230
functional response, predator–prey interaction 227, 236
　type I 236, **237**
　type II 236–237, **237**
　type III **237**, 237–238
　Weasels, rodent prey 238, 242
fungal encephalitis 124–125
fur coat
　colour see specific mustelids (Species Index)
　evolution 48–49
fur coat, winter whitening **312**, 312–321, **313**, **319**
　advantages 320–321
　heat loss reduction 320
　other species in Britain **319**
　see also Stoat (Species Index); Weasel (Species Index)
fur farms 161, **162**, 164–165, 174, 177
　banned/ended in Scotland 164–165, 177
　banned in Britain 164, 177
　banned in Ireland 177
fur trade/use 1, 22, 47, 48
　16th century 53, 54
　American Mink see American Mink (Species Index)
　animals killed, number estimate 53
　for clothing see clothing, fur used for
　European Polecat 50, **65**, 308, 310
　exports from Ireland (1697–1819) 98
　invasive non-native species 148
　legislation/charters 50, 53, 54, 55
　Middle Ages 49–55, **50**
　　decline at end of 54–55
　Pine Marten see Pine Marten (Species Index)
　Renaissance 284
　Sable 49, 289, 290
　status associated 50, 51, 288–291
　Stoat 50, 55
　Stone Marten 49–50
　Tudor period 53, 54, 289–290
　Weasel 50

gait 3
European Polecat 43
Pine Marten 13
Stoat 5, 9
Weasel 5, 33
game birds 64, 66, 71, 99, 150
mustelid predator control 59, 64
Pine Marten impact on Mull 182
protection from Pine Martens 256
Stoats effect 73, 248
see also specific game birds
gamekeepers 64–65, 66–67, 71, 150
density, 1911 68, 71
reduction, 20th century 71, 79
roles 64, 68, 151–152
Stoat/Weasel delivery to New Zealand 152
gamekeeping 63–71, 74, 79
20th century 68, 71, 79, 91
hand-reared game 71, 73
National Gamebag Census data 76–77, **77**, 77–78
Pine Martens, in Ireland 99
shooting of game 64–65, 65–66, **66**, 71
Stoat records 76–77, 78
Weasel records 76–77, **77**, 77–78
generalists (dietary) 14, 163, 220, 236, 246
genetic bottleneck 87, 295, 296
genetic diversity 295–296, 297
George III, King 288
George IV, King 288, 289
George V, King **289**
gestation period
Pine Marten 45
Weasel 33, 46
gin trap 67, **67**, 72, 81, 99
glaciations 18, 31, 41
GPS loggers/tracking, on reintroduced Pine Martens 132–134, *134*, **135**
Greek mythology 277
Greeks, ancient 276, 277, 279, 281, 282
Grey, Sir George 150
gris (fine squirrel skin) 51
grouse
breeding success and predation 257
habitats, and predator–prey interaction 257, 258
removal of Stoats effect 240
grouse shooting 71
guard hair 11, **11**, 48, 320
'gular spots' 5

habitat
evolutionary changes, impact 17
management 104
modelling, for suitability for reintroductions 105
suitability, for reintroductions 104–105
see also specific mustelids (Species Index)
hair
guard 11, **11**, 48, 320
indigestible (Weasels/Stoats) 214
hair samples
Pine Marten surveys 190, **191**, 191–194, **192**, *193*
snagging 190, **191**, **192**, 194
Stoat and Weasel hair 197–198, **198**
hair snares 190
hair traps 190, **192**
hair tubes
Grey Squirrel and Pine Marten detection 262
limitations 197–198
Pine Marten hair 190, **191**, 194
'single use' 197–198
species detected in Wales 199, 199–200
Stoat and Weasel hair 197–198, **198**
modification, Wales **198**, 198–199
Hall, Arnold 251
handling time 236
haplotype 296
Pine Marten 97, 109
harvests, poor (16th century) 55, 56
heat conservation, fur coat 320
heat loss 32
kits 278–279
winter whitening of coat 320
Hebridean Mink Project **165**, 165–167, 175–176
hedgerow removal 309
Henry, Richard 158, 250–251
Henry IV, Henry V, Henry VI (Kings) 51, **51**
Henry VII, King 289
Henry VIII, King 55, **55**, 56, 289
history
ancient beliefs 276–281
ancient Greeks 276, 277, 279, 281, 282
Middle Ages 50, 50–52, **51**, **52**
Paleolithic image 275, **275**
Renaissance *see* Renaissance
small mustelids in UK 18–19
Tudor Period 55, 289–290, **290**
Hopi people 278
'house weasels' 276
humans, negative relationship with animals 57
hunting
for furs *see* fur trade/use
in Middle Ages 52–53
of mustelids as vermin/predators *see* predator (mustelids) control; vermin control

of Pine Martens *see* Pine Marten (*Species Index*)
of rabbits, by ferrets 84, 84–85, **178**, 277
of Stoats 62, 76–77, 78
of Weasels 76–77, **77**
see also specific mustelids (*Species Index*)

ice ages 18
Inner Hebrides
  Pine Martens 181
  Stoats and Weasels 40
interglacials 18, *18*, 31
International Humane Trapping Standards (AIHTS) 72
International Union for Conservation of Nature (IUCN)
  Conservation Translocation Specialist Group 104
  Red List 102, 208, **253**, 295, 308
    Least Concern 295, 308
intra-sexual competition 297–298
introduced non-native predators 107–108
Invasive Alien Species (IAS) Regulation, EU 162
invasive native species 183
invasive non-native species (INNS) 147–184
  Grey Squirrel 147, **259**, 260
  historical aspects 147–148
  IAS Regulation (EU) 162–163
  in New Zealand *see* New Zealand
  number in Britain 147
invasive non-native species (INNS), small mustelids 148–184, 251, 295
  adaptive management 184
  American Mink *see* American Mink (*Species Index*)
  climate change impact 184
  consequences 157–158, 183
  control strategies 176–177
    effectiveness, criteria for 158
    incomplete/localised, failure 76, 161, 177, 180
    'vacuum' effect 169, 170, 175
  costs to UK/Ireland 174, 176, 180, 183
  ethics of control 180, 184
  feral Ferrets 178–180
  Ferret transport/releases, New Zealand 150–151
  genetic diversity 295–296
  Hebridean Mink Project **165**, 166–167, 175–176
  island populations *see* islands; New Zealand
  New Zealand *see* New Zealand
  Pine Martens on Skye and Mull 180–183, 184
  Stoats in New Zealand *see* New Zealand
  Stoats in Orkney 160–161
  as symptom of environmental changes 183–184

Weasels in New Zealand *see* New Zealand
Ireland, fauna 4, 24–26
All-Ireland Squirrel and Pine Marten Survey 264
American Mink 14, 174–177, **176**
  distribution 176, **176**
  fur farms 174, **175**, 177
  habitat favourability model 176
  trapping/control 175–176
  see also American Mink (*Species Index*)
Bank Voles 220
BioBlitz 176
Brehon Laws 293
colonisation by species 25–26, 41
competition between mustelids 27, 41
differences from Britain 25–26, *26*
European Polecat absence 11, 208
European Polecat misidentification 208
ferrets (feral) 11, 178–179
folklore (mustelid) 291–293
Grey Squirrel distribution **260**, 260–261, 262, 268
  decline, Pine Marten recovery 259, 261, 262, 264, 269
  habitat, and Pine Marten recovery 261–262
  increased, despite Pine Martens 262
  Pine Marten density importance 268
  Pine Marten distribution relationship 261–262, 268
  Sheehy's study sites 262, 264, 269
Grey Squirrel response to Pine Marten scent 267–268
hair trapping, from Stoats and Weasels 197
Irish Stoat *see* Irish Stoat (*Species Index*)
Mostela camera traps for Irish Stoats 202–203, **203**
Pine Martens 69, 70, 97–102
  decline 70, 98–99, 100, 101
  densities/population sizes 272, 297
  diet 229, **263**, 299
  dietary niche 299–300
  distribution (1870–1975) 70, 99, **100**
  distribution survey (1978–1980) 100
  distribution survey (2009–2013) 261–262
  distribution surveys (2005, 2019) 102, 264
  folklore about 292–293
  game shooting 70, 99
  genetic diversity, low 102
  habitat loss effect 99, 272
  habitat type 99, **101**, 102, 272, 297, 302
  home range size 298–300, 299
  hunting of (1500s) 98

late Bronze Age 97
legal protection 101–102
Middle Ages 53
origin of 97–98
poisoning by farmers 70, 101
population densities 301
protection on estates 99
recovery (from decline) 101–102, 272
recovery, Grey Squirrel decline 259, 261, 262, 264, 269
as vermin, control 99
woodland increase effect 102
Red Squirrel distribution **260**, 260–261, 264
Pine Marten distribution relationship 261–262, 264, 265
Sheehy's study sites 262, 264
Red Squirrel recovery 259
Red Squirrel response to Pine Marten scent 268
Stoats 4, 9, **25**, 41
folklore about 291–292
Weasel absence 41
Wildcat 292–293
Wildlife Act 1976 73
islands
control of invasive species 184
endemic species, vulnerability 183–184
feral Ferrets 178–179, 180
Irish Stoat populations 39
minimum conditions for mustelid populations 39
size, effect on mustelid populations 39–41
Stoat and Weasel populations 35, 39–41, 40, 183–184
see also New Zealand
Terschelling 38–39
voles, needed for Weasels 41
see also specific islands
Isle of Man, Irish Stoat 39, 40
Isle of Mull *see* Mull, Isle of
Isle of Skye *see* Skye, Isle of
Isle of Wight 41
isotopes, dietary information from 211–212

jaws, Weasel 30
'jiggler' 194, 196

Kielder Forest, England 225, 243–244
King, Carolyn 29, 30, 212–213, 215, 284
Kitchener, Andrew 86
kits and litters *see specific mustelid species (Species Index)*
kiwi species (birds) 158

Klimt, Gustav **65**

Last Glacial Maximum (LGM) 25, 26
Late Miocene 17
legal protection
European Polecat 89, 308
Irish Stoat 73, 220
Pine Marten 64, 92, **92**, 95, 182, 310
Stoat 72
legislation
on furs, Middle Ages 50, 53, 54, 55
vermin control 55
lemmings
in Ireland, extinction 41
see also Lemmings *(Species Index)*
lettice (white Weasel skins) 50
lifespan
delayed implantation association 46
European Polecat 43
Pine Marten 44–45, 91
Stoat and Weasel 34
Lotek Litetrack-30 133
Lotto, Lorenzo 284, **286**
Lovegrove, Roger 63

MacPherson, H.A. (taxidermist) 61, 91
MacPherson, Jenny **296**, **300**, **322**
reintroduction of Pine Martens to Wales **110**, **114**, 116, 118, 119
martens
American Marten *see* American Marten *(Species Index)*
dietary generalists 14, 105
fur for clothing 51
Pine Marten *see* Pine Marten *(Species Index)*
types in Britain pre-1800s 22
'white-breasted' ('house') 22, **23**, 49
'yellow-breasted' ('wood') 22, **23**
Mary I, Queen 290
Mascall, Leonard 60
mating 33, 44, 45–46
delayed implantation and *see* delayed implantation (DI)
see also specific mustelids (Species Index)
McMurtrie, Peter 159–160
McNicol, Catherine (Cat) **126**, 127, 129
melanocyte-stimulating hormone (MSH) 316
mesocarnivores 212
Mesolithic (Middle Stone Age) 19, 47, 70, 181
Messenger, John 95, 96
metabarcoding 205–206, 208
dietary analysis 211

mice 219
  as prey 29, 81
microsatellites (DNA) 86–87, 190
Middle Ages
  fur trade/use for clothing 50, 50–52, **51, 52**
  legislation on furs 50, 53, 54, 55
  open field system 63
Middle Miocene 17
Middle Stone Age (Mesolithic) 19, 47, 70, 181
Middle Vallesian 'crisis' 17
minever (fine squirrel skin) 51
minimum convex polygons (MCPs) 304
mink
  British farm 163
  European 164
  feral 161, 165, **168**, 174
  fur *see* American Mink (*Species Index*)
  Irish farm 174
  see also American Mink (*Species Index*)
'mink raft' **165**, 170–172, **172, 173**
Miocene Epoch 17
mitochondrial DNA 86, 87, 295, 296
monitoring small mustelids *see* surveying and monitoring
Mos, Jeroen 201
Mostela camera traps **201**, 201–203, **203**
  advantages 204
  European Polecat 203, 204, **204**
  Irish Stoat 202–203, **203**
  species detected 203
  Stoat, low detection rate 202
  Weasels **201**, 202
moulting, Weasels 32, 316
movement
  bounding 3, 5, 13
  Stoats 304, 321
  Weasels 5, 6, 321
  see also gait
Mull, Isle of
  American Mink 181
  feral cats 181
  Pine Martens 93, 180–183, 184
    numbers 181, 184
  Stoats and Weasels 39, 40, 40, 41
mustelids *see* small mustelids
mythology and folklore 2, 275–294
  about European Polecats 294
  about Irish Stoats 291–292
  about Pine Martens 292–293, 293–294
  about Weasels 2, 277–284
    childbirth/fertility, role 277–281
    conception by Weasels 2, 279–280, **280**, 281

global folklore 293–294
  as witches 281–284
  in Ireland (mustelid) 291–293
myxomatosis 37, 81, 155, 322
  in Australia 81
  effect on European Polecat diet 221, 223
  effect on Stoat population 37, 78, 81, 155, 214, 215, 248–249, 296
  effect on Weasel population 77
  transmission, and outbreaks 81

naivety, Grey Squirrel susceptible to predation 265–269, 272
National Gamebag Census (NGC)
  European Rabbits 81–82, **82**
  Stoats and Weasels 76–77, **77**, 77–78
Natural Resources Wales **107**, 142
NatureScot (Scottish Natural Heritage) 110, 113, 144–145, 160
Neolithic 47–48
nest(s)
  Stoat, under snow 197
  Weasel 32
nest boxes 255, **255**
  protection from Pine Martens 255–256
New Jersey zoo 137, 138
New Zealand
  feral cats 155, 250
  feral Ferrets 178, 180, 250
  Ferret breeding/release 151, 155
  Ferrets, importing failure 150–151, 271
  offshore populations 157
  food removal as natural enemy 250
  native fauna extinctions 157, 183–184
  offshore island colonisation 156–157, 158, 183
  rabbits, damage from 148–149, **149**, 153, 249
    costs of 153
  rodents, predator (Stoats) impact 250
  Stoats 14, 148, 150, 151–157, 183, 250–251, 297
    cat avoidance 155
    colonisation of islands 156, **156**, 157, 158, 183
    competition and coexistence 154
    consequences of invasion by 157–160
    delivery and costs 152
    diet 218–219, 249
    fur colouring in winter 318, 320
    genetic diversity 295–296, 297
    habitats for 155–156
    hair trapping 197
    home range size 305–306, 306
    impact on birds 150, 155, 158, 218–219, 249

impact on natural fauna 14, 152–153, 155–156, 157–158
impact on rabbits 152, 153, 154, 249, 250–251, 271
impact on rodents 250
importing from Britain 152, 153, 155
predator (Stoat) control 158–160
predator–prey relationships 249, 271
refuge in forests 155–156
Resolution Island 159–160
Secretary Island 159
swimming to islands 156, **156**
trapping and trap types 158–159
Weasels 14, 148, 150, 154, 155, 183, 250
decline in number 156
delivery and costs 152
hair trapping 197
importing from Britain 151–152, 153
on islands 157
nocturnal mustelids 10, 43, **230**, 305
non-native species 183
harmless, failure to establish 183
invasive *see* invasive non-native species (INNS)
risk of being invasive 183
Northern Ireland
citizen survey, squirrel and Pine Marten interactions 269
feral Ferrets on Rathlin island 178–179, 180
Grey Squirrel in Pine Marten diet 264, 265
Grey Squirrel response to Pine Marten scent 267–268
Pine Marten recovery 101–102
numerical response 218–219, 238, 242, 244

occupancy models 189–190
assumptions and limitations 189
citizen science surveys with 207–208
eDNA survey analysis 207
false positives 207
long-term trends, estimates 208
reliability of predictions 207–208
site-specific covariates 189
survey-specific covariates 189–190
Ojibwe people, Wisconsin 277, 293
open field system 63
Orkney Islands 161
Stoats on 160–161
eradication 161
native species threatened by **160**, 160–161
Orkney Native Wildlife Project 161
otters
hunting 1, **54**, **164**

interspecific aggression with mink 164
see also Otter (*Species Index*)
Outer Hebrides 165, **165**
American Mink 165–167, 175–176
Pine Marten extinction 53
wading bird populations 165–166
overharvesting 110, 310–311
Ovid 279–280
Ownes, Clifford **276**

Paleolithic image of Weasel 275, **275**
Parmigianino 284, **286**
Pennant, Thomas 22
Perrault, Charles 52
persecution of mustelids 55–63
for fur/pelts *see* fur trade/use
as predators of game 64, 65–66, **66**, 66–71
see also predator (mustelids) control
as vermin *see* vermin control
see also trapping
Pests Act (1954) 72
pets, Weasels as 276, **276**
pheasant shoots 71, 73
phylogenetic tree 15, **16**
Physiologus 280–281
Pine Marten Recovery Project **126**, 126–127
plague, outbreaks 55–56
Pleistocene 17, 18, 19, 30, 31, 148
Pliny 276, 277–278
Plutarch 280
poaching 64
Poland
bats as prey 256, **256**
Białowieża National Park in 19, 246, 252
fossil Weasels 31
Pine Marten diet 229, **256**
Pine Marten home range size 298, 299, 300
rodent numbers determined by food supply 246
rodent population, predation effect 246, 252
Stoat decline after mink invasion 42
Weasel home range size 303
Weasels and Stoats 41
'polecat', use of term (abusive) 1
polecats *see* European Polecat (*Species Index*)
population (human), 16th century 55–56
population bottlenecks 295
stoats 295–296, 297
population cycles, prey *see* prey population cycles
population sizes
difficulty estimating 297
see also specific mustelids (*Species Index*)

poultry
  mink killing 167
  Pine Martens taking 181
  polecats killing 58
  Stoats killing 62, 161
  Weasels killing 282
'powdering' 290
Powell, Roger 10, 111
predation 235–240
  impact, studies 239–240
  mortality (compensatory/additive) 235–236, 250
  study of 238–239
predator(s), mustelids as 1, 3
  ambush 321
  body shape importance 5
  specialist or generalist 14, 163, 212, 220, 236, 246
  Stoat and Weasel comparison 28–31, 76
predator (mustelid) control 66–71
  20th century 71, 72, 76–78, **77**
  21st century 72–73, 76–78, **77**
  ethical and welfare issues 180, 184
  future threat 310–311
  partial/localised, ineffective 76, 161, 177, 180
  Pine Martens 69–70, 91
  polecats 70–71, 79
  Weasels and Stoats 67–69, **68**, 72, 73–74, 76–78
  see also Stoat (*Species Index*); vermin control; Weasel (*Species Index*)
predator-deflection marks 9–10
predator–prey interactions 235–258
  climate effect 257, 258
  consumption rate 236, 237
  daily food requirements (of predators) 239, 239
  efficiency of predators 236
  European Polecat and Water Vole 248
  fertility and mortality affecting 250–251, 257
  food removal as natural enemy of prey 246, 250–251
  functional response 227, 236, 242, 248
    types 236–238, **237**, 252
  human-caused habitat change effect 255–256
  numerical response 218–219, 238, 242, 244, 248
  Pine Martens 252–258, 271–272
    bats **256**, 256–257
    birds 252, 254–255
    Black Grouse 257, 258
    coexistence with bird species 258
    game birds 256, 257, 258
    model, of impact on birds 254–255
    Red Squirrel **253**, 253–254, **262**
  predation mortality

additive 235–236, 250
  compensatory 235
predator control of prey 235, 236, 238–239
predator effect on prey number *see* prey population cycles
predator removal, benefit for birds 257–258
prey density effect 236, **237**, 237–238
rodent populations *see* rodent prey
spatial variation, landscape level 255
specialist predator hypothesis 244
species' vulnerability, factors affecting 254
Stoats 247–250
  coadaptation 249–250
  failure to control rabbits 152–153, 247, 249, 250, 271
  Field Voles 248
  impossible to reduce prey numbers 247, 250
  on islands 249–250
  lemmings 247, **247**
  in New Zealand 249–250, 251
  rabbits 248–249
  rodents 247, 250
  Water Voles 248
study of predation, methods 238–239
Weasels 240–247
  effect on prey population cycles 240–242, 243
  functional response 236, 238, 242
  habitat effect on impact 240–241
  impossible to reduce prey number 247
  Kielder Forest, voles 243–244, 245
  northern habitats 242–243, 244–245, 245–246
  numerical response 238, 242
  temperate habitats 243
  voles/lemmings 241–243, 243–244, 245, 246
prey 1
  death, reasons 235, 238
  driving predator abundance 250
  evolutionary diversification 17
  mustelid predators *see specific mustelids* (*Species Index*)
  naivety, Grey Squirrel susceptibility 265–269, 272
  nutritional value 212–213
prey population cycles
  predator effect on 235, 236, 238, 240–242, 243
    Pine Marten effect 252
    specialist predator effect 240–241, 241–242, 244
    Stoats effect 247–250
  rabbits, Stoats effect 248–249, 250
    failure to control 152–153, 247, 249, 250, 271

GENERAL INDEX · 367

rodents, in New Zealand 250
voles and lemmings 241–243, 246
  food supply (winter) affecting 246
  Kielder Forest 243–244, 245
  specialist and generalist predators 246
  Stoats effect 247–248
  vole body size (Chitty effect) 242–243
  Water Vole 248
  Weasels affecting 241–243, 243–244, 245, 246
Protection of Animals Act (1911) 72

Q-methodology 127

Raasay, stoats and weasels on 39, 40, **40**
rabbit(s) 80
  as American Mink prey 41–42, **42**
  benefit of predator culling for 79–80
  breeding rate, Stoat failure to control 152–153, 247, 249, 250, 271
  commercial importance 58–59
  decline, RHD virus 221–222, 249, 250, 322
  eradication/decline postwar 80
  as European Polecat prey **42**, 43, 58–59, 80, 81, 221, **221**
  ferrets used for hunting 84, 84–85, **178**
  as food source (for humans) 58–59, 80
  killing/snaring 80
  myxomatosis *see* myxomatosis
  National Gamebag Census 81–82, **82**
  in New Zealand *see* New Zealand
  population increase (1930s) 79–80
  population increase (1960s–2000s) 81–82
  as Stoat prey *see* Stoat (*Species Index*)
  as Weasel prey 214–215, 218
rabbit haemorrhagic disease (RHD) virus 221–222, 249, 250, 322
radio collar, on Pine Martens **299**
  removal 121–122
  translocated Pine Martens 115, 116, **117**, 118, 120, **121**, 127, 145, **262**
radio tracking
  Pine Martens **300**
    home range size data 298–299, **299**
    Stoat home range size 304–305
    Weasel home range size 304
reinforcement 103
  Pine Martens in Wales 106
reintroduction(s) 2, 104, 310
  community, importance 141–142
  definition 103
  disease risk 108
  habitat suitability assessment for 104–105, **105**

impact on prey populations, concerns 252
as last resort 146
males *vs* females, model 111
modelling, for martens 111–112
native species coexisting with prey species 107–108
non-native predators, impact on prey 107
overharvesting of donor populations 310
Pine Martens *see below*
prey species decline consideration 108
release process, soft or hard release 115–116, 132
successful, criteria 111–112, 151
reintroduction, Pine Martens to England 2, 97, 105, 274
  Forest of Dean 141, 144–146, **146**
  radio tracking, released martens 145
  roadkill concerns 144, 145
  trapping in Scotland 144–145
  Grey Squirrel management and 272–273
reintroduction, Pine Martens to Wales 103–135, 273, 274
  aim, targets, objectives 110–111, 143, 272
  behaviour after release 134
  behavioural assessment before release 131–132, **132**
  breeding success 118, **119**, 119–122, **121**, **122**, 123
  budget/costs 143–144
  Cambrian Mountains, woodland 105–106
  camera trap, monitoring 138, **139**, 140
  captured (donor) animals, assessment **114**, 114–115
  community, importance 141–142, 144
  concerns over predation by 107–108, 252
  den boxes (artificial) **119**, 120
  distribution survey 138–139, 140, 144
  distributional range expansion 139–141, **140**
  donor source, finding 109–111, 143
  establishment and rate of spread 122, 124, 126, 143
  failure, causes 112–113, 143
  fungal infection 124–125
  fur/bib colourings 125, 137, **137**, 194, **195**, 196
  genetic diversity of donor population 109–110
  Grey Squirrel management and 272–273
  habitat suitability for releases 105, **105**, 129, **129**
  home range habitats **129**, 129–130, 133
  home range size **128**, 128–129
  identification/recognition 125, 137, **137**, **195**, 196
  location sites 105–106, **106**, 120

long-distance dispersal 130–132, 136–138
long-term monitoring 138–139, **139**, 143, 144
males *vs* females, number 111–112
mating 118–124
microchipping before 115, 136
monitoring 125–127, **126**, 132–134, 134–135, 143
mortality, and causes of 124–125
number required for release 109, 112, 116, 118, 122
personality of animals 130–132, 135
planning 111–112
post-release movements 127–130, **128**, 143
pre-release pens 116, **117**, **128**, **131**, **135**
public opinion survey 106
radio collar removal 121–122, 134–135
radio telemetry/GPS equipment 132–134, 134, **135**
radio tracking after release 116, **117**, 118, 120, 121, 127
recaptures 134–135
release process (soft release) 115–116, **117**, 132
release sites 106, **107**, 116, 129, **129**, 143
release timing 113, 116, 118, 119, 125
research and adaptive management 126–127
risk assessment to donor population 109
Savannah marten (PM38) 136–138, **137**
scat examination/surveys 131, **132**, 139
Sid Vicious, and Miss Piggy 130, 131, 140
social feasibility (people) 106, 141, 142
spread to Forest of Dean 141, 143–144
stress in animals 112–113, 118–119, 131
stress measurement (fGCMs) 127, 131, 132, **132**
stress reduction methods 112–113
success, measuring 142–144
time to settle in area 128, **128**
translocation (process) 115, **115**
trapping/capture and timing 113, **114**
volunteers helping 142, 144
Welsh-born kits 120–122, **121**, **122**
reintroduction biology 104
Renaissance
  fur use for clothing 284, **285**, **286**
  Weasels in 284–288
Resolution Island, New Zealand 158, 159–160
Richard II, King 84–85
roadkill 105, 125, 145, 220, 233–234
  Irish Stoat 220
  Pine Martens 105, 112, 125, 145, 207, 274
  polecats 309, 310
Robe of State 288
rodent prey 14, 28
  of European Polecat 163, 221, **224**, 231, 232

evolution aspects 17–18
population cycles 241–246, 250
  food supply effect 246, 250–251
  predator effect 240, 241, 242–244, **245**
  Stoat effect in New Zealand 250
  Stoat/Weasel failure to reduce 247, 250
reproduction rate 250
  of Stoats 29, 34, 214–215, 218, 233
  of Weasels *see* Weasel (Species Index)
rodenticides, mustelid susceptibility 231–234
  anticoagulant 231–232, 233
  second-generation anticoagulant (SGARs) 232, 233
royalty, wearing of ermine 51, 52, **52**, 288–290, **289**, **290**

sable, fur from 49, 289, 290
Savannah, Georgia, Pine Marten found 136–138, **137**
Scalpay, stoats and weasels on 39, 40, 40
scats
  American Mink 186
  analysis for diet study 209–211
    digestibility differences 211
    frequency of occurrence (FO) 210
    misidentification 210–211
    prey DNA/molecular analysis 211
    relative frequency of occurrence (RFO) 210
    sampling error and bias 210
  distinguishing, difficulties 186
  European Polecat 186
  Pine Marten *see* Pine Martens (*Species Index*)
  Red Fox 188, 209–210
  Stoat 186, **186**
  for surveying/monitoring 186
  Weasel 186, **186**, 212
scent 10, 11
  of scats 186, 187–188
scent glands 1, 2–3, 10
Scotland 216
  American Mink 164–165
    control projects 173–174
  Black Grouse numbers 258
  European Polecat
    population number 307–308
    as vermin, control 68
  fur farming ban 164–165
  gin trap use 72
  Pine Martens
    18th–20th centuries 69, 91–92, **92**
    1994 survey 92
    detection probability, survey transects 190

diet 225–226
distribution/numbers 69, 91–94, **92**
as donor source for reintroductions 109–111
expansion to southern Scotland 92–93, 274
Galloway Forest survey 191–194, *193*
game bird populations and 258
home range size 298, *299*
impact on Grey and Red Squirrels 269
in Middle Ages 53, 55
overestimate of population size 297
population density 191–193, 297, 300–301
pre-/post-legal protection, distribution 92, **92**
surveying/monitoring methods 190
Stoats
home range size 305–306, *306*
as vermin, control 68
winter whitening 312, **312**, *313*
Weasels
home range size 303
as vermin, control 68
Scottish Invasive Species Initiative 173
Scottish Mink Initiative 173
Scottish Natural Heritage (NatureScot) 110, 113, 144–145, 160
second-generation anticoagulant rodenticides (SGARs) 232, 233, 234
Secretary Island, New Zealand 159
Segar, William 284, **285**, 290, **290**
sexual dimorphism
European Polecat 221, 222
Irish Stoat 26, 220
Stoat and Weasel 26, 28, 28–29, **217**, 217–218
sexual symbolism 284, **285**
Shakespeare 282
sheep farming, rabbits impact in New Zealand 149, **149**
shooting, of game *see* gamekeeping
shotguns 66
Simpson, George Gaylord 15
Sinclair, John 68
sizes 3–4, **4**
cold climate implications 31–32
evolution 17–18
small mustelids 28
variation with latitude (Bergmann's Rule) 220, 228
see also specific mustelids (Species Index)
skull
fossil, Pine Martens 24
Mustela robusta (fossil) 19
Pine Marten *vs* Stone Marten 20
Weasel 5, 30

skunks 15
Skye, Isle of
Pine Martens 93, 180–183
Stoats and Weasels 39, 40, *40*
small mustelids 3–5, **4**
characteristics 1
distinguishing between species 5
distribution 14
history in British Isles 18–19
species included, and sizes 3–4, **4**
species number in Britain/Ireland 25
see also specific topics (e.g. body shape)
snares, hair 190
snaring
Pine Martens 91, 99, 100
rabbits 59, 80
snow cover
decline, impact on Stoat fur colour 311, 313–314, 321
impact on Weasel fur colour 314–316
see also fur coat, colouring of
specialist predator 14, 163, 218, 236
regional, polecats 163, 221
Stoats 28–31, 218
Weasels 28–31, 212, 218, 246
specialist predator hypothesis 244
species richness, invasive species and 184
Spring Traps Approval Order 72
squirrel furs (coats) 51
squirrelpox virus 147
stink badgers 15
Stoat-skin purse 292
stoats *see* Stoat (*Mustela erminea*) (*Species index*)
stomach contents, examination 209, 212, 214
Strabo (Greek writer) 84
Strachan, Mark 6
Strachan, Rob 186, 188
stress in animals
capture-mark-release-recapture 185–186
measurement methods 127, 131, 132, **132**
Pine Marten reintroductions 112–113, 118–119, 131
reduction, methods 112–113
strychnine poisoning 70, 101
surveying and monitoring 185–208
'absent' species, false negative results 188–189
camera traps *see* camera traps
difficulties 297
eDNA analysis *see* environmental DNA (eDNA) analysis
European Polecat 203, 204, **204**, 208
1960s–2010s national survey **83**, 83–85

2014–2015 national survey 83, **83**, 89, **89**
  in Britain, citizen scientist approach 309
  misidentification, citizen science 208
  Mostela camera trap 203, 204, **204**
  rodenticides in carcasses 232–233
field signs (tracks/scats) 186
  see also scats
hair detection *see* hair samples; hair tubes
Irish Stoat 202–203, **203**
live trapping 185–186
  vs other methods 204, 205
  see also trapping
metabarcoding 205–206
Mostela camera trap use *see* Mostela camera traps
occupancy models *see* occupancy models
Pine Martens 187–196, 205
  abundance estimation methods 196
  camera traps (trail cameras) 194, 205
  detection probability, transect size 190
  difficulties 297
  eDNA sample 207
  false negative results 188–189
  genotyping (scats) 192–193
  hair samples/sampling 190, **191**, 191–194, **192**, 193
  live trapping 191–194, 193
  method comparison, Galloway Forest 191–194, 193
  method effectiveness and locality 194
  occupancy models 189–190
  problems associated with scats 188–189
  scats 94, 97, 139, 187–189, 190, 193, 193, 194
  sex bias 193
Stoats 196–208
  camera traps ('Mostela') 196–197, **201**, 202
  footprint detection 197
  hair trapping, DNA analysis 197–201
  hair tube modification **198**, 198–199
  hair tubes, low detection 197–199, **198**
  low detection in camera traps 202
  Wales 198–199, 199
  traditional methods with eDNA 207, 208
Weasels 196–208
  camera traps ('Mostela') 196–197, **201**, 201–202, 203
  footprint detection 197
  hair trapping, DNA analysis 197–201, 199
  hair tube modification **198**, 198–199
  Wales 198–199, 199, 200, **200**
Sweden
  European Polecat distribution 311

Stoat home range size 304
Weasel fur colour in winter 315–316
'sweet mart' 14, 61

tail
  length, comparison between species 28
  predator-deflection marks 9–10
  sizes and colours 5, **6**, 9, **9**, 10, 13
  see also specific mustelids (Species Index)
teeth
  Pine Marten 114, **114**
  Weasel 30
temperature
  heat loss and, Weasels 32, 278–279
  minimum for fur whitening in winter 316
  range, for stoats 26
territorial nature 108, 297–298, 304
Terschelling, Netherlands 38–39
Thompson, William 99
threats (current/future) 310–311
Tomlinson, Alexandra 108, 114
trail cameras *see* camera traps
translocations *see* conservation translocation
trap(s)
  alarms 172, 174
  for American Mink 168–169, 170–173, 174
  cage 113, 168, 170
  camera *see* camera traps
  Fenn 'humane' trap 72
  gin traps 67, **67**, 72, 81, 99
  on mink rafts 170–173, **172**, **173**
  self-resetting Stoat 158
  tunnel 73, **73**, 168–169, 202
trapping
  in 19th century 66–67, **67**
  20th and 21st century restrictions 72
  AIHTS restrictions 72–73
  American Mink *see* American Mink (*Species Index*)
  Grey Squirrel control 274
  Irish Stoat 202–203, **203**
  in Middle Ages 52–53
  Pests Act (1954) 72
  Pine Martens 60–61, **62**, 91, 99, 100
    for reintroductions 113, **114**
  for predator control, future threat associated 310–311
  Stoat *see* Stoat (*Species Index*)
  for surveying/monitoring 185–186
    behavioural traits affecting 193–194
    benefits 185
    camera traps *vs* 204

GENERAL INDEX · 371

limitations 185–186, 205
scats/hair samples comparison 191–194, *193*
Weasel *see* Weasel (*Species Index*)
tree climbing 13
Pine Martens 13, 14, 43, **44**, **48**, **98**, 264, **301**
tunnel traps 73, **73**, 168–169, 202
Turner, Peter 99

UFOs (unidentified furry objects) 95–97

'vacuum effect' 169, 170, 175
vair (Baltic squirrel fur) 52
van Maanen, Erwin 194, 196
Veale, Andrew 159
Velander, Kathy 94
Vermin Acts 56–57, 60
vermin control 53, 54, 55–63, 64
16th century (Tudor period) 55–57, 60
19th century 60, 64, 68
20th century 68–69
community participation 57
by gamekeepers 64, 68–69
killing methods 60, **60**
legislation 55, 56
payments for 57, 59, 60, 61, 62, 63
reasons for (16th century) 55–56
vermin types 56–57
see also specific mustelids (*Species Index*)
Veronese, Paolo 284, **286**
VHF radio collars 133, 134, **135**
Vincent Wildlife Trust 86, 88, **89**, 92, 95, 97
Camera Trap Loan Scheme 138
European Polecat survey 204
Mostela camera trap trial 203, 204
Pine Marten reintroductions 104, 116, 118, 126, 142
Forest of Dean 143, 145
Pine Marten sighting appeal 95–97
rodenticides in polecat carcasses 232
voles
population cycles 242–243
food supply (winter) affecting 246
predators (Weasels) affecting 242–243, 244, 245
Weasel island populations and 41
see also specific voles (*Species Index*)
volunteers
American Mink control 173
citizen science *see* citizen science
reintroduction of Pine Martens to Wales 142, 144
von Waldheim, Johann Gotthelf Fischer 15

Wales 216
European Polecat
distribution **69**, 71, 83, **83**, 89, 164, 223
population number 307
Pine Martens 94, 96, 103
Black Grouse population 257
failure to detect (1990s–2010s) 103
habitat suitability evaluation 104
persistence/survival/records 94, 96, **96**, 103
reintroductions *see* reintroduction, Pine Martens to Wales
rodenticides in polecat carcasses 233
Stoats, sites for 201
Weasel sites 200, **200**, 307
warfarin 232
water-based eDNA metabarcoding 205–206
'weasel words' 282
weasels
names and colloquial use of 4
way of life 31–33
see also Weasel (*Mustela nivalis*) (*Species index*)
weight, comparison between species 28
whiskers, isotope data 211–212
'white-breasted' ('house') martens 22, **23**, 49
wild species
domestic gene introgression 86, 88–89
hybridisation of polecats and ferrets 84, 85, 86, 87, 88
Wildlife and Countryside Act 89, 92
witches, weasels as 281–284
Wodehouse, P.G. 71
Wolstonian glaciation 18
woodland 216
19th century cover extent 63
bird decline and 255
fragmentation, impact on Pine Martens 310–311
Pine Martens in 13, 22, **48**, **110**, **129**, 225, 301–302, 321
donor source areas 109–111, **110**
home range size 298
reintroduction areas 105–106
Weasels in 216, 303
diet in 215–216
World War, First 71, 91
World War, Second 71, 80

'yellow-breasted' ('wood') martens 22, **23**

zibellini (flea furs) 284

Picture Credits

All photographs by the author or in the public domain and sourced from Wikimedia Commons or Internet Archive with the exception of those listed below. References are to figure numbers.

1: Robert Cruikshanks. 3: Clive Craik. 4: Pat Morris. 5: Mark Strachan. 6: John Dellow. 7: Henry Schofield. 8: Vincent Wildlife Trust (VWT). 9: Robert Cruikshanks. 10, 11: Pat Morris. 14. Robert Cruikshanks. 16: Jean Michel Bompar. 17: Robert Cruikshanks. 18: Ruth Hanniffy. 19: Carl Morrow/Alamy Stock Photo. 20: ARIE VAN 'T RIET/SCIENCE PHOTO LIBRARY. 21: Robert Cruikshanks. 23: Frank Greenaway. 24: McPhoto/Rolfes/Alamy Stock Photo. 25: BIOSPHOTO/Alamy Stock Photo. 26: Mark Lockett. 28: DavidEdwards8/Shutterstock. 29: mammalpix/Alamy Stock Photo. 30: Paul Hobson/Alamy Stock Photo. 32: Jason Hornblow. 33: Pat Morris. 35: The Print Collector/Alamy Stock Photo. 36: World History Archive/Alamy Stock Photo. 37: © Leonard de Selva/Bridgeman Images. 38: Chronicle/Alamy Stock Photo. 39: blickwinkel/Alamy Stock Photo. 40: imageBROKER.com GmbH & Co. KG/Alamy Stock Photo. 44: whitemay/Getty Images. 45: ARTGEN/Alamy Stock Photo. 46: The Royal Collection of the United Kingdom. 47, 48: Pat Morris. 50: Stephan Schramm/Alamy Stock Photo. 51: Neil Henderson/Alamy Stock Photo. 53: Joel Walley/Alamy Stock Photo. 54: Pat Morris. 56: The Museum of Modern Art, New York/Scala, Florence. 58: Pat Morris. 61: Johnny Birks. 64: VWT/Judy Mackenzie. 66: Robert Cruikshanks. 67: VWT. 69: Ruth Hanniffy. 71, 74: Henry Schofield. 75a: Lizzie Croose. 75b: VWT. 76: Robert Cruikshanks. 77: VWT. 79: Henry Schofield. 81. Nick Upton. 84: Robert Cruikshanks. 85, 88: Nick Upton. 90: Nick Upton/naturepl.com. 92: VWT. 97: Papers Past, National Library of New Zealand. 98: Mark Lockett. 99: Pat Morris. 100: Gary Clarke. 101: Minden Pictures/Alamy Stock Photo. 102: FLPA/Alamy Stock Photo. 104: John Dellow. 105: Pat Morris. 106: Vincent Lowe/Alamy Stock Photo. 107: Jerome Murray - CC/Alamy Stock Photo. 108, 111, 112: John Dellow. 115: Johnny Birks. 116: Lizzie Croose. 117: Pat Morris. 118: Lizzie Croose. 119: Ruth Hanniffy. 120b: Keziah Hobson. 121: Josie Bridges. 124: VWT. 125: Ruth Hanniffy. 126: VWT. 127: Henry Schofield. 129: Joel Walley/Alamy Stock Photo. 130: Brian Hewitt/Alamy Stock Photo. 132: robin chittenden/Alamy Stock Photo. 133: Dermot Breen. 134: Nature Picture Library/Alamy Stock Photo. 137: Buiten-Beeld/Alamy Stock Photo. 138: Terry Whittaker/naturepl.com. 140: Joe Blossom/Alamy Stock Photo. 142: DAVID ORMEROD/Alamy Stock Photo. 143: BIOSPHOTO/Alamy Stock Photo. 144: Ernie Janes/Alamy Stock Photo. 146: Dennis Jacobsen/Shutterstock. 149: Chris Grady/Alamy Stock Photo. 151: Richard Newton/Alamy Stock Photo. 152: Henry Schofield. 153: UniquePhotoArts/Shutterstock. 155: seawhisper/Shutterstock. 156: Jason Hornblow. 158a: Jim Cumming/Shutterstock. 158b: Gaertner/Alamy Stock Photo. 160: Protasov AN/Shutterstock. 161: The Bradshaw Foundation. 162: Pat Morris. 163: Bodleian Libraries, University of Oxford. 164: Chalon-sur-Saône. Bibliothèque municipale. 165a: Biblioteque Nationale de France. 165b: The British Library. 166: Frank Greenaway. 167b, 168c: The Walters Art Museum, Baltimore. 170: PA Images/Alamy Stock Photo. 172: Petr Muckstein/Shutterstock. 173: Henry Schofield. 174: Nick Upton/naturepl.com. 175: Lizzie Croose. 176: Colin Smith. 177: VWT/Moss Taylor. 178: Lillian Tveit/Shutterstock. 179: outdoorsman/Shutterstock. 181a: Richard Seeley/Shutterstock. 181b: John Navajo/Shutterstock. 182a: Victor Tyakht/Shutterstock. 182b: Burlet Florent/Shutterstock. 184: Hugo Fourdin. 185a: Paul A Carpenter/Shutterstock. 185b: Karen Miller Photography/Shutterstock. 186: Colin Smith.

The New Naturalist Library

1. *Butterflies* — E. B. Ford
2. *British Game* — B. Vesey-Fitzgerald
3. *London's Natural History* — R. S. R. Fitter
4. *Britain's Structure and Scenery* — L. Dudley Stamp
5. *Wild Flowers* — J. Gilmour & M. Walters
6. *The Highlands & Islands* — F. Fraser Darling & J. M. Boyd
7. *Mushrooms & Toadstools* — J. Ramsbottom
8. *Insect Natural History* — A. D. Imms
9. *A Country Parish* — A. W. Boyd
10. *British Plant Life* — W. B. Turrill
11. *Mountains & Moorlands* — W. H. Pearsall
12. *The Sea Shore* — C. M. Yonge
13. *Snowdonia* — F. J. North, B. Campbell & R. Scott
14. *The Art of Botanical Illustration* — W. Blunt
15. *Life in Lakes & Rivers* — T. T. Macan & E. B. Worthington
16. *Wild Flowers of Chalk & Limestone* — J. E. Lousley
17. *Birds & Men* — E. M. Nicholson
18. *A Natural History of Man in Britain* — H. J. Fleure & M. Davies
19. *Wild Orchids of Britain* — V. S. Summerhayes
20. *The British Amphibians & Reptiles* — M. Smith
21. *British Mammals* — L. Harrison Matthews
22. *Climate and the British Scene* — G. Manley
23. *An Angler's Entomology* — J. R. Harris
24. *Flowers of the Coast* — I. Hepburn
25. *The Sea Coast* — J. A. Steers
26. *The Weald* — S. W. Wooldridge & F. Goldring
27. *Dartmoor* — L. A. Harvey & D. St Leger Gordon
28. *Sea Birds* — J. Fisher & R. M. Lockley
29. *The World of the Honeybee* — C. G. Butler
30. *Moths* — E. B. Ford
31. *Man and the Land* — L. Dudley Stamp
32. *Trees, Woods and Man* — H. L. Edlin
33. *Mountain Flowers* — J. Raven & M. Walters
34. *The Open Sea: I. The World of Plankton* — A. Hardy
35. *The World of the Soil* — E. J. Russell
36. *Insect Migration* — C. B. Williams
37. *The Open Sea: II. Fish & Fisheries* — A. Hardy
38. *The World of Spiders* — W. S. Bristowe
39. *The Folklore of Birds* — E. A. Armstrong
40. *Bumblebees* — J. B. Free & C. G. Butler
41. *Dragonflies* — P. S. Corbet, C. Longfield & N. W. Moore
42. *Fossils* — H. H. Swinnerton
43. *Weeds & Aliens* — E. Salisbury
44. *The Peak District* — K. C. Edwards
45. *The Common Lands of England & Wales* — L. Dudley Stamp & W. G. Hoskins
46. *The Broads* — E. A. Ellis
47. *The Snowdonia National Park* — W. M. Condry
48. *Grass and Grasslands* — I. Moore
49. *Nature Conservation in Britain* — L. Dudley Stamp
50. *Pesticides and Pollution* — K. Mellanby
51. *Man & Birds* — R. K. Murton
52. *Woodland Birds* — E. Simms
53. *The Lake District* — W. H. Pearsall & W. Pennington
54. *The Pollination of Flowers* — M. Proctor & P. Yeo
55. *Finches* — I. Newton
56. *Pedigree: Words from Nature* — S. Potter & L. Sargent
57. *British Seals* — H. R. Hewer
58. *Hedges* — E. Pollard, M. D. Hooper & N. W. Moore
59. *Ants* — M. V. Brian
60. *British Birds of Prey* — L. Brown
61. *Inheritance and Natural History* — R. J. Berry
62. *British Tits* — C. Perrins
63. *British Thrushes* — E. Simms
64. *The Natural History of Shetland* — R. J. Berry & J. L. Johnston
65. *Waders* — W. G. Hale
66. *The Natural History of Wales* — W. M. Condry
67. *Farming and Wildlife* — K. Mellanby
68. *Mammals in the British Isles* — L. Harrison Matthews
69. *Reptiles and Amphibians in Britain* — D. Frazer
70. *The Natural History of Orkney* — R. J. Berry

71. *British Warblers* — E. Simms
72. *Heathlands* — N. R. Webb
73. *The New Forest* — C. R. Tubbs
74. *Ferns* — C. N. Page
75. *Freshwater Fish* — P. S. Maitland & R. N. Campbell
76. *The Hebrides* — J. M. Boyd & I. L. Boyd
77. *The Soil* — B. Davis, N. Walker, D. Ball & A. Fitter
78. *British Larks, Pipits & Wagtails* — E. Simms
79. *Caves & Cave Life* — P. Chapman
80. *Wild & Garden Plants* — M. Walters
81. *Ladybirds* — M. E. N. Majerus
82. *The New Naturalists* — P. Marren
83. *The Natural History of Pollination* — M. Proctor, P. Yeo & A. Lack
84. *Ireland: A Natural History* — D. Cabot
85. *Plant Disease* — D. Ingram & N. Robertson
86. *Lichens* — Oliver Gilbert
87. *Amphibians and Reptiles* — T. Beebee & R. Griffiths
88. *Loch Lomondside* — J. Mitchell
89. *The Broads* — B. Moss
90. *Moths* — M. Majerus
91. *Nature Conservation* — P. Marren
92. *Lakeland* — D. Ratcliffe
93. *British Bats* — John Altringham
94. *Seashore* — Peter Hayward
95. *Northumberland* — Angus Lunn
96. *Fungi* — Brian Spooner & Peter Roberts
97. *Mosses & Liverworts* — Nick Hodgetts & Ron Porley
98. *Bumblebees* — Ted Benton
99. *Gower* — Jonathan Mullard
100. *Woodlands* — Oliver Rackham
101. *Galloway and the Borders* — Derek Ratcliffe
102. *Garden Natural History* — Stefan Buczacki
103. *The Isles of Scilly* — Rosemary Parslow
104. *A History of Ornithology* — Peter Bircham
105. *Wye Valley* — George Peterken
106. *Dragonflies* — Philip Corbet & Stephen Brooks
107. *Grouse* — Adam Watson & Robert Moss
108. *Southern England* — Peter Friend
109. *Islands* — R. J. Berry
110. *Wildfowl* — David Cabot
111. *Dartmoor* — Ian Mercer
112. *Books and Naturalists* — David E. Allen
113. *Bird Migration* — Ian Newton
114. *Badger* — Timothy J. Roper
115. *Climate and Weather* — John Kington
116. *Plant Pests* — David V. Alford
117. *Plant Galls* — Margaret Redfern
118. *Marches* — Andrew Allott
119. *Scotland* — Peter Friend
120. *Grasshoppers & Crickets* — Ted Benton
121. *Partridges* — G. R. (Dick) Potts
122. *Vegetation of Britain & Ireland* — Michael Proctor
123. *Terns* — David Cabot & Ian Nisbet
124. *Bird Populations* — Ian Newton
125. *Owls* — Mike Toms
126. *Brecon Beacons* — Jonathan Mullard
127. *Nature in Towns and Cities* — David Goode
128. *Lakes, Loughs and Lochs* — Brian Moss
129. *Alien Plants* — Clive A. Stace and Michael J. Crawley
130. *Yorkshire Dales* — John Lee
131. *Shallow Seas* — Peter J. Hayward
132. *Falcons* — Richard Sale
133. *Slugs and Snails* — Robert Cameron
134. *Early Humans* — Nicholas Ashton
135. *Farming and Birds* — Ian Newton
136. *Beetles* — Richard Jones
137. *Hedgehog* — Pat Morris
138. *The Burren* — David Cabot & Roger Goodwillie
139. *Gulls* — John C. Coulson
140. *Garden Birds* — Mike Toms
141. *Pembrokeshire* — Jonathan Mullard
142. *Uplands and Birds* — Ian Newton
143. *Ecology and Natural History* — David M. Wilkinson
144. *Peak District* — Penny Anderson
145. *Trees* — Peter A. Thomas
146. *Solitary Bees* — Ted Benton & Nick Owens
147. *Shieldbugs* — Richard Jones
148. *Ponds, Pools and Puddles* — Jeremy Biggs & Penny Williams

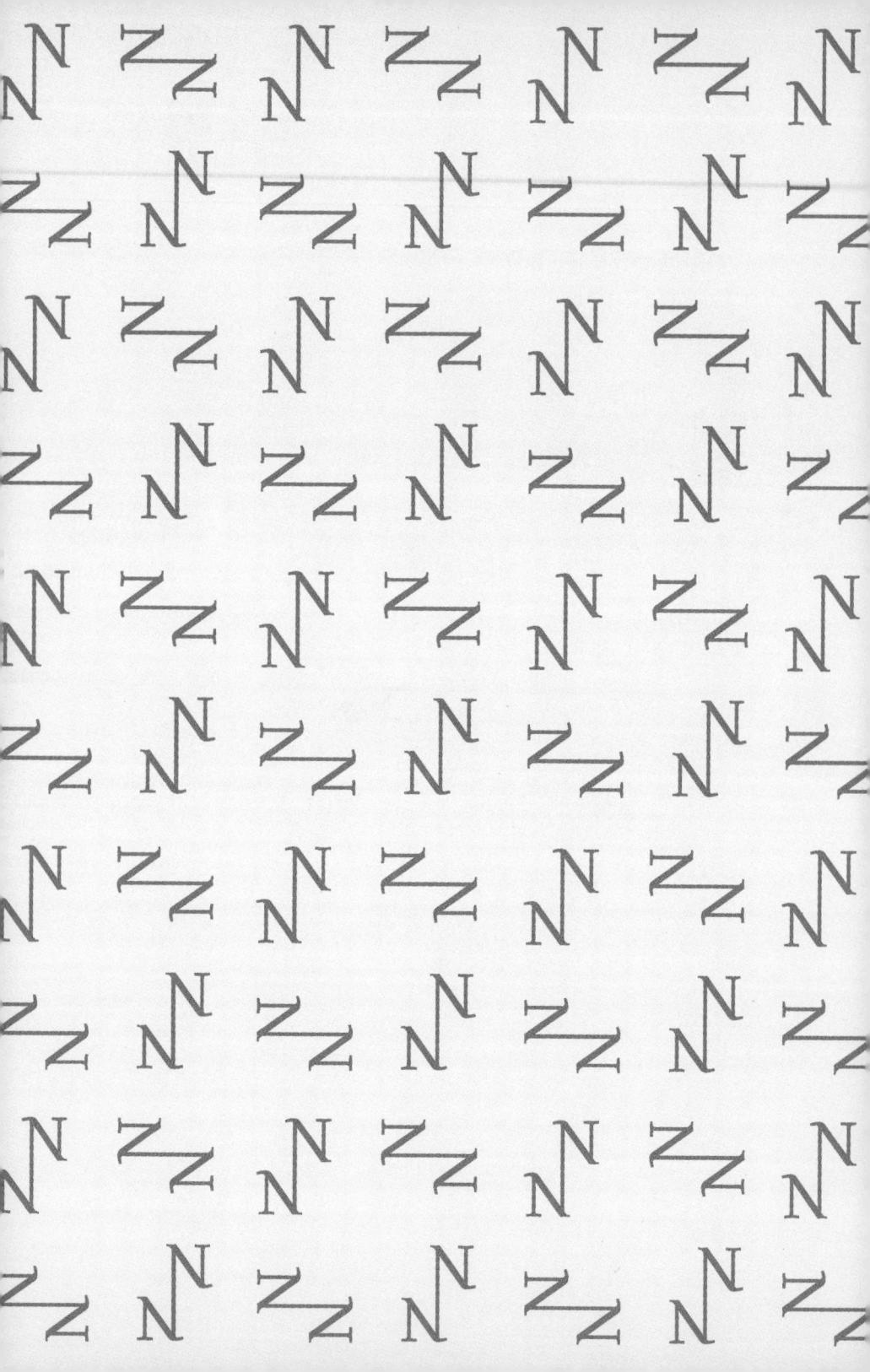